Survey of CHEMICAL SPILLS COUNTERMEASURES

Survey of
CHEMICAL SPILLS COUNTERMEASURES

edited by

MERV FINGAS
JENNIFER CHARLES
P.G. LANGILLE
L.B. SOLSBERG

LEWIS PUBLISHERS
Boca Raton Boston London New York Washington, D.C.

Library of Congress Cataloging-in-Publication Data

A survey of chemical spill countermeasures / P.G. Langille . . . [et al.] ; prepared by Emergencies Science Division, Environmental Technology Advancement Directorate, Environmental Protection Service, Environment Canada.
      p. cm.
   "October, 1997."
   Includes bibliographical references and indexes.
   ISBN 0-56670-313-1 (alk. paper)
   1. Chemical spills--Management. I. Langille, P. G. II. Canada. Emergencies Science Division.
TD196.C45S87 1997
628.5'2—dc21
for Library of Congress   97-40553
                                                                                                                      CIP

   This book contains information obtained from authentic and highly regarded sources. Reprinted material is quoted with permission, and sources are indicated. A wide variety of references are listed. Reasonable efforts have been made to publish reliable data and information, but the author and the publisher cannot assume responsibility for the validity of all materials or for the consequences of their use.

   Neither this book nor any part may be reproduced or transmitted in any form or by any means, electronic or mechanical, including photocopying, microfilming, and recording, or by any information storage or retrieval system, without prior permission in writing from the publisher.

   The consent of CRC Press LLC does not extend to copying for general distribution, for promotion, for creating new works, or for resale. Specific permission must be obtained in writing from CRC Press LLC for such copying.

   Direct all inquiries to CRC Press LLC, 2000 Corporate Blvd., N.W., Boca Raton, Florida 33431.

   Trademark Notice: Product or corporate names may be trademarks or registered trademarks, and are used only for identification and explanation, without intent to infringe.

   This book has been prepared by the Emergencies Science Division of Environment Canada and approved for publication. Approval does not necessarily signify that the contents reflect the views and policies of Environment Canada. Mention of trade names or commericial products does no necessarily constitute recommendation or endorsement for use.

© 1998 by CRC Press LLC

No claim to original U.S. Government works
International Standard Book Number 1-56670-313-1
Library of Congress Card Number 97-40553
Printed in the United States of America 1 2 3 4 5 6 7 8 9 0
Printed on acid-free paper

Abstract

This book is a survey of commercially available equipment, processes, and agents (collectively termed countermeasures) for controlling chemical spills. It includes only those products that have been evaluated, have proven performance, or were judged to have potential merit for chemical spill response. It is an updated and extensively revised version of the original "Survey of Chemical Spill Countermeasures", which was published in 1986. This revised and updated version contains data made available to Environment Canada's Emergencies Science Division up to April, 1996.

This book is intended as a reference for those responsible for purchasing, evaluating, or implementing countermeasures for chemical spills. It could also be used by government and industry to establish priorities for related research and development. It is not intended as a comprehensive spill response field manual.

To assist the reader in locating information, two Indexes are included at the back of the book. The first Index lists the various countermeasures by product name, and the second lists them by the name of the manufacturer or distributor.

Acknowledgments

This book was originally compiled by Phil Langille and Laurie Solsberg of Counterspil Research Incorporated of West Vancouver, British Columbia under contract to Environment Canada. The report was edited and revised by Jennifer Charles and reviewed by Merv Fingas both of the Emergencies Science Division at the Environmental Technology Centre in Ottawa. Sections of the report were also reviewed by Harry Whittaker and the following other members of the Emergencies Engineering Division whose assistance is gratefully acknowledged: Caroline Ladanowski, Anne Legault, Brian Mansfield, and David Cooper.

The graphics in this book were prepared by Lauren Forgie of Post-modern Designs. Some graphics were also done by S. Brenninkmeyer of Les Services Cartographiques 2+1 Incorporated.

The manufacturers, distributors, and developers of spill control technology are acknowledged for providing technical data for this survey both in the form of personal communications and brochures, reports, and papers, as well as for reviewing the writeups on their products. Extensive use was also made of file literature maintained by Environment Canada, including conference proceedings, manufacturers' sales information, and technical journals.

Introduction

This book is a survey of commercially available equipment, processes, and agents (collectively termed countermeasures) developed primarily for controlling chemical spills. Due to the broad scope and related time constraints of such a survey, listings are limited primarily to North American technologies. In addition to listings on individual products and pieces of equipment, several more comprehensive technology reviews are included that may list contact information for products not reviewed. These are called General Listings.

The products are categorized in various sections according to the primary response operation for which they are used. The following are the categories and the subsections within them.

Containment	Leak Mitigation
	Containment on Land
	Containment on Water
	Containment in/under Water
	Vapour Control
Removal	Removal from Land
	Removal from Surface Water
	Removal from Subsurface Soils
	Sorbents
	Vacuum Systems
	Dredging
Temporary Storage	Flexible Containers
	Rigid Wall Containers
	Liners
Transfer	Transfer of Liquids
	Transfer of Gases
	Transfer Hoses
Treatment/Disposal	Spill Treating Agents
	Liquid-Solid Separation Systems
	Fundamental Water Treatment Processes
	Aqueous Treatment Systems
	Vapour Treatment Systems
	Solids Treatment Systems

Within these sections, the products are listed in alphabetical order and within each listing, data are organized under the following headings: Description, Operating Principle, Status of Development and Usage, Technical Specifications (*Physical, Operational, Power/Fuel Requirements, and Transportability*), Performance, Contact Information, Other Data, and References. Entries are numbered to facilitate use of the survey as a catalogue as well as for easier cross-referencing. Two indexes are provided at the back of the report. The first index lists all products by name. The second index lists the manufacturers and distributors of the various products.

The information in these listings is only a partial representation of the manufacturers' capabilities and equipment offerings. Neither does this survey attempt to include all products or manufacturers of such products, but is a cross section of those products that are available commercially. It includes only those countermeasures that have been evaluated, have proven performance, or have potential merit for chemical spill response.

The survey does not include cleanup measures that are intended primarily for oil spills and that have not been used or tested for specific chemicals; new ideas or concepts that have not been tested or built as prototypes; concepts or equipment that have not been used or tested in the past five years; equipment or materials that are specific to small-scale spillage, such as in laboratories, kits, or assemblages, or supplies and equipment that are not unique to the spill response industry; processes or equipment that have not been used or tested for applications to spills, although they are applicable to the same chemicals for a fixed application; basic equipment such as shovels and brooms; and spill-treating agents containing surfactants and biological agents such as bacteria.

Abbreviations and Acronyms

°C	degrees Celsius
°F	degrees Fahrenheit
Å	angstrom unit
BTEX	benzene, toluene, ethylbenzene, xylenes
bhp	brake horsepower
BOD	biological oxygen demand
Btu	British thermal unit
Btuh	British thermal unit per hour
bu	bushel
cfh	cubic foot, feet per hour
cfm	cubic foot, feet per minute
cm	centimetre
COD	chemical oxygen demand
cP	centipoise (dynamic viscosity)
cSt	centistokes (kinematic viscosity)
DNAPLs	dense, non-aqueous phase liquids
FRP	fibreglass-reinforced plastic
ft	foot, feet
ft^2	square foot, feet
ft^3	cubic foot, feet
g	force of gravity
g	gram
gal	U.S. gallon (= 3.79 L)
gpd	U.S. gallons per day
gph	U.S. gallons per hour
gpm	U.S. gallons per minute
h	hour
ha	hectare
HDPE	high density polyethylene
HEPA	high efficiency particulate air
hp	horsepower
Hz	Hertz (cycles per second)
I.D.	inside diameter
in H_2O	inches of water
in Hg	inches of mercury
in	inch
in WC	inches of water column
kg	kilogram
kJ	kilojoules
km	kilometre
kPa	kilopascal
kW	kilowatt
L	litre
lb	pound
LDPE	low density polyethylene
LEL	lower explosive limit
m	metre
M	molar
m^2	square metre
m^3	cubic metre
mi	mile
min	minute
mm	millimetre
N·m	Newton-metre
NO_x	nitrogen oxide(s)
NPT	national pipe thread
O.D.	outside diameter

oz	ounce
PAHs	poly aromatic hydrocarbons
PCBs	polychlorinated biphenyls
pH	logarithm of the reciprocal of the hydrogen ion concentration
ppb (wt)	parts per billion by weight
ppb	parts per billion
pph	pounds per hour (1 pph = 0.216 cfm)
ppm	parts per million
ppm (wt)	parts per million by weight
psi	pound-force per square inch
psi(g)	pound-force per square inch gauge
PVC	polyvinyl chloride
qt	quart
rph	revolutions per hour
rpm	revolutions per minute
s	second
scfm	cubic feet per minute at standard temperature and pressure
SVC	semi volatile compound
SVOCs	semi volatile organic compounds
t	tonne
TDS	total dissolved solids
THC	total hydrocarbons
TLV	threshold limit value
TOD	total oxygen demand
TPH	total petroleum hydrocarbon
TSS	total suspended solids
μm	micrometre
U.S. EPA	United States Environmental Protection Agency
UV	ultraviolet (wavelength of light)
V	Volt
VAC	Volts, alternating current
VDC	Volts, direct current
VOCs	volatile organic compounds
W	watt
WC	water column
WCG	water column gauge

Table of Contents

1 Containment

1.1 Leak Mitigation
- 1. Acid Leak Control Kit 1
- 2. Cherne Petro Plugs 2
- 3. Chlorine Institute Emergency Kits 3
- 4. Devcon Zip Patch & Magic Bond 4
- 5. Edwards & Cromwell Response Kits 5
- 6. Furon (Bunnell) Teflon Flange and Valve Shields 7
- 7. Holmatro Vacuum Leak Sealing Pad 8
- 8. ILC Dover DrumRoll 10
- 9. Link-Seal Pipe Penetration Seals 11
- 10. Matheson MESA (Mobile Emergency Scrubbing Apparatus) 13
- 11. Milsheff Spray-Stops Valve and Flange Covers 15
- 12. Paratech Inflatable Leak Sealing Systems 16
- 13. Petersen Inflatable Pipeline Stopper Plugs 18
- 14. PLIDCO Pipeline Repair Fittings 20
- 15. Plug N'Dike Products and Plug Rugs 22
- 16. Vetter Systems Leak Sealing & Pipe Sealing Systems 24

1.2 Containment on Land
- 17. ATL Port-A-Berm 27
- 18. Bentonite Soil Sealing Systems 29
- 19. Chem-Tainer Polyethylene Spill Containment Pallet 32
- 20. Clark Spilstopper 33
- 21. Contain-It Plus Polyethylene Spill Containment Pallet 35
- 22. Froth-Pak Portable Foam System 36
- 23. Hi Tech Berm 38
- 24. Poly-Spillpallet 6000 Polyethylene Spill Containment Pallet 39
- 25. Sewer Guard Watertight Manhole Inserts 40
- 26. Shields Welded Steel Spill Containment Pallet 42
- 27. Syntechnics Track Collector Pan System 43
- 28. Trouble Shooter Portable Containment Berms 44

1.3 Containment on Water
- 29. Spill Containment/Deflection Barriers - General Listing 47

1.4 Containment In/Under Water
- 30. Portadam Barrier System 51

1.5 Vapour Control
- 31. Chemically Active Covers - General Listing 53
- 32. Induced Air Movement - General Listing 56
- 33. Coppus Portable Ventilators 58
- 34. Inert Gas Blankets - General Listing 60
- 35. Mechanical Covers - General Listing 63
- 36. Airotech EcoSpheres 65
- 37. Sprung Instant Structures 66
- 38. Vapour Suppressing Foams - General Listing 68

2 Removal

2.1 Removal from Land
- 39. Enviropack Emergency Spill Cleanup Kits 73

2.2 Removal from Surface Water
- 40. Mechanical Skimmers - General Listing 75

2.3 Removal from Groundwater
- 41. CEE Product Recovery Systems 77
- 42. Dynamic Petro-Belt Hydrocarbon Skimmer 79
- 43. ESI Mobile Recovery Trailer 80
- 44. Keck Product Recovery System 82
- 45. NEPPCO PetroPurge Pump Systems 84
- 46. ORS Scavenger Hydrocarbon Recovery Systems 86
- 47. PetroTrap and SkimRite Free Product Skimmers 89
- 48. Protec Recip Pump and Rotary Pump 91
- 49. Timco Isomega Recovery Pump 93

2.4 Removal from Subsurface Soils
- 50. Carbtrol Regenerative and Multi-phase Extraction Systems 95
- 51. EG&G Rotron Regenerative Blowers for Environmental Remediation ... 96
- 52. Global Vacuum Extraction and Vapour Liquid Separator Modules 98
- 53. Horizontal Directional Drilling for Trenchless Remediation 100
- 54. Hrubout® In-situ Soil Processor 101
- 55. Lamson Centrifugal Exhausters for Environmental Remediation 103
- 56. M-D Rotary Positive Displacement Blowers 104
- 57. NEPPCO SoilPurge Soil Vapour Treatment System 105
- 58. Pego Soil Vapour Extraction Units 106
- 59. RETOX (Regenerative Thermal Oxidizer) 107
- 60. Roots Blowers and Vacuum Pumps for Environmental Remediation ... 109

2.5 Removal - Sorbents
- 61. Spill Sorbents - General Listing 111

2.6 Vacuum Systems
- 62. Suction Systems - General Listing 113
- 63. Max Vac Vacuum Systems 115
- 64. Minuteman International Vacuum Systems 116
- 65. Nilfisk Vacuum Systems 118
- 66. Powervac Oil Recovery System 120
- 67. Ro-Vac Vacuum System 121
- 68. Tornado Vacuum Systems 123
- 69. Trans-Vac Oil Recovery Units 125
- 70. Vactagon Vacuum Systems 127
- 71. Vac-U-Max Vacuum Systems 129

2.7 Dredging
- 72. Dredging Equipment - General Listing 131
- 73. Crisafulli Dredges 140
- 74. H&H Dredges 142
- 75. Mud Cat Model 370 PDP Dredge 144
- 76. Pit Hog Auger Dredges 145
- 77. Trac Pump Dredging System 147
- 78. VMI Mini Dredges and Piranha Auger Dredge Attachment 148

3 Temporary Storage

3.1 Flexible Containers
- 79. General Listing 151
- 80. Canflex Sea Slug and Open-Top Reservoir 155
- 81. Helios Flexible Bulk Containers 157
- 82. Port-A-Tank 158
- 83. Terra Tank 159

3.2	**Rigid Wall Containers**	161
	84. General Listing	161
	85. Custom Metalcraft Transportable Storage Systems	166
	86. Enviropack Polyethylene Overpacks/Salvage Drums	167
	87. Greif Containers - Steel Overpacks/Salvage Drums	168
	88. ModuTank Modular Storage Tanks	170
	89. Skolnik Steel Overpacks/Salvage Drums	173
3.3	**Liners**	
	90. Geomembrane Liners - General Listing	175
	91. CDF Drum Liners	184
	92. Hazardous Waste Container Bag Liners	185
	93. Permalon Drum Liners	186
	94. Waste Container Bag Liners	187
4	**Transfer**	
4.1	**Transfer of Liquids**	
	95. Pumps - General Listing	189
	96. Amflow-Lift Peristaltic Hose Pump (formerly Mastr-Pump)	192
	97. Flygt Submersible Pumps	194
	98. Gilkes Portable Emergency Turbo Pump	196
	99. Lutz Drum Pumps	197
	100. Megator Sliding-Shoe Pump	199
	101. Sala Roll Pump	201
	102. Solar-powered Pumping Systems	202
	103. Warren Rupp SandPIPER Pumps	203
4.2	**Transfer of Gases**	
	104. Vacuum Pumps for Chemicals - General Listing	205
	105. Busch Vacuum Pumps for Chemicals	206
	106. Gast Vacuum Pumps	209
4.3	**Transfer Hoses**	
	107. Acid, Chemical Hoses - General Listing	211
	108. Flexaust Hose and Duct	214
	109. Gates Industrial, Acid-Chemical Hoses	216
	110. Goodall Chemical Hoses	218
5	**Treatment/Disposal**	
5.1	**Spill Treating Agents**	
	111. Ansul Spill Treatment Kits and Applicators	221
	112. CAPSUR®, MEXTRACT®, and PENTAGONE® Spill Agents	224
	113. Cartier Spill Response Kits	226
	114. Epoleon Odor-Neutralizing Agents	228
	115. Foul-Up Hazardous Wet-Waste Solidifying Agent	230
	116. Mercon Spill-treating Kits	231
	117. Nochar Solidifying Agents	232
5.2	**Liquid-Solid Separation Systems**	
	118. Alfa-Laval Centrifuges	233
	119. Andritz Mobile Dewatering Systems	235
	120. Denver Sand-Scrubber™ System	236
	121. Flo Trend Hydroclone Centrifugal Separator, Vibratory Screen, and Container Filter	237
	122. Kason Drum Sifter	239
	123. Lavin Centrifuges	241
5.3	**Fundamental Water Treatment Processes**	
	124. General Listing	243

5.4 Aqueous Treatment Systems

- 125. 3L Filters Oil/Water Separators ... 251
- 126. Advanced Oxidation Water Treatment Systems ... 252
- 127. Air Plastics - Air Stripping Towers ... 257
- 128. Algasorb Process ... 258
- 129. Andco Mobile Treatment System ... 260
- 130. BioAcceleration Treatment System ... 262
- 131. Breeze Air Stripping System ... 264
- 132. Brinecell Oxidizers ... 265
- 133. Calgon Activated Carbon Adsorption Treatment Systems ... 267
- 134. Carbtrol L-1, L-4, and L-5 Activated Carbon Adsorption Canisters ... 269
- 135. Carbtrol Diffused Air Strippers ... 271
- 136. Cyanide Destruction System ... 273
- 137. Environment Canada's Reverse Osmosis Units ... 274
- 138. Environment Canada's Steam Stripping Units ... 278
- 139. Environment Canada's Microfiltration/Ultrafiltration Unit ... 281
- 140. ESI Cascade LP500 Series Low Profile Air Stripper ... 284
- 141. GLE Slant Rib Coalescing (SCR) Oil/Water Separator ... 286
- 142. Hazelton Maxi-Strip Systems ... 287
- 143. Hudson CS Series Coalescing Oil/Water Separators ... 289
- 144. Hydro-Stripper Air Stripper ... 290
- 145. Koch Tubular Ultrafiltration (UF) System ... 291
- 146. Liquid-Miser Activated Carbon Adsorption Canisters ... 293
- 147. ORS LO-PRO Air Strippers ... 295
- 148. Osmonics Crossflow Membrane Filtration Systems ... 297
- 149. PACE Oil/Water Separators ... 299
- 150. PACT® Batch Wastewater Treatment System ... 301
- 151. ShallowTray® Low Profile Air Strippers ... 303
- 152. TriWaste Micro-flo Mobile Water Treatment System ... 305
- 153. Water Scrub Units - Activated Carbon Adsorption Canisters ... 306
- 154. ZenoGem Process ... 309

5.5 Vapour Treatment Systems

- 155. Calgon VentSorb & Sweetstreet Activated Carbon Adsorption Canisters ... 311
- 156. Global Vapour Treatment Module - Chloro-Cat Catalytic Oxidizer ... 313
- 157. John Zink Flaring Systems ... 315
- 158. NAO Trailer-mounted Emergency Flaring Systems and Mobile VOC Wagon ... 316
- 159. Odor-Miser Activated Carbon Adsorption Canisters ... 319
- 160. Torvex Catalytic Oxidation Systems ... 320
- 161. Tub Scrub Units - Activated Carbon Adsorption Canisters ... 321

5.6 Solids Treatment Systems

- 162. Andersen Mobile Soil Decontamination System ... 323
- 163. Alberta Taciuk Process (ATP) System ... 325
- 164. Bennett MKIII Transportable Rotary Incinerator ... 326
- 165. Brulé Incinerators ... 327
- 166. Calciner-based Thermal Processing Systems ... 328
- 167. CleanSoils Thermal Desorber ... 329
- 168. Coreco Rotary Incinerators ... 331
- 169. Heyl & Patterson Transportable Soil Remediation Unit ... 332
- 170. Porcupine Processor ... 333
- 171. Trecan Solid Waste Incinerators ... 334
- 172. Vulcanus 400 Mobile Incinerator ... 336

INDEX - By Product Name ... 337

INDEX - By Manufacturer/Distributor ... 339

Section 1.1

Containment - Leak Mitigation

ACID LEAK CONTROL KIT

No. 1

DESCRIPTION

A kit designed to patch slits or jagged holes in non-pressurized tanks or drums without special surface preparation.

OPERATING PRINCIPLE

Controls leaks and vapours quickly with acid-resistant sealing tape. A magnetic backup plate assembly can be placed over the sealant to secure the patch. The patch can also be secured with a webbed strap.

STATUS OF DEVELOPMENT AND USAGE

Commercially available product in general use.

TECHNICAL SPECIFICATIONS

Physical

Each kit contains one 5 cm wide × 9 m long (2 in × 30 ft) roll of acid-resistant isobutylene sealant, one magnetic backup plate assembly, one 5 cm wide × 2.4 m long (2 in × 8 ft) webbed belting, and a pair of Nitrile gloves, packaged in a 7.5 L (2 gal) pail. Two kits are provided in a carton, with an approximate shipping weight of 4.5 kg (10 lb). Replacement kit refills are also available which contain two rolls of sealant.

Operational

Plugs slits or jagged holes on non-pressurized vessels without special surface preparation.

Power/Fuel Requirements

None

Transportability

Person-portable

CONTACT INFORMATION

Manufacturer

PND Corporation
14320 N.E. 21st Street
Suite 6
Bellevue, WA 98007
U.S.A.

Telephone (206) 562-7252
Facsimile   (206) 562-7254

Canadian Distributor

Can-Ross Environmental Services Ltd.
2270 South Service Road West
Oakville, Ontario
L6L 5M9

Telephone (905) 847-7190
Facsimile   (905) 847-7175

OTHER DATA

See No. 15 in this section for information on **other Plug n'Dike products**.

REFERENCES

(1) Distributor's Literature.

CHERNE PETRO PLUGS

DESCRIPTION

Chemical-resistant inflatable plugs for stopping puncture leaks or plugging pipes.

OPERATING PRINCIPLE

Plugs are inserted into the pipe opening and expanded using air to compress the sealing material against the opening.

STATUS OF DEVELOPMENT AND USAGE

Commercially available products generally marketed to the plumbing industry. Petro Plugs are manufactured specifically for applications with chemical and petroleum products.

TECHNICAL SPECIFICATIONS

Physical

Muni-Ball Plugs

The following are the standard sizes of Muni-Ball Plugs. Custom sizes are also available.

Sizes	mm	100	150 to 200	200 to 250	250 to 300	300 to 350	350 to 450	400 to 600	600 to 900
	(in)	4	6 to 8	8 to 10	10 to 12	12 to 15	15 to 18	16 to 24	24 to 36
Usage Range	mm	87 to 112	119 to 217	172 to 268	229 to 320	280 to 396	356 to 477	375 to 635	540 to 953
	(in)	3.4 to 4.8	4.7 to 8.6	6.8 to 11	9 to 13	11 to 16	14 to 19	15 to 25	21 to 38

Operational

Muni-Ball Plugs

Maximum Allowable Backpressure   Air   68 kPa (10 psi)
                                                               Water   11 m head (35 ft head)

Power/Fuel Requirements

Muni-Ball plugs require a source of compressed air for inflation.

Transportability

Person-portable

CONTACT INFORMATION

Manufacturer

Cherne Industries Incorporated
5700 Lincoln Drive
Minneapolis, MN 55436-1695
U.S.A.

Telephone (612) 933-5501
Canada Toll Free (800) 843-7584
Facsimile (612) 938-6601

REFERENCES

(1) Manufacturer's Literature.

CHLORINE INSTITUTE EMERGENCY KITS

DESCRIPTION

Kits for stopping leaks in chlorine cylinders, one-ton containers, and tank cars and tank trucks.

OPERATING PRINCIPLE

Kits "A" and "B" contain devices and tools to stop leaks at the valve, fusible plug, and in the sidewalls of 45- and 68-kg (100- and 150-lb) chlorine cylinders conforming to specification 3A480, and chlorine one-ton containers conforming to specification 106A500X, respectively. Kit "C" stops leaks at the safety valve or angle valves of chlorine tank cars and tank trucks that conform to specifications 105A500W and MC331.

STATUS OF DEVELOPMENT AND USAGE

Commercially available product in general use. Kits are also available for SO_2, NH_3, AHF, and AHCl.

TECHNICAL SPECIFICATIONS

Physical - Kits are contained in polyethylene boxes 89 x 34 x 35 cm (35 x 13.5 x 14 in) in size.

Kit Type	Gross Weight	Net Weight
Kit A	49 kg (109 lb)	45 kg (99 lb)
Kit B	51 kg (113 lb)	46 kg (103 lb)
Kit C	76 kg (168 lb)	72 kg (158 lb)

Operational - Kits will service the following chlorine containers. These three kits are not suitable for use on chlorine barge tanks, but such kits can be made if drawings are supplied. Instructional videos and personal protection equipment are available.

Kit Type	Applicable Containers
Kit A	DOT 3A480 45 kg (100 lb) and 68 kg (150 lb) cylinders
Kit B	DOT 106A500 X 907 kg (2,000 lb) container
Kit C	Tank car and tank trucks, MC 331

Transportability - The polyethylene boxes containing the kits can be carried by two people.

Manufacturer

Indian Springs Specialty Products Inc.
2095 W. Genesee Road
P.O. Box 118
Baldwinsville, NY 13027

Telephone (315) 635-6243
Facsimile (315) 635-7473

Information

The Chlorine Institute Incorporated
2001 L Street, N.W.
Suite 506
Washington D.C. 20036

Telephone (202) 775-2790
Facsimile (202) 223-7225

REFERENCES

(1) Manufacturer's Literature.

DEVCON ZIP PATCH & MAGIC BOND

DESCRIPTION

Devcon Zip Patch is a fast-setting patch kit for effecting permanent waterproof emergency field repairs to many surfaces. **Magic Bond** is a hand-kneadable epoxy putty for permanently patching holes in metal, concrete, fibreglass, and ceramics.

OPERATING PRINCIPLE

The **Zip Patch** is applied to the leak and smoothed with a plastic putty knife to form a waterproof seal within minutes. **Magic Bond** putty is kneaded with fingers for about one minute until the blue colour disappears, applied, and smoothed out.

STATUS OF DEVELOPMENT AND USAGE

Commercially available products in general use.

TECHNICAL SPECIFICATIONS

Physical - The **Zip Patch** is 10 x 23 cm (4 x 9 in) in size. **Magic Bond Putty** is packaged in a re-sealable plastic tube containing one 110-g (4-oz) stick of putty. It forms a compressive strength of 82,700 kPa (12,000 psi) ASTM D 695. Both products have a minimum shelf stability of 6 months at 24°C (75°F).

	Zip Patch	Magic Bond
Adhesive Tensile Shear (ASTM D 1002)	31,700 kPa (4,600 psi)	4,800 kPa (750 psi)
Tensile Strength (ASTM D 638)	69,600 kPa (10,100 psi)	16,900 kPa (2,450 psi)
Flexural Strength (ASTM D 790)	131,000 kPa (19,000 psi)	29,500 kPa (4,280 psi)
Cure Shrinkage (ASTM D 2566)	0.0010 in/in	0.0030 in/in
Cured Hardness (ASTM D 2240)	75 Shore	75 Shore

Operational - **Zip Patch** can be used for field repairs on iron, steel, stainless steel, titanium, fibreglass, aluminum, concrete, ceramics, acrylics, composites, PVC, and most other plastic surfaces, but not on non-stick surfaces such as Teflon. It requires little surface preparation and can be used on wet or oily surfaces. It tolerates rain or salt water during application as moisture in patching area is displaced. Hardens to a tough, durable, waterproof finish in minutes. Maximum operating temperature is 93°C (200°F). **Magic Bond Putty** hardens in 10 minutes and cures to full strength in 24 hours. It can be used on wet surfaces and applied under water, in either fresh or salt water. It will not yellow when exposed to ultraviolet light and is resistant to hydrocarbons, ketones, alcohols, esters, halocarbons, aqueous salt solutions, and dilute acids and bases. Maximum operating temperature is 120°C (250°F).

Manufacturer

ITW Devcon Environmental Products
30 Endicott Street
Danvers, MA 01923, U.S.A.

Telephone (508) 777-1100
Facsimile (508) 774-0516

Canadian Distributor

ITW Devcon Environmental Products
Box 240
Coombs, British Columbia, V0R 1M0

Telephone (604) 248-4851
Facsimile (604) 248-4852

REFERENCES

(1) Manufacturer's Literature

1.1 Containment - Leak Mitigation

EDWARDS AND CROMWELL RESPONSE KITS No. 5

DESCRIPTION

Several different types of pre-packaged kits for temporarily plugging or patching leaks of liquid or bulk hazardous materials. Series A, E, D, A-2, and AE (Universal Series) are for plugging or patching small to medium holes in any type of low-pressure vessel. Series C (Pipe Plugger Kit) is for plugging knocked-off pipes up to 101 mm (4 in) in diameter. Series C-2 (External Pipe Patch Kit) is for patching low-pressure lines from 13 to 101 mm (0.5 to 4 in) in diameter. Series F-AS (Roll-over Kit) is designed for patching large holes in vessels when it is not possible to secure a patch with cables or chains.

OPERATING PRINCIPLE

Series A, E, D, A-2, and AE - Leaks are patched using tire plugs, hot or cold hose tape, screw patches, lead wool, epoxy putty, wood wedges with felt blankets, surface plugs, "T" patches, or ladder patches.

Series C - The plugs are inserted into the end of the pipe and compressed by tightening a bolt that passes through the plug. The 25 to 101 mm (1 to 4 in) sizes have pressure relief vents to help insert and seal the plug under pressure. When the plug is secured in place, these vents can be crimped off for a final seal.

Series C-2 - Metal compression bandages, combined with soft or hard Neoprene or rubber bandage liner materials, are placed over the leaking pipe and tightened around the pipe using the tools contained in the kit.

Series F-AS - A variety of large "T" bolts are used to support a rigid frame/patch assembly applied directly onto the hole. Compression bolts are then threaded through the rigid frame and tightened to make the patch conform to the vessel and control the leak. The "F-AS" patch can also be used on the ends or corners of vessels.

STATUS OF DEVELOPMENT AND USAGE

Commercially available product in general use.

TECHNICAL SPECIFICATIONS

Physical

Series A, E, D, A-2, and AE - Universal Kits are contained in steel carrying cases and are available in various configurations, all of which also contain barricade tape and basic hand tools (standard or non-sparking).

Series C - These kits contain 10 standard pipe size plugs from 19 to 101 mm (0.75 to 4 in), a wooden plug, and all necessary tools and hardware. Kit C-1 is an optional version of Kit C that includes a valve and pipe-off set for the 38 to 101 mm (1.5 to 4 in) sizes. This allows the vented flow to be shut off with the valve, piped to a recovery vessel, or flared. Each kit is contained in a steel carrying case.

Series C-2 - This kit contains 10 standard pipe size, external pipe bandages covering nominal pipe sizes from 13 to 101 mm (0.5 to 4 in). Additional soft and hard neoprene bandage liner material increases the versatility of the bandages. The kit includes all necessary hand tools (standard or non-sparking) and is contained in a steel carrying case.

Series F-AS - This kit contains one patch 330 x 584 mm (13 x 23 in), one stainless steel support frame, four stainless steel "T" bolts, and all necessary tools and hardware. The kit is contained in a steel carrying case.

Operational

Kits are designed for temporary leak mitigation or containment only. Series A, E, D, A-2 and AE can be applied to drums, cans, tank trucks and cars, or any other low-pressure vessel. Series C can be applied to knocked-off gas meters and tank truck discharge valves, and Series C-2 to all low-pressure lines from 13 to 101 mm (0.3 to 4 in) in diameter. Series F-AS is for large leaks in vessels when cables or chains cannot be fixed around the vessel, or on the ends or corners of vessels.

Power/Fuel Requirements

None

Transportability

Steel boxes containing kits are person-portable.

CONTACT INFORMATION

Manufacturer

Edwards and Cromwell Mfg., Incorporated
4301 Jeffrey Drive
Baton Rouge, LA 70816
U.S.A.

Telephone (504) 292-3377
Facsimile (504) 292-7519

OTHER DATA

A descriptive video of all Edwards and Cromwell Hazardous Material Response Kits is available from the manufacturer free-of-charge. Training and a training video are also available from the manufacturer.

REFERENCES

(1) Manufacturer's Literature.

1.1 Containment - Leak Mitigation

FURON (BUNNELL) TEFLON FLANGE AND VALVE SHIELDS

No. 6

DESCRIPTION

Designed to protect operating personnel from the hazards of corrosive or high temperature leaks at flanged joints or valve packings.

OPERATING PRINCIPLE

Made of 100% Teflon FEP (fluorinated ethylene propylene co-polymer), the shields resist most severe chemical and temperature environments. They can be quickly attached or removed and deflect and contain leaks, such as drops, sprays, or streams, to reduce the impacts of such leaks. Their transparent design allows visual inspection of flanges, gaskets, or valve packings, and tightening of flange bolts with shield in place, thus reducing maintenance time.

STATUS OF DEVELOPMENT AND USAGE

Commercially available product in general use.

TECHNICAL SPECIFICATIONS

Physical

Material	Teflon FEP
Colour	Transparent
Sizes	Standard sizes are available to fit virtually all valves and ASA flanges
Tensile strength	18,000 to 22,000 kPa (2,600 to 3,200 psi)
Melt index	0.95 to 1.40

Operational

Resistant to acids, caustics, and most other industrial chemicals. Can be attached or removed quickly. The flange covers can be shrunk-fit if desired. The shields are UV-stabilized to protect against deterioration caused by sunlight.

Power/Fuel Requirements

None.

Transportability

Person-portable, compact, and lightweight.

CONTACT INFORMATION

Manufacturer

Furon/Bunnell Plastics
1-295 Harmony Road
Mickleton, NJ 08056
U.S.A.

Telephone (609) 423-6630
Facsimile (609) 423-7072

REFERENCES

(1) Distributor's Literature.

HOLMATRO VACUUM LEAK SEALING PAD   No. 7

DESCRIPTION

A leak-sealing and drainage control system that can be secured to a tank by a vacuum generated by a vacuum pump. The liquid inside the tank can also be removed through a second chamber within the sealing pad.

OPERATING PRINCIPLE

The vacu-pad is positioned over a leak with the valve open and vacuum pump operating. The vacuum pump empties the vacuum compartment and removes the liquid through a hose. When sufficient vacuum has been created in the sealing compartment to seal the leak, the valve is closed, securing the vacu-pad to the tank. The liquid in the vacuum compartment is automatically removed through the hose.

STATUS OF DEVELOPMENT AND USAGE

Commercially available product in general use.

TECHNICAL SPECIFICATIONS

Physical

The Vacu-pad is equipped with PVC fittings, gauge- and air-operated venturi-type vacuum pump, and a built-in check valve. All materials are acid-resistant and non-sparking.

Dimensions

Sealing pad	60 × 39 cm (24 x 15 in)
Thickness	3.6 cm (1.4 in)
Sealing compartment	25 × 5 cm (10 x 2 in)
Hose connection	4 cm (1.6 in)

Operational

The Vacu-pad seals leaks within seconds after application. After sealing the leak, the vacuum is maintained by a built-in check valve. A gauge is used to monitor the vacuum pressure. The Vacu-pad rapidly provides a complete seal or allows controlled off-loading on flat and curved surfaces.

Max. air pressure	600 kPa (86 psi)
Air consumption	300 L/min (at 600 kPa)

Power/Fuel Requirements

An air compressor is required to operate the vacuum pump.

Transportability

Person-portable

1.1 Containment - Leak Mitigation

CONTACT INFORMATION

Manufacturer

Holmatro Industrial Equipment
P.O. Box 33
4940 aa Raamsdonksveer
Holland

Telephone 31-162-589200 (dial "0" and ask operator for appropriate country and routing codes)
Facsimile 31-162-522482

Holmatro Inc.
1110 Benfield Blvd.
Millersfield, MD 21108
U.S.A.

Telephone (410) 907-6633
Facsimile (410) 907-1638

Holmatro Asia
Representative Office
Block 373, #02-128
Bukit Batok St. 31
Singapore 2365

Telephone 65-5643923
Facsimile   65-5613507

REFERENCES

(1) Manufacturer's Literature.

ILC DOVER DRUMROLL

No. 8

DESCRIPTION

A chemical-resistant, inflatable bladder that instantly stops leaks in 208-L (55-gal) drums.

OPERATING PRINCIPLE

In the event of a drum leak, the DrumRoll is wrapped and secured around the drum and inflated with the attached CO_2 cartridge to immediately seal the leak.

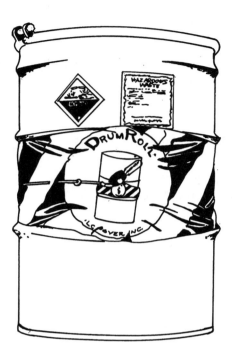

STATUS OF DEVELOPMENT AND USAGE

Commercially available product in general use.

TECHNICAL SPECIFICATIONS

Physical

The DrumRoll is stored in a compact, rugged plastic tube. The bladder is Teflon-coated to resist most chemicals.

Operational

Activated by a CO_2 cartridge, the DrumRoll can easily be installed by one person to quickly stop a drum leak. It is fastened and activated on the opposite side of the leak allowing it to be applied and activated safely. The DrumRoll can also be used on overpacks. It stops leaks up to 18×23 cm (7×9 in) in size. The effective operating temperature of the DrumRoll is from -18 to $60°C$ (0 to $140°F$).

Power/Fuel Requirements

The CO_2 cartridge comes with the DrumRoll.

Transportability

Person-portable

CONTACT INFORMATION

Manufacturer

ILC Dover Incorporated
P.O. Box 266
Frederica, DE 19946
U.S.A.

Telephone (302) 335-3911
Facsimile   (302) 335-0762

REFERENCES

(1) Manufacturer's Literature.

1.1 Containment - Leak Mitigation

LINK-SEAL PIPE PENETRATION SEALS
No. 9

DESCRIPTION

A modular mechanical plug for sealing the annular space between pipes and the holes or casings through which they pass. Suitable for sealing pipes passing through containment berms or casings at road crossings, or double-containment piping.

OPERATING PRINCIPLE

The Link-Seal consists of interlocking synthetic rubber links shaped to continuously fill the annular space between the pipe and the wall opening. The links are loosely assembled with bolts to form a continuous rubber belt around the pipe with a pressure plate under each bolt head and nut. The seal assembly is positioned around the pipe in the wall opening, and the bolts are tightened, causing the rubber to expand and provide a complete seal between the pipe and the wall opening.

STATUS OF DEVELOPMENT AND USAGE

Commercially available product in general use.

TECHNICAL SPECIFICATIONS

Physical

Link-Seal comes in eight different thicknesses to fit various sizes of annular spaces. They have effectively sealed pipes from 0.6 cm to 3 m (0.3 in to 10 ft) in diameter, although there is no maximum limit to the diameter that can be handled. Available in a number of service designations including Standard Service (Insulating Type), Corrosive Service, Oil-resistant Service, High- or Low-temperature Service, and Fire-rated Service. The following lists the type of material used for the various components of the Link-Seal.

Standard Service (Insulating Type)
- Pressure Plate   Glass-reinforced nylon plastic
- Bolt and Nut   Zinc-galvanized, low carbon steel
- Sealing Element   EPDM rubber

Corrosive Service
- Pressure Plate   Glass-reinforced nylon plastic
- Bolt and Nut   18-8 stainless steel (316)
- Sealing Element   EPDM rubber

Oil-resistant Service
- Pressure Plate   Zinc-galvanized, low carbon steel
- Bolt and Nut   Zinc-galvanized, low carbon steel
- Sealing Element   Nitrile rubber

High- or Low-temperature Service
- Pressure Plate   Zinc-galvanized, low carbon steel
- Bolt and Nut   Zinc-galvanized, low carbon steel
- Sealing Element   Silicone rubber

Fire-rated Service
- Pressure Plate   Zinc-galvanized, low carbon steel
- Bolt and Nut   Zinc-galvanized, low carbon steel
- Sealing Element   Silicone rubber

Operational

Rated for 12 m (40 ft) of head or 140 kPa (20 psi) pressures, Link-Seal provides electrical insulation between the pipe and the wall, reducing the chances of cathodic reaction between these two members. Link-Seal compensates for a considerable amount of both angular misalignment and eccentricity, while still forming an efficient seal.

Power/Fuel Requirements

None

Transportability

Person-portable, compact, and lightweight.

PERFORMANCE

Link-Seal has been used in such diverse applications as a pipeline in Arabia, a pneumatic drug-distribution system in a modern New York hospital, and a fire-protection system in a British nuclear power plant.

CONTACT INFORMATION

Manufacturer

Thunderline/Link Seal
A Division of PSI
19500 Victor Parkway
Suite 275
Livonia, MI 48152
U.S.A.

Telephone (313) 432-9700
Facsimile (313) 432-9704

OTHER DATA

The specifications provided here represent only the most popular options. Many additional variations are available, such as different elastomer for the sealing element (Neoprene, SBR, natural rubber) or different surface treatments for the low carbon steel bolt and nut (cadmium plating).

REFERENCES

(1) Manufacturer's Literature.

1.1 Containment - Leak Mitigation

MATHESON MESA (MOBILE EMERGENCY SCRUBBING APPARATUS)

No. 10

DESCRIPTION

A mobile scrubber system for containing and mitigating leaks in gas cylinders.

OPERATING PRINCIPLE

If a gas cylinder leaks, the main module of the MESA is brought to the location and the cylinder is placed inside the cabinet. Air is drawn through the base of the cabinet, swept up past the leaking cylinder, and exhausted to the dry scrubber where it passes through two chemisorbents. The chemisorption process irreversibly removes the hazardous gas from the stream. Once the leak is contained and being treated, it can be repaired under controlled conditions. The cylinder and valve are reached through an access port in the cabinet door. If the leak cannot be repaired, the wet scrubber module is brought to the site, connected to the main module, and the remaining gas in the cylinder is drained and scrubbed, using the unit's two wet scrubbers and one dry scrubber.

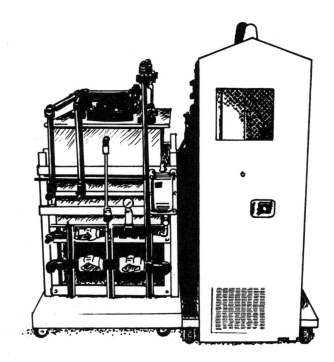

STATUS OF DEVELOPMENT AND USAGE

Commercially available product in general use.

TECHNICAL SPECIFICATIONS

Physical

Size   Dry Module - 2 × 0.7 × 1.9 m (6.7 × 2.6 × 6.2 ft)
           Wet Module - 1.4 × 0.8 × 1.6 m (4.6 × 2.7 × 5.2 ft)

Weight   Total System, including dry module, wet module, and dry chemisorbent vessel - 860 kg (1,900 lb).

Cabinet construction features 11-gauge, cold-rolled steel with continuously welded seams. Gas-tight gaskets fit around the door to ensure a positive seal. A recessed door handle prevents accidental opening and snagging. The door latches at three points.

Operational

A visual indicator changes colour when the adsorbent in the dry scrubber is spent. The adsorbent comes in a vessel that is readily transportable [208-L (55-gal) drum in a 320-L (85-gal) overpack], allowing for easy waste disposal and replacement. MESA is equipped with a series of CGA connections to accommodate a wide variety of compressed gas cylinders. The appropriate connection is used to hook the cylinder up to a gas control panel, located in the cabinet, via a flexible hose. A keypad on the control panel is used to enter the cylinder size and gas mixture. An integral microprocessor calculates and controls the batch times for cylinder draining and purging sequences. Pneumatically actuated valves and a mass flow transducer facilitate this fully automated process. Based on the gas mixture to be drained, the microprocessor provides a recipe for preparing one of four elementary wet scrubber solutions. An integral pump facilitates loading, mixing, and eventual emptying of the vessel.

MESA is designed to accommodate the following gases:

Ammonia	Hydrogen sulphide
Arsenic trichloride	Monomethylamine
Arsine	Nitrogen dioxide
Boron trichloride	Nitric oxide
Boron trifluoride	Phosgene
Chlorine	Phosphine
Diborane	Phosphorus pentafluoride
Dichlorosilane	Silicon tetrachloride
Cimethylamine	Silicon tetrafluoride
Hydrogen bromide	Sulphur dioxide
Hydrogen chloride	Sulphur tetrafluoride
Hydrogen fluoride	Tungsten hexafluoride
Hydrogen selenide	

Power/Fuel Requirements

Dry Module   120 VAC, 60 Hz, 13 Amps
Wet Module   120 VAC, 60 Hz, 12 Amps

Consumables include dry chemisorbent and wet scrubber chemicals.

Transportability

MESA will fit through a standard 0.9 × 2 m (3 × 6.7 ft) doorway, and can be transported by pickup truck or other emergency response vehicle. Maneuvering the system is facilitated by heavy-duty caster wheels. Once in position, a foot brake locks the system in place.

PERFORMANCE

MESA can reduce toxic gas concentrations in the treated effluents to less than Threshold Limit Value (TLV) levels.

CONTACT INFORMATION

Manufacturer

Matheson Gas Products Canada
530 Watson Street East
P.O. Box 89
Whitby, Ontario
L1N 5R9

Telephone (905) 668-3397
Facsimile (905) 668-6937
Canada Toll Free (800) 263-2620

Distributors

Matheson Gas Products

Edmonton (403) 471-4036
Toronto   (416) 798-7079
Sarnia   (519) 332-8100
Ottawa   (613) 526-0208

Les Gas Speciaux MEGS
Northern Abcor

Montreal (514) 956-7503
Sudbury   (705) 673-4848

OTHER DATA

Purchase of a Mobile Emergency Scrubbing Apparatus System includes up to two days of training at the customer's facility. Wet scrubber chemicals are not included with the purchase of the unit, as these are basic chemicals (NaOH, H_2SO_4, NaOCL, NH_3, H_2O) that can be purchased from a local supplier. Replacement dry adsorbent drums are available from the manufacturer.

REFERENCES

(1) Manufacturer's Literature.

1.1 Containment - Leak Mitigation

MILSHEFF SPRAY-STOPS VALVE AND FLANGE COVERS

No. 11

DESCRIPTION

Vinyl covers for valves and flanges that indicate leakage by means of a colour change. Covers also deflect and contain leaks, such as drops, sprays, or streams, to reduce the impact of such leaks.

OPERATING PRINCIPLE

Made of woven glass fabric, coated with chemical-resistant vinyl or Teflon. The vinyl fabric is coated with a special formulation of indicators and polymers that cause it to change colour when contacted by an acid or caustic when the inside of the cover is exposed to the leak. The covers are secured with tie strings and can be attached or removed in seconds.

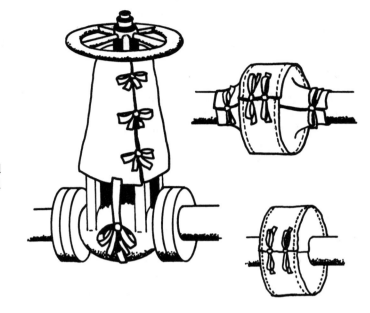

STATUS OF DEVELOPMENT AND USAGE

Commercially available product in general use.

TECHNICAL SPECIFICATIONS

Physical

Material	Glass weave fabric, coated with chemical-resistant vinyl or Teflon
Colour	International orange for vinyl, beige for Teflon
Sizes	Standard sizes are available to fit virtually all valves and ASA flanges. Specific sizes can be supplied on special order. Pump and expansion-joint covers are also available on special order.

Operational and Transportability

These covers are resistant to acids, caustics, and most other industrial chemicals and are sun-, rain-, and mostly fume-proof. They have a contoured shape to provide proper fit and can be attached or removed quickly. Moderate-temperature vinyl covers can be used at up to 104°C (220°F). High-temperature Teflon covers can be used at up to 260°C (500°F).

These covers are person-portable, compact, and lightweight.

PERFORMANCE

The manufacturer will provide a list of reagents and their effect on the covers, including the reaction, the colour change, and the reaction time. The manufacturer will also provide samples of the material for customer testing and evaluation.

Manufacturer

Milsheff Inc.
2472 East Main Street
Bridgeport, CT 06610
U.S.A.

Telephone (203) 366-8834
Facsimile (203) 367-8073

Canadian Distributor

Industrial Plastics Canada
P.O. Box 93
Fort Erie, Ontario
L2A 5M6

Telephone (905) 871-0412
Facsimile (905) 871-6494

REFERENCES

(1) Manufacturer's Literature

PARATECH INFLATABLE LEAK SEALING SYSTEMS

No. 12

DESCRIPTION

Chemical-resistant, pneumatically inflatable, reinforced Neoprene bags that completely seal leaks in pipes, storage tanks, tank cars, and tank trucks.

OPERATING PRINCIPLE

The bag or bandage attaches to leaking vessels by means of ratcheting nylon belts and is then filled with air to effectively seal smooth or rough surfaces.

STATUS OF DEVELOPMENT AND USAGE

Commercially available product in general use.

TECHNICAL SPECIFICATIONS

Physical

These systems are sold as kits and are made of chemical-resistant, reinforced Neoprene. The following are detailed specifications for various kits.

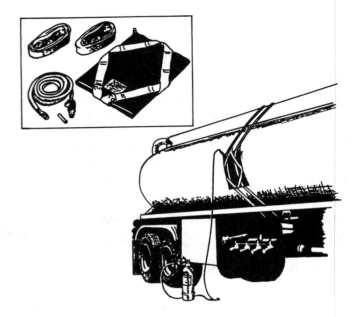

Tank, Pipe, and Drum Sealing Kit, Model 900
1   15 × 15 cm (6 × 6 in) leak-sealing bag
1   15 × 30 cm (6 × 12 in) leak-sealing bag
1   25 × 25 cm (10 × 10 in) leak-sealing bag
1   38 × 53 cm (15 × 21 in) leak-sealing bag
1   60 × 60 cm (24 × 24 in) leak-sealing bag
1   harness for 38 × 53 cm (15 × 21 in) bag
1   harness for 60 × 60 cm (24 × 24 in) bag
2   1.2 m (4 ft) wraparound Velcro strap
3   2.4 m (8 ft) wraparound Velcro strap
1   9 m (30 ft) orange ratchet belt
1   6 m (20 ft) orange extension belt
1   9 m (30 ft) yellow ratchet belt
1   6 m (20 ft) yellow extension belt 2
1   single controller with bypass valve
1   5 m (16 ft) orange hose

Tank Sealing Kit, Model 300
1   38 × 53 cm (15 × 21 in) leak-sealing bag
1   60 × 60 cm (24 × 24 in) leak-sealing bag
1   harness for 38 × 53 cm (15 × 21 in) bag
1   harness for 60 × 60 cm (24 × 24 in) bag
1   9 m (30 ft) orange ratchet belt
1   6 m (20 ft) orange extension belt
1   9 m (30 ft) yellow ratchet belt
1   6 m (20 ft) yellow extension belt
1   single controller with bypass valve
1   5 m (16 ft) orange hose

Tank Sealing Kit, Model 200
1   60 × 60 cm (24 × 24 in) leak-sealing bag
1   harness for 60 × 60 cm (24 × 24 in) bag
1   9 m (30 ft) orange ratchet belt
1   9 m (30 ft) yellow ratchet belt
1   single controller with bypass valve
1   5 m (16 ft) orange hose

Tank Sealing Kit, Model 100
1   38 × 53 cm (15 × 21 in) leak-sealing bag
1   harness for 38 × 53 cm (15 × 21 in) bag
1   9 m (30 ft) orange ratchet belt
1   9 m (30 ft) yellow ratchet belt
1   single controller with bypass valve
1   5 m (16 ft) orange hose

Pipe and Drum 3-Bag Sealing Kit
1   15 × 15 cm (6 × 6 in) leak-sealing bag
1   15 × 30 cm (6 × 12 in) leak-sealing bag
1   25 × 25 cm (10 × 10 in) leak-sealing bag
2   1.2 m (4 ft) wraparound Velcro strap
3   2.4 m (8 ft) wraparound Velcro strap
1   single controller with bypass valve
1   5 m (16 ft) orange hose

1.1 Containment - Leak Mitigation

Operational

The inflatable leak sealers work effectively on both smooth and rough surfaces. They apply up to 150 kPa (22 psi) pressure over the damaged area to stem flow, seal the leak, and reduce contamination. A special harness design allows the bag to slide into position easily, even after the straps are in place. Additional strapping can be added to secure the bag. A 4.8 m (16 ft) air hose and control valve allow responders to work at a safe distance from the hazard zone.

Power/Fuel Requirements

A compressed air source is required, such as a compressor, compressed air storage cylinders, truck air brakes, or a hand or foot pump.

Transportability

Person-portable

CONTACT INFORMATION

Manufacturer

Paratech Incorporated
1025 Lambert Road
Frankford, IL 60423-7000
U.S.A.

Telephone (815) 469-3911
Facsimile (815) 469-7748

Canadian Distributors

Region	Distributor	Region	Distributor
Newfoundland	Bren-Ker Industrial Supplier Box 1298 Marytown, Newfoundland A0E 2M0 Telephone (709) 279-2238 Facsimile (709) 279-3000	Quebec	C.M.P. Mayer Fire Equipment 1999 Nobel Unit 23 Ste. Julie, Quebec J3E 1Z7 Telephone (514) 922-3691 Facsimile (514) 922-3693
Maritime Provinces	Polaris Fire & Safety 263 Hillside Drive Boutilier's Point, Nova Scotia B0J 1G0 Telephone (902) 826-2185 Facsimile (902) 826-2186	Southern Ontario	Class A Fire Equipment P.O. Box 626 Madoc, Ontario K0K 2K0 Telephone (613) 473-4490 Facsimile (613) 473-5319
Northern Ontario	Superior Safety Equipment P.O. Box 1150 Thunder Bay, Ontario P7C 4X9 Telephone (807) 623-2797 Facsimile (807) 623-4027	Western Provinces	Fleck Brothers Fire & Safety 4084 McConnel Court Burnaby, British Columbia V5A 3N7 Telephone (604) 420-3535 Facsimile (604) 421-8803

OTHER DATA

Paratech supplies a manual foot pump and 370 L (13 ft^3), 1,400 L (50 ft^3), and 2,300 L (80 ft^3) working air cylinders with valves and gauges, for use with all their air bags. Paratech also manufactures tapered hardwood and neoprene plugs used to control leaks of both liquid and vapour.

REFERENCES

(1) Manufacturer's Literature.

PETERSEN INFLATABLE PIPELINE STOPPER PLUGS

No. 13

DESCRIPTION

Multi-sized, pneumatically inflatable bags that provide a complete seal in pipes from 10 to 168 cm (4 to 66 in) in size.

OPERATING PRINCIPLE

The Pipeline Stopper Plugs are inserted into the pipe opening and inflated with air to stop flow in the pipe.

STATUS OF DEVELOPMENT AND USAGE

Commercially available product in general use, which is frequently used to facilitate sewer repairs. Special configurations also developed for unique applications.

TECHNICAL SPECIFICATIONS

Physical

These plugs will seal pipes ranging from 10 to 168 cm (4 to 66 in) in size. The following are detailed specifications for the various models.

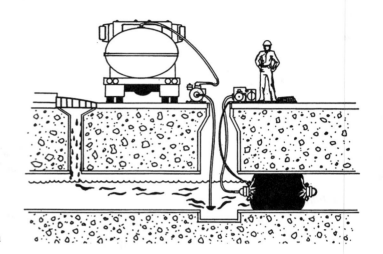

Model	Pipe Size (I.D.)	Deflated Length	Deflated Diameter	Weight
12-3041	10 cm (4 in)	66 cm (26 in)	6 cm (2.5 in)	2 kg (4 lb)
12-3042	10 cm (4 in)	66 cm (26 in)	8 cm (3.2 in)	2 kg (4 lb)
12-3043	10 cm (4 in)	66 cm (26 in)	13 cm (5 in)	2.7 kg (6 lb)
12-3061	15 cm (6 in)	74 cm (29 in)	6 cm (2.5 in)	2.3 kg (5 lb)
12-3062	15 cm (6 in)	74 cm (29 in)	8.9 cm (3.5 in)	2.7 kg (6 lb)
12-3063	15 cm (6 in)	74 cm (29 in)	13 cm (5 in)	4 kg (8 lb)
12-3081	20 cm (8 in)	81 cm (32 in)	7 cm (2.7 in)	2.7 kg (6 lb)
12-3082	20 cm (8 in)	81 cm (32 in)	8.9 cm (3.5 in)	3.2 kg (7 lb)
12-3083	20 cm (8 in)	81 cm (32 in)	13 cm (5 in)	3.6 kg (8 lb)
12-4102	25 cm (10 in)	102 cm (40 in)	11 cm (4.5 in)	4 kg (9 lb)
12-4103	25 cm (10 in)	102 cm (40 in)	13 cm (5 in)	5 kg (11 lb)
12-4122	30 cm (12 in)	122 cm (48 in)	9.4 cm (3.7 in)	6 kg (14 lb)
12-4123	30 cm (12 in)	122 cm (48 in)	13 cm (5 in)	6 kg (14 lb)
12-4153	38 cm (15 in)	152 cm (60 in)	13 cm (5 in)	8 kg (18 lb)
12-4183	46 cm (18 in)	183 cm (72 in)	13 cm (5 in)	14 kg (30 lb)
12-4203	51 cm (20 in)	193 cm (76 in)	13 cm (5 in)	14 kg (31 lb)
12-4223	56 cm (22 in)	203 cm (80 in)	13 cm (5 in)	14.5 kg (32 lb)
12-4243	61 cm (24 in)	213 cm (84 in)	13 cm (5 in)	16 kg (35 lb)
12-4263	66 cm (26 in)	224 cm (88 in)	13 cm (5 in)	17 kg (38 lb)
12-4283	71 cm (28 in)	234 cm (92 in)	13.2 cm (5.2 in)	20 kg (44 lb)
12-4303	76 cm (30 in)	244 cm (96 in)	13.5 cm (5.3 in)	21 kg (47 lb)
12-4333	84 cm (33 in)	259 cm (102 in)	14 cm (5.5 in)	24 kg (52 lb)
12-4363	91 cm (36 in)	274 cm (108 in)	14.5 cm (5.7 in)	25 kg (56 lb)

1.1 Containment - Leak Mitigation

Model	Pipe Size (I.D.)	Deflated Length	Deflated Diameter	Weight
12-4423	107 cm (42 in)	305 cm (120 in)	15 cm (6 in)	30 kg (65 lb)
12-4483	122 cm (48 in)	335 cm (132 in)	16 cm (6.2 in)	32 kg (71 lb)
12-4543	137 cm (54 in)	370 cm (144 in)	18 cm (7 in)	38 kg (83 lb)
12-4603	152 cm (60 in)	396 cm (156 in)	19 cm (7.6 in)	42 kg (92 lb)
12-4663	168 cm (66 in)	427 cm (168 in)	20 cm (8 in)	46 kg (101 lb)

Operational

The pneumatic sealing plugs can rapidly provide a complete seal in pipes sized from 10 to 168 cm (4 to 66 in). Maximum allowable back pressure varies from 42 m (140 ft) head for the 168 cm (66 in) Stopper Plug to 3.7 m (12 ft) head for the 10 cm (4 in) Stopper Plug.

Power/Fuel Requirements

A source of water or compressed air is required, i.e., a compressor, compressed air storage cylinder, or a hand pump.

Transportability

Person-portable

PERFORMANCE

The manufacturer will provide step-by-step documentation and photos showing how a major industrial firm controls sewer emergencies using Petersen Inflatable Pipeline Stopper Plugs, in the event of oil or chemical spills.

CONTACT INFORMATION

Manufacturer

Petersen Products Company
419 Wheeler Avenue
Fredonia, WI 53021
U.S.A.

Telephone (414) 692-2416
Facsimile (414) 692-2418

REFERENCES

(1)   Manufacturer's Literature.

PLIDCO PIPELINE REPAIR FITTINGS

DESCRIPTION

A range of fittings for repairing pipelines carrying oil, gas, petrochemicals, and other products, some of which can be applied while the pipeline remains in operation.

OPERATING PRINCIPLE

Flange+Repair Ring - Halves of the fitting are placed around the leaking flanges, the side bolts are tightened, and a sealant is injected through button-head fittings into the space between the flanges. Patented steel girder rings hold packing in place during installation.

Shear+Plug - The seal-off is accomplished with a metal-to-metal seal and is not restricted to the temperature limitations of elastomers. As no shell cutter is used to enter the pipe, no metal chips fall into the line to cause possible downstream problems. Instead, a shear blade cuts a one-piece section out of the pipe and deposits it into the receptacle housing for easy, complete removal.

Split+Sleeve - Halves of the fitting are placed around a ruptured pipe and securely fastened over the leak. Patented steel girder rings hold packing in place. The sleeves can be seal-welded for a permanent repair if necessary.

Smith+Clamp with Companion Weld+Cap - A repair fitting that locates and stops pit holes. When a permanent repair is specified, the PLIDCO Weld+Cap can be welded in place over the Smith+Clamp while the pipeline is in operation. A pilot pin guides the pointed cone into the pit hole. Pressure is applied by force screw, not by drawing bolts on the opposite side. This forces the cone point into the hole, stopping the leak. Exerting pressure directly on the cone eliminates the danger of a badly damaged pipe caving in.

STATUS OF DEVELOPMENT AND USAGE

Commercially available product in general use.

TECHNICAL SPECIFICATIONS

Physical

Flange+Repair Kit - Stocked in sizes for ANSI flanges 3.8 to 31 cm (1.5 to 12 in), these kits can also be supplied for bonnet flanges, heat exchangers, and other special applications. Sealants, compatible with fluid, temperature, and pressure, can be recommended for individual applications.

Shear+Plug - The standard Shear+Plug is available for pipes from 3.8 to 31 cm (1.5 to 12 in) in size. Plugs for larger sizes and higher pressures are available on request.

Split+Sleeve - Stocked in sizes 3.8 to 122 cm (1.5 to 48 in). Other sizes, lengths, vents, and hinges are available. Approximate weights range from 13 kg (29 lb) for the 3.8 cm (1.5 in) fitting to 1,000 kg (2,230 lb) for the 122 cm (48 in) fitting. Buna-N packing is standard. Other packings are available for individual applications.

Smith+Clamp with Companion Weld+Cap - These are stocked in sizes 3.8 to 122 cm (1.5 to 48 in). Other sizes available by special order. Buna-N packing is standard. Other packings available on application. Weld+Cap fittings are available from stock in 10 to 122 cm (4 to 48 in) pipe sizes, contoured and beveled for welding.

Operational

Flange+Repair Ring - Working pressure depends on the type of sealant, which in turn depends on the type of material that is leaking.

Shear+Plug - The standard PLIDCO Shear+Plug is rated at 10,200 kPa (1,480 psi) pressure and is designed using applicable codes. Higher pressure and temperature ratings [up to 538°C (1,000 °F)] are available using special alloy steel.

Split+Sleeve - Standard working pressure is 6,900 kPa (1,000 psi). Higher working pressures are available.

Smith+Clamp with Companion Weld+Cap - The PLIDCO Weld+Cap is designed for working pressures up to 13,790 kPa (2,000 psi), and meets Transport Canada requirements as well as the latest U.S. specifications for unfired pressure vessels.

Power/Fuel Requirements

None

Transportability

Smaller fittings are person-portable [an 18-cm fitting weighs 45 kg (100 lb)], while larger fittings require lifting equipment and a vehicle for transportation.

PERFORMANCE

PLIDCO Leak Repair Fittings are widely used for making on-stream repairs to pipelines carrying oil, gas, petrochemicals, and other products. For example, when a backhoe damaged a natural gas pipeline, closing a major highway and causing evacuation of the area, a 25 x 76 cm (10 × 30 in) PLIDCO Split+Sleeve was used to repair the pipeline rupture safely and quickly without shutdown.

CONTACT INFORMATION

Manufacturer

PLIDCO The Pipe Line Development Company
870 Canterbury Road
Cleveland, OH 44145
U.S.A.

Telephone (216) 871-5700
Facsimile (216) 871-9577

REFERENCES

(1) Manufacturer's Literature.

PLUG N'DIKE PRODUCTS AND PLUG RUGS

No. 15

DESCRIPTION

Plug N'Dike is a nontoxic, granular material that mixes with water and forms a seal to control flammable and hazardous leaks and spills. It is a combination of a high water absorption, cornstarch copolymer and a bentonite base. Plug N'Dike products are a first response method to seal punctures and control spills.

OPERATING PRINCIPLE

The combination of a high water absorption polymer in a bentonite base absorbs over 300 times its own weight in water and produces a sticky paste that adheres to dirty, greasy, or rusty surfaces to plug leaks. It can also be used to build dikes to divert or contain spills, or to plug storm sewers and drains.

STATUS OF DEVELOPMENT AND USAGE

Commercially available product in general use.

TECHNICAL SPECIFICATIONS

Physical

Plug N'Dike - Dry granular is available in one 19-L (5-gal) pail or four 3.8-L (1 gal) bottles. **Premixed paste** is available in fifteen 0.6-kg (1.3 lb) containers per carton and three 3.8-L (1-gal) pails. Disposable gloves are provided in all kits. The density of Plug N'Dike is approximately 1,040 kg/m^3 (65 lb/ft^3).

Leak Lock Kits - A ready-to-use kit containing 0.6 kg (1.3 lb) of premixed Plug N'Dike, epoxy/hardener, gloves, sorbent, packaged in a molded plastic case, 4 kits per carton, or as a cover patch kit only which includes epoxy/hardener, gloves and sorbent, packaged in a ziplock bag, 4 kits per carton.

Plug N'Patch - A kit containing one 7.6-L (2-gal) plastic pail of granular Plug N'Dike, epoxy/hardener, sorbent and abrasive pads, and disposable gloves, packaged in a 23-L (6-gal) pail.

Plug Rugs - A fabric-reinforced Plug N'Dike pad used to seal leaks and prevent spills from entering drains and storm sewers. Available in three sizes: 20 × 20 cm (8 × 8 in), twelve mats per carton, 40 × 40 cm (16 × 16 in), four per carton, and 40 × 60 cm (16 × 24 in), four per carton.

Operational

Plug N' Dike plugs leaks with up to 2.1 m (7 ft) of hydrostatic head. It can plug leaks on dirty, greasy, or rusty surfaces while material is flowing over it. This product cannot be used, however, with strong acids or with some water-soluble chemicals. Unless an epoxy patch (available separately) is used, Plug N' Dike will eventually dry and crack, with dryout time depending on weather conditions. Plug N' Dike has a one-year shelf life.

Power/Fuel Requirements

Water is required to mix the granular form of Plug N' Dike.

Transportability

Person-portable

PERFORMANCE

State Highway, New York - A tractor-trailer overturned resulting in a 25 × 10 cm (10 x 4 in) tear along a seam. Leaking gasoline was stopped using a 60 cm (24 in) Plug Rug (see Other Data). Environmental damage and cleanup costs were minimized and 28,770 L (7,600 gal) of gasoline were salvaged.

1.1 Containment - Leak Mitigation

Fire Department, Ohio - A vehicle's gas tank sustained a 5 × 1.3 cm (2 x 0.5 in) puncture in an accident. Even though the tank was dirty and greasy, Plug N' Dike stopped the leak immediately.

Fire Department - A truck carrying drums of chemicals jumped an embankment, spilling chemicals into a creek. Plug N' Dike was used to dike downstream and along the sides of the creek to contain the contaminated water.

CONTACT INFORMATION

Manufacturer

PND Corporation
14320 N.E. 21st Street
Suite 6
Bellevue, WA 98007
U.S.A.

Telephone (206) 562-7252
Facsimile (206) 562-7254

Canadian Distributor

Can-Ross Environmental Services Ltd.
2270 South Service Road West
Oakville, Ontario
L6L 5M9

Telephone (905) 847-7190
Facsimile (905) 847-7175

REFERENCES

(1) Manufacturer's Literature.

VETTER SYSTEMS
LEAK SEALING & PIPE SEALING SYSTEMS

DESCRIPTION

Leak Sealing Bags and Leak Bandages are pneumatically inflatable bags that provide complete seals to stop leaks in pipes, storage tanks, tank cars, and tank trucks. **Leak Drainage Bags** similarly stop leaks while allowing the responder to control the outflow and empty the damaged tank.

Pipe Sealing and Flo-Thru Plugs are pneumatically inflatable bags that provide either a complete seal or controlled bypass pumping in pipes ranging from 19 mm to 2.4 m (0.8 to 8 ft) in size.

OPERATING PRINCIPLE

Leak Sealing Bags, Leak Bandages, and Leak Drainage Bags attach to leaking vessels with ratcheting nylon belts. The bandage or bag is then filled with air to create an effective seal on round or "squarish" tanks, smooth or rough surfaces, and steel or plastic.

Blind Sealing Plugs or Flo-Thru Plugs are inserted into the pipe opening and inflated with air to stop flow from the pipe. Blind Sealing Plugs will completely seal the pipe, while Flo-Thru Plugs allow for bypass pumping or flow.

STATUS OF DEVELOPMENT AND USAGE

Commercially available products in general use. Blind Sealing Plugs and Flo-Thru Plugs are frequently used to facilitate sewer repairs.

TECHNICAL SPECIFICATIONS

Physical

The **Leak Sealing Bags, Leak Drainage Bags**, and **Leak Bandages** are made of tough multi-layer material with reinforced nylon fabric, and additional steel reinforcement on the edges. The following are detailed specifications for the various sealing bags.

Model	Size	Vessel Size (diameter)	Weight
Leak Sealing Bag			
LD 50/30 BS	69 × 30 cm (27 × 12 in)	≥ 48 cm (19 in)	7 kg (15 lb)
LD 50/30 S BS	61 × 29 cm (24 × 11.8 in)	≥ 48 cm (19 in)	4 kg (9 lb)
Leak Drainage Bag			
DLD 50/30 BS	62 × 31 cm (24.4 × 12 in)	≥ 48 cm (19 in)	8 kg (17 lb)
Leak Sealing Bandage			
LB 5-20 BS	99 × 21 cm (39 × 8.3 in)	5 to 20 cm (2 to 7.9 in)	2 kg (5 lb)
LB 20-48 BS	1.8 m × 21 cm (5.8 ft × 8.3 in)	20 to 48 cm (7.9 to 19 in)	3.6 kg (8 lb)

Multi-size Blind Sealing Plugs seal pipes ranging from 25 mm to 1.2 m (1 in to 4 ft) in size. Most of these plugs have tiedowns, hose connectors, and a 9.5 mm (3/8 in) check coupler that allows quick inflation and deflation with a large diameter air hose. Blind Sealing Plugs operate at 240 kPa (35 psi) and achieve approximately 70 kPa (10 psi) back pressure. The BS 20/48 model operates at 150 kPa (22 psi). The following are detailed specifications for the multi-size Blind Plugs.

1.1 Containment - Leak Mitigation

Model	Pipe Size	Dimensions	Weight
BS 1/1.5	25 to 38 mm (1 to 1.5 in)	2.5 × 12 cm (1 × 4.8 in)	0.5 kg (1 lb)
BS 1.5/3	38 to 76 mm (1.5 to 3 in)	3.8 × 20 cm (1.5 × 8 in)	0.5 kg (1 lb)
BS 3/5	76 to 127 mm (3 to 5 in)	7.6 × 13 cm (3 × 5 in)	0.5 kg (1 lb)
BS 4/8	102 to 203 mm (4 to 8 in)	8.4 × 53 cm (3.3 × 21 in)	1 kg (2 lb)
BS 6/12	152 to 305 mm (6 to 12 in)	15 × 53 cm (6 × 21 in)	2 kg (4 lb)
BS 8/16	203 to 406 mm (8 to 16 in)	19 × 55 cm (7.5 × 21.6 in)	4 kg (9 lb)
BS 12/20	305 to 508 mm (12 to 20 in)	30 × 56 cm (12 × 22 in)	9 kg (19 lb)
BS 12/24	305 to 610 mm (12 to 24 in)	30 × 79 cm (12 × 31 in)	6 kg (14 lb)
BS 20/32	508 to 813 mm (20 to 32 in)	46 cm × 1 m (18 in × 3.5 ft)	16 kg (35 lb)
BS 20/48	508 mm to 1.2 m (20 to 48 in)	48 × 97 cm (19 × 38 in)	37 kg (82 lb)

Pipes from 1 m to 2.4 m (3.5 to 8 ft) can be sealed with the **single-size Blind Sealing Plugs**. The large diameter, single-size sealing plugs operate at 50 kPa (7 psi) and can be inflated with a large-diameter hose connected to an air compressor or compressed air storage cylinders. The following are specifications for the single-size plugs.

Model	Pipe Size	Dimensions
BS 42	1 m (3.3 ft)	1 × 1.3 m (3.5 × 4.3 ft)
BS 48	1.2 m (4 ft)	1.2 × 1.3 m (4 × 4.3 ft)
BS 54	1.4 m (4.5 ft)	1.4 × 1.7 m (4.5 × 5.7 ft)
BS 60	1.5 m (5 ft)	1.5 × 2.0 m (5 × 6.6 ft)
BS 72	1.8 m (6 ft)	1.8 × 2.4 m (6 × 7.9 ft)
BS 96	2.4 m (8 ft)	2.4 × 3.3 m (8 × 11 ft)

The **Flo-Thru Plugs** are all multi-sized, with four plug sizes to seal pipes from 76 mm to 1.2 m (3 in to 4 ft) in diameter. The following are detailed specifications for the multi-size Flow-Thru Plugs.

Model	Pipe Size	Dimensions	Weight
FT 3/6	76 to 152 mm (3 to 6 in)	7.1 × 30 cm (2.8 × 12 in)	0.6 kg (1.4 lb)
FT 4/8 *	102 to 203 mm (4 to 8 in)	9.9 × 69 cm (3.9 × 27 in)	7 kg (15 lb)
FT 6/12	152 to 305 mm (6 to 12 in)	15 × 66 cm (5.9 × 26 in)	5 kg (11 lb)
FT 8/20 *	203 to 508 mm (8 to 20 in)	19 × 61 cm (7.5 × 24 in)	14 kg (30 lb)
FT 12/24	305 to 610 mm (12 to 24 in)	30 × 91 cm (12 × 36 in)	11 kg (24 lb)
FT 20/30 *	508 to 813 mm (20 to 36 in)	46 × 71 cm (18 × 28 in)	32 kg (70 lb)
FT 20/48 *	508 mm to 1.2 m (20 in to 4 ft)	46 × 94 mm (18 × 37 in)	43 kg (94 lb)

Vetter Systems also custom manufactures Sealing and Flo-Thru plugs for oval pipes and pipe sizes not listed above.

Operational

The **Leak Sealing Bags, Leak Drainage Bags**, and **Leak Bandages** are resistant to most acids and chemical solutions. An optional acid protection cover gives the bag extra resistance to highly aggressive media. These products can be used in temperatures of up to 100 °C (212 °F) for brief periods of time, or at 60 °C (140 °F) for extended periods of time. They operate at inflation pressures of 150 kPa (22 psi) and can safely hold back pressures of 140 kPa (20 psi). They can rapidly provide a complete seal or allow controlled off-loading.

The **pneumatic sealing plugs** can rapidly provide a complete seal or allow controlled bypass pumping or flow in pipes from 19 mm to 2.4 m (0.8 in to 8 ft) in size.

Power/Fuel Requirements

A source of compressed air is required, i.e., a compressor, compressed air storage cylinders, truck air brakes, or a hand or foot pump.

Transportability

Person-portable

PERFORMANCE

A Vetter Sealing Bag was successfully used to stop liquid chlorine gas leaking from a rail car in the derailment in Mississauga, Ontario in 1979. After five days of attempting to plug the leak, a Vetter Leak Sealing bag was secured on the rail car with steel chains and timber beams. Upon inflation, it stopped the escape of the chlorine gas.

CONTACT INFORMATION

Manufacturer

Vetter Systems, Inc.
1005 International Drive
Oakdale, PA 15071-9223
U.S.A.

Telephone (412) 695-3100
Facsimile (412) 695-3232

OTHER DATA

Vetter Systems also produces the **Gully Sealing Kit** - a complete, self-contained kit consisting of a pneumatically inflatable bag that provides a complete seal in pipes from 305 to 508 mm (12 to 20 in) in size. The collapsible bag is placed inside the pipe, secured, and inflated using its own CO_2 supply, regulator, 3.6 m (5 ft) tethering line, and 3.6 m (5 ft) inflation hose.

REFERENCES

(1)   Manufacturer's Literature.

Section 1.2

Containment On Land

1.2 Containment - On Land

ATL PORT-A-BERM   No. 17

DESCRIPTION

An inflatable containment berm designed to contain leaking tanks, drums, pails, and other storage vessels, to capture discharges from decontamination operations, or to act as secondary containment for bulk transfers or for isolating hazardous wastes.

OPERATING PRINCIPLE

Four separate air chamber tubes form the inflatable berm or dike. These chambers are attached to the liner using liner-retention loops and then filled with air to provide a quickly deployed containment system.

STATUS OF DEVELOPMENT AND USAGE

Commercially available product in general use.

TECHNICAL SPECIFICATIONS

Physical

Custom sizes are available in heights up to 1.3 m (4.2 ft) and lengths up to 30 m (100 ft) to contain a 1,135,620 L (300,000 gal) spill. The following are detailed specifications for the ten standard sizes.

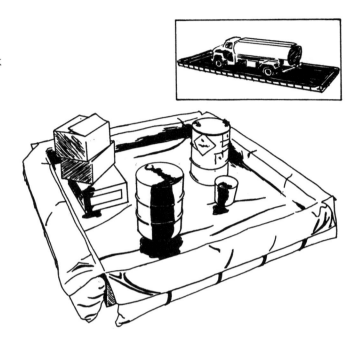

Size	Draft	Maximum Capacity	Required Air Volume	Liner	Empty Weight
3 × 3 m (10 × 10 ft)	43 cm (17 in)	2,800 L (750 gal)	2.0 m^3 (70 ft^3)	single	45 kg (100 lb)
3.9 × 3.9 m (13 × 13 ft)	43 cm (17 in)	6,800 L (1,800 gal)	2.3 m^3 (80 ft^3)	single	70 kg (150 lb)
5.5 × 5.5 m (18 × 18 ft)	43 cm (17 in)	12,000 L (3,400 gal)	3.7 m^3 (130 ft^3)	single	90 kg (200 lb)
7.3 × 7.3 m (24 × 24 ft)	43 cm (17 in)	23,000 L (6,100 gal)	4.8 m^3 (170 ft^3)	single	125 kg (280 lb)
9.8 × 9.8 m (32 × 32 ft)	43 cm (17 in)	40,800 L (10,800 gal)	6.5 m^3 (230 ft^3)	single	200 kg (440 lb)
9.8 × 9.8 m (32 × 32 ft)	86 cm (34 in)	81,700 L (21,600 gal)	26.0 m^3 (920 ft^3)	single	265 kg (580 lb)
22.5 × 10.4 m (74 × 34 ft)	86 cm (34 in)	202,100 L (53,400 gal)	43 m^3 (1,500 ft^3)	single	535 kg (1,180 lb)
13.7 × 4.9 m (45 × 16 ft)	43 cm (17 in)	28,700 L (7,600 gal)	6.2 m^3 (220 ft^3)	double	260 kg (570 lb)
19.8 × 4.9 m (65 × 16 ft)	43 cm (17 in)	41,600 L (11,000 gal)	8.2 m^3 (290 ft^3)	double	360 kg (800 lb)
15.2 × 6.7 m (50 × 22 ft)	43 cm (17 in)	43,900 L (11,600 gal)	7.4 m^3 (260 ft^3)	double	345 kg (760 lb)

Operational

Port-A-Berms can be set up on asphalt, concrete, sand, or soil if the surface is well groomed and level, and are fully effective from -18 to 60 °C (0 to 140 °F). To minimize fabric stresses, however, they should be packed and unpacked at 5 to 50 °C (40 to 120°F). In high wind environments, one third of each tube should be filled with water and the balance with air or nitrogen. Each tube is equipped with a 17 kPa (2.5 psi) relief valve to prevent over-pressurization. If sharp or abrasive equipment is to be put in the berm, a protective liner (optional) or panels of plywood should be laid out inside. The following products have been tested by the manufacturer and found to be chemically compatible with the Port-A-Berm.

A = little or no effect   *B = minor to moderate effect*   *C = severe effect*

Acetic acid (5%)	B	Calcium hydroxide	A
Ammonium phosphate	A	Chlorine solution (20%)	A
Animal oils	A	Ammonium hydroxide (concentrated)	A
Aqua regia	C	Corn oil	A
ASTM Fuel A and B	A	Crude oil	A
Benzene	B	Diesel fuel	A
Calcium chloride solution	A	Ethyl acetate	C

Ethanol	A	Phenol formaldehyde	B
Furfural	C	Phosphoric acid (50%)	A
Gasoline	B	Phthalate plasticizer	C
Glycerine	A	Potassium chloride	A
Hydraulic fluid (mineral)	A	Potassium sulphate	A
Hydrochloric acid (50%)	A	Salt water (15%)	A
Hydrofluoric acid (50%)	A	Sea water	A
Hydrofluorosilicic acid (30%)	A	Sodium acetate solution	A
Isopropyl alcohol	A	Sodium bisulphite solution	A
JP-4 Jet fuel	A	Sodium hydroxide (60%)	A
Kerosene	A	Sodium phosphate	A
Linseed oil	A	Sulphuric acid (50%)	A
Magnesium chloride	A	Tanic acid (50%)	A
Magnesium hydroxide	A	Toluene	B
Methyl ethyl ketone	C	Transformer oil	A
Mineral spirits	A	Turpentine	A
Motor oil	A	Urea formaldehyde	A
Naphtha	A	Vegetable oil	A
Nitric acid (5%)	B	Water (93°C, 200°F)	A
Nitric acid (50%)	C	Xylene	B
Perchloroethylene	C	Zinc chloride	A
Phenol	C		

Spills collected within a Port-A-Berm should be promptly neutralized or transferred to permanent containers. The Port-A-Berm is a temporary holding medium to afford hours or a few days of containment. Chemical resistance data are based on an exposure limit of seven days duration. Material samples are available to customers for immersion testing.

Power/Fuel Requirements

A high-volume, low-pressure, non-sparking air source is required. As an option, the manufacturer supplies a 110 VAC powered blower that meets these requirements.

Transportability

Port-A-Berms collapse into a compact bundle that is one percent of their full volume. Their weight will determine the applicable transport mode (see Technical Specifications - Physical).

PERFORMANCE

Port-A-Berms are used by the U.S. Army, Navy, Marine Corps, and thousands of industrial plants throughout the world.

CONTACT INFORMATION

Manufacturer

Aero Tec Laboratories Incorporated
Spear Road Industrial Park
Ramsey, NJ 07446
U.S.A.

Telephone (201) 825-1400
Facsimile (201) 825-1962

Canadian Representative

Patlon Industries Ltd.
5502 Timberlea Boulevard
Mississauga, Ontario
L4W 2T7

Telephone (905) 624-5572
Facsimile (905) 624-0975

OTHER DATA

A similar system is marketed in the United States by the Eldred Corporation, P.O. Box 3611, Rock Island, IL 61204-3611, U.S.A., Telephone (309) 788-6944, Facsimile (309) 788-7925.

REFERENCES

(1) Manufacturer's Literature.

BENTONITE SOIL-SEALING SYSTEMS — No. 18

DESCRIPTION

Bentonite is a natural clay mineral with a molecular structure that permits water absorption of up to 15 times its dry weight. It can be used to construct cutoff walls for preventing lateral migration of pollutants in soil surrounding effluent storage lagoons, landfills, and tailings storage ponds.

OPERATING PRINCIPLE

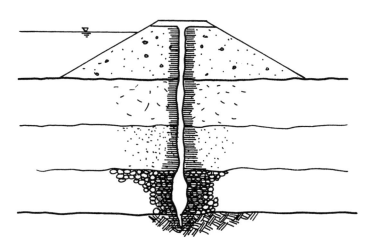

Bentonite's water-absorbing quality makes it useful in preventing the migration of fluids. Under confined conditions, as in a cutoff wall, the wetted, swelled bentonite particles are forced against each other and into voids between soil particles, forming a barrier to the passage of many fluids. Cutoff walls are constructed using either a slurry trench technique or the vibrating beam technique. For the slurry technique, a bentonite slurry is added to a narrow trench while the trench is being excavated, and acts to support the trench until it is filled with a suitable material. This creates a homogeneous, low-permeability wall that restricts the flow of water and hazardous materials through the soil. The vibrating beam method uses pressure injection to construct a cutoff wall, eliminating the need to excavate. Bentonite liners and covers can be constructed with the slurry technique, with pure bentonite, or with a soil-bentonite mixture.

STATUS OF DEVELOPMENT AND USAGE

Commercially available product that requires significant time for implementation and is therefore suitable only for long-term projects. More than 1,000 slurry trench cutoffs have been constructed in North America in the last 20 years.

TECHNICAL SPECIFICATIONS

Physical

Specially formulated bentonite products are available in powder (with 75 to 85% passing 200 mesh) and granular form, a fine, free-flowing material with an average sieve analysis of 65 to 85% passing between 40 and 160 mesh, and with less than 10% finer than 200 mesh. The mineralogical properties must be known in order to understand how bentonite works as a fluid barrier. The following are the mineralogical properties for a typical Canadian bentonite.

Composition (%)

Montmorillonite	79
Illite	9.5
Quartz	5
Plagiaoclase	3
Gypsum	2
Carbonate	1.5

Specific Surface (m²/g)

Specific Surface	631

Adsorbed Cations (% of charge)

Sodium (Na^+)	49.6
Potassium (K^+)	0.7
Calcium (Ca^{2+})	42.2
Magnesium (Mg^{2+})	7.5

Cation Exchange Capacity

CEC (meq/100 g)	82

Operational

General - Bentonite lowers soil permeability in direct proportion to the amount added, the type added, and the uniformity of blending when mixed with soils. It is activated by wetting, which is a reversible process. It can be dried, re-swelled, and also restored to its pre-application condition. Due to bentonite's "bound-water" condition, bentonite particles can move without breaking the seal. Therefore, ground movements that would rupture some liners may not affect a bentonite barrier. In extreme cases where ground movement would rupture the seal or a foreign object would penetrate it, bentonite particles will migrate with the seepage to effect a resealing action. Chemical attack, which may come from the adjacent soil, groundwater, or contaminants, is a concern when using bentonite barriers as it usually alters the sealing properties of the bentonite by ion exchange.

Cutoff Walls - Several variations of the slurry wall technique have been developed using different material to construct the barrier. These variations include the soil-bentonite slurry trench cutoff, the cement-bentonite cutoff, the plastic-concrete cutoff, and the concrete diaphragm wall. Each of these cutoff walls have different permeability, strength, and flexibility characteristics. The technique chosen depends on the performance required, the layout of the site, and the physical environment in which the wall is to be placed. The permeability of a well designed soil-bentonite cutoff wall can vary from 10^{-6} to 10^{-8} cm/s. Cold temperatures are not conducive to slurry wall construction, as frozen backfill may cause discontinuities in the wall upon thawing.

Liners - Bentonite liners can be constructed with a slurry, pure bentonite (called a membrane liner), or with a soil-bentonite mixture. The slurry technique is used to seal leaking reservoirs or dugouts. The membrane liner, which consists of pure bentonite, is spread over the sides and base of an impoundment to form a complete membrane. The soil-bentonite liner consists of a homogeneous mixture of soil and bentonite which is used in the same manner as a membrane liner. Liners are usually constructed with a permeability specification of 10^{-7} cm/s or less. With a head of one, a soil liner with this permeability will release 315,000 L/ha/year (32,850 gal/acre/yr).

Power/Fuel Requirements

None

Transportability

Heavy equipment is required to apply a bentonite soil-sealing system. The specific requirements vary with site-specific design criteria.

PERFORMANCE

A bentonite soil-sealing system must be properly designed and constructed to be effective. Factors and processes to be considered in the design include soil and slurry composition factors, environmental factors, construction factors, and chemical and physical processes. The effectiveness of bentonite in retarding the flow of many chemicals is still not fully understood. Recent measurements and case histories indicate that the hydraulic conductivity of field installations is often significantly greater than predicted by laboratory design measurements. Possible causes of failure include liners that are too thin, improper construction, inadequate construction quality assurance, desiccation and freezing, and chemical attack.

The hydraulic conductivity of the clay is greatly influenced by the chemical composition of the contaminant to be contained. Studies have shown that when compacted clay soils are exposed to concentrated organics, hydraulic conductivity increases by 1 to 4 orders of magnitude. Clays stabilized with lime and cement have a much smaller increase in hydraulic conductivity when exposed to concentrated organics. Further studies are being conducted.

As it appears that some liquid will inevitably flow through clay barriers, research is also being done on retaining the organics within the clay, and thus "filtering" the fluid. These "designer" clays have great potential for use in spill response. Experiments have shown that saturation of the clay with a quaternary ammonium cation can effectively filter organics from the leachate. The long-term stability of these compounds in the soil environment, however, is still not certain. Although these treated clays have been reported to be 10 to 30 times more effective than natural soil in adsorbing volatile organic compounds on a unit-weight basis, no field data are available to document the actual amount of organics these clays retain.

1.2 Containment - On Land

Case Studies

Use of Bentonite in a Nuclear Fuel Waste Disposal Vault - In Canada, nuclear fuel waste is disposed of in a vault, deep underground in plutonic rock. Protective barriers have been designed to isolate the waste and minimize the leaching and transport of radionuclides away from the disposal vault. The barriers include a corrosion-resistant container, buffer and backfill materials, and borehole and shaft seal, all of which supplement the natural barrier provided by the surrounding rock. Atomic Energy of Canada Limited is investigating the potential of using a compacted mixture of bentonite and sand for the buffer material.

Slurry Trench Cutoff Wall for Mercury Containment - In October 1978, high levels of mercury (up to 89 ppm) were discovered in the sediments at the confluence of Thunder Creek and the Moose Jaw River within the city limits of Moose Jaw, Saskatchewan. The contaminated sediments were subsequently removed and placed in a storage area secured by constructing a drainage system and a cutoff wall around the site, and placing a cap liner over the deposit. The drainage system was intended to intercept groundwater flowing towards the river and divert it around the storage site. The cutoff wall was intended to isolate the contaminants from Thunder Creek.

The slurry trench cutoff, which was probably the most important part of the mercury-containment system, had the following functions: to control the rate of seepage of groundwater through the wall to the river; to restrict the flow of water into the bank from the river during periods of high flows along the creek; to act as a barrier to the transportation of contaminated sediment by the groundwater; and to extend far enough back from the creek to stop the river from flowing through the sediments behind and parallel to the cutoff wall during periods of high water level.

The cutoff wall was constructed with a minimum width of 0.6 m (2 ft) over a length of 685 m (2,250 ft), and extended 0.6 m (2 ft) down into the shale bedrock aquiclude to a maximum depth of 5.5 m (18 ft). The slurry wall was backfilled and then capped with a 0.5 m (1.7 ft) thick clay cap, which was to provide fill for any settlement. The containment system took four months to construct and has performed satisfactorily since 1981.

Vibrating Beam Cutoff Wall for PCB Spill Containment - A cutoff wall was constructed using the vibrating beam technique to contain a PCB spill at a plant site in Regina, Saskatchewan. The wall was constructed to a depth of 10 m (33 ft) and was completed in four months. Performance data are available from Ground Engineering Ltd.

CONTACT INFORMATION

Manufacturer

Canadian Clay Products
Environmental Control Division
P.O. Box 70
Wilcox, Saskatchewan
S0G 5E0

Telephone (306) 732-2085
Facsimile (306) 732-2085

REFERENCES

(1) Brown, K.W. and J.C. Thomas, "New Technology for Liners", in: *Proceedings of a Conference on the Prevention and Treatment of Groundwater and Soil Contamination in Petroleum Exploration and Production*, Info-Tech, Calgary, Alberta (May, 1989).
(2) Buettner, W., "Bentonite Seepage Control Barriers", in: *Proceedings of the First Canadian Engineering Technology Seminar on the Use of Bentonite for Civil Engineering Applications*, Ground Engineering Limited, Regina, Saskatchewan (March, 1985).
(3) Dixon, D.A. and M.N. Gray, "The Use of Bentonite in a Nuclear Fuel Waste Disposal Vault", in: *Proceedings of the First Canadian Engineering Technology Seminar on the Use of Bentonite for Civil Engineering Applications*, Ground Engineering Limited, Regina, Saskatchewan (March, 1985).
(4) Manufacturer's Literature.

CHEM-TAINER POLYETHYLENE SPILL CONTAINMENT PALLET

No. 19

DESCRIPTION

Polyethylene pallets for storing chemical drums, designed so that hazardous spills can be safely contained in the pallet basin. Pallets hold either two or four 208-L (55-gal) drums.

OPERATING PRINCIPLE

Two or four drums are supported on an open-grate, molded deck above the sump. Any leaks from the containers will drain down through the deck and be contained in the sump. Suitable for forklift handling.

STATUS OF DEVELOPMENT AND USAGE

Commercially available product in general use.

TECHNICAL SPECIFICATIONS

Physical

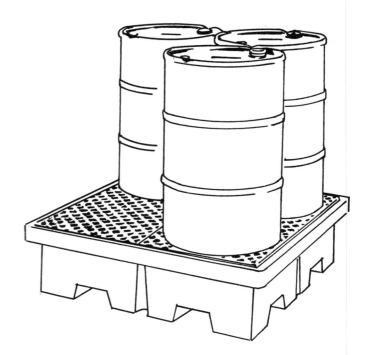

The 2-drum pallet is 1.4 × 0.7 × 0.4 m (4.6 × 2.3 × 1.4 ft) and the 4-drum pallet is 1.4 × 1.4 × 0.3 m (4.5 × 4.5 × 1.2 ft).

Operational

Chem-Tainer pallets can hold most 208-L (55-gal) drums, side by side, with their approximate width of 1.2 m (4 ft). The pallets have a load capacity of 1,135 k (2,500 lb) and are fully nestable when not in use. An outdoor storage cover is available for both sizes of pallets to protect drums and pallet sump from weather.

Power/Fuel Requirements

None

Transportability

Chem-Tainer pallets have four-way forklift access for moving fully loaded pallets. These pallets are designed for in-plant use, not for transport. Empty pallets are person-portable.

CONTACT INFORMATION

Manufacturer

Chem-Tainer Industries Inc.
361 Neptune Ave
West Babylon, NY 11704
U.S.A.

Telephone (516) 661-8300.
Toll-free   (800) 275-2436
Facsimile (516) 661-8209

OTHER DATA

Chem-Tainer also manufactures a range of other spill containment products, including **Double Wall Bulk Storage Tanks** [3785-L (1000-gal) capacity], **Containment Basins** [380 to 1,890-L (100 to 500-gal) capacity], a **Security Spill Containment Vessel,** a Spill Basin for containing spills while dispensing chemicals, a **Single Drum Containment Vessel**, a **Drum Bib** to catch spills and drips from pumps, and a **Drum Funnel** which eliminates overspills while pouring, as well as **Tank Accessories and Fittings**. Chem-Tainer also manufactures a wide range of **Bulk Storage Tanks** and **Open Top Tanks**.

REFERENCES

(1) Manufacturer's Literature.

CLARK SPILSTOPPER No. 20

DESCRIPTION

A specially engineered mat designed to seal off gravity flow of hazardous materials into openings. The mat consists of a polyurethane elastomer core, laminated to abrasion-resistant covers.

OPERATING PRINCIPLE

The flexible Spilstopper mat conforms to drains, grates, or manholes to provide an effective seal on gravel, sand, grass, asphalt, concrete, or pavement.

STATUS OF DEVELOPMENT AND USAGE

Commercially available product in general use.

TECHNICAL SPECIFICATIONS

Physical

Material	Polyurethane rubber and elastomer
Colour	High visibility orange
Weight	Approximately 11 kg/m^2 (2.3 lb/ft^2)
Thickness	13 mm (0.5 in)
Sizes	**Standard**
	0.6 × 0.6 m (2 × 2 ft)
	0.8 × 0.8 m (2.5 × 2.5 ft)
	0.9 × 0.9 m (3 × 3 ft)
	1 × 1 m (3.5 × 3.5 ft)

Custom
Up to 1 m (3.5 ft) wide, in any length

Operational

This cover is impermeable to fluids and its physical characteristics are unchanged between -29 and 82 °C (-20 to 180 °F). It conforms to nearly any shape and can be stretched to almost twice its length and still recover completely. A 13-mm (0.5-in) diameter sphere placed under a mat produces no perceivable distortion 13 mm (0.5 in) from the edge of the sphere. The cover is not damaged by pedestrian traffic and should be stored in the shade when not in use to avoid discolouration. The cover is inert to water, petroleum products, and most caustics. It can be washed with nonabrasive detergents or petroleum solvent cleaners.

According to the manufacturer, certain concentrations of the following products have been found to be chemically compatible with the Spilstopper.

Acetone	Freon
Aluminum salts	Gasoline
Ammonia	Glycerine
Animal oils	Glycol ether
Barium salts	Hydrochloric acid (med. conc.)
Benzyl alcohol	Hydrogen peroxide
Boric acid solution	Kerosene
Butane	Linseed oil
Butanol	Methylene chloride
Calcium	Naphtha
Copper salts	Nitrogen oxides
* Cyclohexanone	Oil, mineral and vegetable
Formaldehyde	Sodium
Sulphuric acid (med. conc.)	Turpentine

* Tetrahydrofuran (THF)	Uric acid
Triethylamine	Water
Tropylene glycol	

* suitable for once-only use; outer material destroyed

Power/Fuel Requirements

None

Transportability

Person-portable @ 11 kg/m^2 (2.3 lb/ft^2).

CONTACT INFORMATION

Manufacturer

Clark Products Company
916 West 25th Street
Norfolk, VA 23517
U.S.A.

Telephone (804) 625-5917
Facsimile   (804) 625-7651

OTHER DATA

Manufacturer provides a chemical compatibility list as a guide only and suggests pre-use trials.

Clark Products also manufacture **Spildike**, a portable, nonabsorbent barrier for containing indoor spills. The flexible polyurethane barrier comes in 3-m (10-ft) sections which can be joined to form longer sections. Spildike is inert to water, petroleum, and many caustics and can be cleaned with soap and water.

REFERENCES

(1) Manufacturer's Literature.

CONTAIN-IT PLUS
POLYETHYLENE SPILL CONTAINMENT PALLET

No. 21

DESCRIPTION

Portable polyethylene pallet with a 318-L (85-gal) sump that holds 150% of a standard drum's capacity.

OPERATING PRINCIPLE

Designed to hold four drums, either with or without a pallet. Internal structural inserts support the load above the sump, and any leaks drain down and are contained in the sump. The extra wide 1.3 m (4.3 ft) design provides more than 5 cm (2 in) of space between a standard 1.2 m (4 ft) pallet and the inside wall of the tray, thus providing an extra margin of safety.

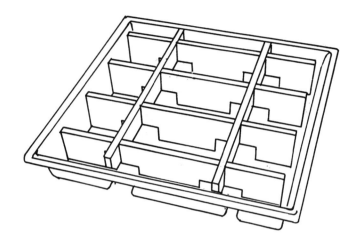

STATUS OF DEVELOPMENT AND USAGE

Commercially available product in general use.

TECHNICAL SPECIFICATIONS

Physical

Size - 1.4 × 1.4 × 0.3 m (4.5 × 4.5 × 1 ft)
 - Holds up to four 208-L (55-gal) drums
Capacity - 318 L (85 gal); 150% containment of a 208-L (55-gal) drum
Material - 100% high density polyethylene with UV barrier

Operational

Holds a static weight of 2,700 kg (6,000 lb) and accepts most 208-L (55-gal) drums, side by side, as well as drums on wooden pallets. An internal structure supports the weight, not the tray walls. The open design facilitates inspection and cleanup of the pallet. The pallets are fully nestable when not in use, which saves shipping costs and storage space. All-season poly covers with hold-down clips are available to protect the pallet from weather and heavy-duty weather covers with five plys of material (three plys of polysheet laminated around two plys of nylon mesh) are also available.

Transportability

Contain-It Plus pallets have two-way forklift access for moving fully-loaded pallets. Empty pallets are person-portable.

PERFORMANCE

In 1992, Chevron USA tested platform containment systems for use in its new HazMat facility in El Segundo, California. Eight 280-kg (610-lb) drums were placed on two wooden pallets [a total weight of 2,270 kg (5,000 lb)], double stacked, and moved under full load four times a day. The Contain-It Plus was the only system that passed the test. Chevron subsequently placed a substantial order of the pallets. Potential customers can contact the manufacturer for information on the resistance of the polyethylene material to a specific chemical.

CONTACT INFORMATION

Manufacturer

Containment Corporation
10889 Portal Drive
Los Alamitos, CA 90720
U.S.A.

Telephone (714) 821-6570
Toll-free   (800) 235-7421
Facsimile   (714) 821-9949

REFERENCES

(1) Manufacturer's Literature.

FROTH-PAK PORTABLE FOAM SYSTEM

No. 22

DESCRIPTION

A self-contained, portable two-component polyurethane foam-dispensing system that can be used to contain spills of hazardous chemicals on dry ground surfaces. The components of the foam system are contained in two cylinders in a common carton, with applicator hoses and a nozzle head.

OPERATING PRINCIPLE

Each Froth-Pak kit is factory-pressurized with nitrogen. When the trigger is activated, the two components are released under pressure and mixed at the nozzle to produce a polyurethane foam which is applied to the spill area to form a berm and contain the spill.

STATUS OF DEVELOPMENT AND USAGE

Commercial product designed and marketed primarily for construction/insulation applications.

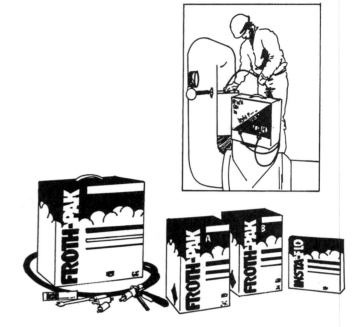

TECHNICAL SPECIFICATIONS

Physical

Froth-Pak kits are available in a variety of sizes and densities in convenient, factory pressurized dispensers. Kits are supplied with gun-hose assembly and standard spray nozzles. Various foam formulations (slow rise, low density, high compression, and flame spread rated) are also available. The foams are two-component polyurethane foams. Component "A" is a Polymeric Isocyanate; component "B" is a Polyether Polyol containing additives, catalyst, and nitrogen as the blowing agent. Froth-Pak foam systems contain **no** fluorocarbons. The density of the foam in most kits is 28 kg/m^3 (1.8 lb/ft^3), although other densities are available. All kits include complete information on operation, troubleshooting, yield, application tips, quality control, curing, storage, disposal, and first aid. The following are specifications for the standard formulation foam kits.

Kit Model No.	Yield	Kit Contains	Ordering Information
Froth-Pak 12	0.03 m^3 (1 ft^3)	One "A" can, one "B" can, two spray nozzles with 0.6 m (2 ft) of hose and one pinch clamp per box	12 units/case 16 kg (36 lb)/case 24 cases/pallet
Froth-Pak 110	0.23 m^3 (8 ft^3)	One "A" can, one "B" can, five spray nozzles, 1.5 m (5 ft) gun-hose assembly, wrench, and Vaseline packet	12 kg (28 lb)/unit 30 units/pallet
Froth-Pak 180	0.47 m^3 (17 ft^3)	One "A" can, one "B" can, ten spray nozzles, 2.3 m (7.5 ft) gun-hose assembly, wrench, and Vaseline packet	17 kg (38 lb)/unit 30 units/pallet
Froth-Pak 600	1.4 m^3 (50 ft^3)	Same as above, but with 3 m (15 ft) gun-hose assembly	50 kg (113 lb)/set 2 cartons/set 12 sets/pallet

1.2 Containment - On Land

Operational

Optimally applied at 24°C (75°F). Minimum application temperature is 2°C (35°F). Expands and solidifies in seconds, if the temperature is high enough to activate the catalyst and initiate foam reaction. Not suitable for blocking a flowing stream or containing spills on wet or vegetated surfaces. Once formed, polyurethane foams are reasonably inert to most other chemicals. One exception is methanol, which does not react, but tends to be absorbed by the foam.

Power/Fuel Requirements

None

Transportability

Completely self-contained and person-portable.

PERFORMANCE

The use of polyurethane foams as physical barriers to hazardous chemical spills was studied by the U.S. EPA in 1973. On dry, firm surfaces, such as concrete and asphalt, at temperatures above -1 to 2°C (30 to 35°F), tests showed that 1.2-m (4-ft) high barriers can easily be constructed that can contain at least a 0.9 m (3 ft) head of water for a period of days. Cold and/or wet surfaces (water, solvents, etc.) will impair the quality of the initial deposit of foam, resulting in poor bonding or adhesion of the foam to the surface. Vegetated surfaces prevent the foam from penetrating and impede the formation of a good seal. Efforts to block or divert already flowing streams were unsuccessful.

CONTACT INFORMATION

Manufacturer

Insta-Foam Products, Inc.
1500 Cedarwood Drive
Joliet, IL 60435-3187
U.S.A.

Telephone (815) 741-6800
Toll-free (800) 800-3626
Facsimile (815) 741-6822

Canadian Representative

Insta-Foam Products, Inc.
P.O. Box 21
Etobicoke, Ontario
M9C 4V2

Telephone (416) 622-6844/45
Facsimile (416) 626-2818

Local distributors retail the company's products throughout Canada and the United States.

REFERENCES

(1) Manufacturer's Literature.
(2) Friel, J.V., *et al.*, "Control of Hazardous Chemical Spills by Physical Barriers", U.S. EPA, EPA-R2-73-185 (PB-221-493) (March, 1973).

HI TECH BERM

No. 23

DESCRIPTION

A modular wall system made of medium-density polyethylene with interlocking swivel connectors, so various shapes of dike systems can be built. When lined, it holds toxic, acidic, or basic fluids or solids.

OPERATING PRINCIPLE

Interlocking panels are connected to produce an instant berm or temporary reservoir.

STATUS OF DEVELOPMENT AND USAGE

Commercially available product in general use.

TECHNICAL SPECIFICATIONS

Physical

Panels are of medium-density polyethylene, 1.2 m (4 ft) high, 1.2 cm (4 ft) wide, and 10 cm (4 in) thick.

Operational

Easily assembled. Straight or curved panels and two-way or four-way connectors are available. Connections must be caulked if used without a liner. Interlocking panels can be stacked. Liners can be incorporated into the berm and secured with a liner cap. For additional stability and strength, panels can be filled with water, sand, or other material, or the berm's base can be imbedded into the soil.

Power/Fuel Requirements

None

Transportability

The Hi Tech Berm is easy to transport as it is modular and lightweight.

PERFORMANCE

High Tech Berms have been used in a wide variety of applications including oil and chemical spills. High Tech Berms Inc. has entered into a contract with the University of Saskatchewan, Civil Engineering Department, to evaluate the following properties of the Hi Tech Berm: tensile strength, load deflection relationships, and shear strength of the panel connections; possible fatigue effects due to repeated loads; stiffness modules of the panels; ability of bracing plates to increase strength of panel; required depth of imbedment and methods of confining base for maximum strength.

CONTACT INFORMATION

Manufacturer

Hi Tech Berms (1993) Inc.
#8, 3111 Millar Avenue
Saskatoon, Saskatchewan
S7K 6N3

Telephone (306) 934-4549
Facsimile (306) 244-1715

Canadian Representative

Wes-Can Erosion Control
#8, 3111 Miller Avenue
Saskatoon, Saskatchewan
S7K 6N3

Toll-free (800) 268-5111

REFERENCES

(1) Manufacturer's Literature.

POLY-SPILLPALLET 6000
POLYETHYLENE SPILL CONTAINMENT PALLET

No. 24

DESCRIPTION

A portable polyethylene pallet with built-in sump capable of containing 110% of a standard drum's capacity.

OPERATING PRINCIPLE

Holds up to four drums supported on an open-grate, molded deck above the sump. Any leaks from the containers drain down through the deck and are contained in the sump. Suitable for forklift handling.

STATUS OF DEVELOPMENT AND USAGE

Commercially available product in general use.

TECHNICAL SPECIFICATIONS

Physical

Size	1.2 × 1.2 × 0.4 m (4 × 4 × 1.4 ft) Holds up to four 208-L (55-gal) drums
Capacity	265 L (70 gal); 110% containment of a 208-L (55-gal) drum
Material	100% medium density polyethylene with UV inhibitor
Weight	40 kg (94 lb)

Operational

The Poly-Spillpallet 6000 accepts most 208-L (55-gal) drums, side by side and has a 2,720 kg (6,000 lb) static load capacity. The design allows stacking of two fully loaded pallets and they are also fully nestable when not in use. The sump sidewalls are translucent, allowing visual inspection of fluid accumulation in sump. Options include horizontal drum rack, drain fittings, drum loading ramp, caster/frame assembly, and an outdoor "Bell" cover to protect the pallet from weather and prevent the sump from filling with rain or snow.

Power/Fuel Requirements

None

Transportability

Poly-Spillpallets have four-way forklift access for moving fully loaded pallets. Empty pallets are person-portable.

CONTACT INFORMATION

Manufacturer

ENPAC Corporation
P.O. Box 1100
Chardon, OH 44024-5100
U.S.A.

Telephone (216) 286-9222
Toll-free   (800) 993-6722
Facsimile (216) 286-9297/(800) 993-6722

REFERENCES

(1)   Manufacturer's Literature.

SEWER GUARD
WATERTIGHT MANHOLE INSERTS

No. 25

DESCRIPTION

Polyethylene manhole inserts for reducing sewage treatment costs by preventing rainwater and surface water runoff from entering the collection system. The Sewer Guard inserts sit directly under the manhole cover, transforming the standard cover into a watertight unit. The units are equipped with automatic gas relief and vacuum relief valves to prevent hazardous buildup of pressure.

OPERATING PRINCIPLE

A correctly sized insert is placed under the manhole cover and rests on the same rim as the cover. A closed-cell, crossed-linked polyethylene gasket conforms to irregularities in the manhole frame and forms a tight, virtually leakproof seal. No special tools are needed to install and there are no other installation costs, either for existing or newly constructed manholes.

STATUS OF DEVELOPMENT AND USAGE

Commercially available product in general use.

TECHNICAL SPECIFICATIONS

Physical

Two models of Sewer Guard inserts are available: the Sewer Guard, which is a bowl-shaped insert with two valves, and the Sewer Guard SD which is a shallow dish insert. Sewer Guards are manufactured to fit any manhole frame and cover and are shipped fully assembled. Exact field measurements of existing manhole frames must be provided when ordering. Special instructions for correctly measuring the manhole frame are available from the manufacturer. The Sewer Guard and each of its components, which include the valve bodies, the valve plugs, the valve springs, and the gasket, are manufactured with corrosion-proof material.

Operational

Sewer Guard inserts can completely seal off manholes very quickly. Made of specially formulated plastic polymers, the inserts will not corrode and cannot be damaged by sewer gases or road oils. They are also highly resistant to road traffic and subfreezing temperatures.

Sewer Guard - This insert's bowl shape is designed to protect the valves if the manhole cover is flipped during removal. The closed-cell, cross-linked polyethylene gasket, heat-welded under the lip, conforms to most irregularities in the manhole frame, and forms a tight, virtually leakproof seal. The spring-loaded gas relief valve, which is designed to relieve gas pressure buildup, is automatically activated when the gas pressure differential in the manhole reaches approximately 3.5 kPa (.5 psi). This prevents dangerous accumulation of sewer gases. The spring-loaded vacuum relief valve, which is designed to relieve the vacuum pressure buildup, is automatically activated when the vacuum pressure differential reaches approximately 16 kPa (2.3 psi). Both valves are self-cleaning and are used as handles for easy removal.

Sewer Guard SD - This insert is a shallow dish which is ideal for larger utility castings. It is equipped with one valve - a gas relief valve, flush-mounted so that the manhole cover can slide over it. This self-cleaning valve is automatically activated when the gas pressure differential in the manhole reaches approximately 3.5 kPa (.5 psi). The insert can be easily removed with a non-corrosive strap handle.

Power/Fuel Requirements

None

Transportability

Person-portable

PERFORMANCE

Years of testing and improvement have resulted in a design linked to performance. The bowl or shallow pan shape traps dirt and debris, the gasket properly seats and seals the inserts, and the high density polyethylene construction material prevents corrosion and damage by sewer gases or road oils. Sewer Guard inserts have also proven highly resistant to traffic wear and subfreezing temperatures.

CONTACT INFORMATION

Manufacturer

Fosroc Inc.
Construction Division
150 Carley Court
Georgetown, KY 40324
U.S.A.

Telephone (502) 863-6800
         1-800-645-3954 (Technical Service Line)
Facsimile (502) 863-4010

REFERENCES

(1) Manufacturer's Literature.

SHIELDS WELDED STEEL SPILL CONTAINMENT PALLET

No. 26

DESCRIPTION

A portable, welded steel construction pallet with built-in sump to provide containment of leaks from 245 to 2,350 L (65 to 620 gal).

OPERATING PRINCIPLE

Sized to hold from 1 to 40 drums, these pallets accept both palletized and non-palletized loads. A fibreglass-floor grating deck holds the load above the sump, and any leaks from the containers drain down and are contained in the sump. Suitable for forklift handling.

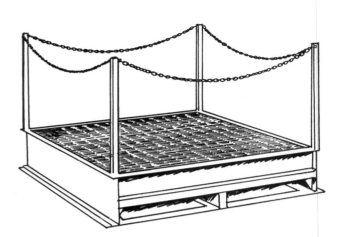

STATUS OF DEVELOPMENT AND USAGE

Commercially available product in general use.

TECHNICAL SPECIFICATIONS

Physical

Constructed of 12-gauge, non-commercial grade A-570 steel, with two-part, anti-corrosive epoxy and polyurethane coatings. The following are specifications for the seven standard sizes.

Model	Size	Weight	Drums stored	Sump capacity
50	0.9 × 0.9 × 0.5 m (3 × 3 × 1.6 ft)	120 kg (260 lb)	1	245 L (65 gal)
100	1.7 × 0.7 × 0.5 m (5.8 × 2.4 × 1.5 ft)	140 kg (315 lb)	2	320 L (85 gal)
308	1.7 × 1.7 × 0.3 m (5.8 × 5.6 × 1 ft)	310 kg (685 lb)	5 to 8	455 L (120 gal)
816	3.1 × 1.7 × 0.3 m (10 × 5.6 × 1 ft)	490 kg (1,080 lb)	11 to 16	910 L (240 gal)
924	4.8 × 1.7 × 0.3 m (16 × 5.6 × 1 ft)	720 kg (1,590 lb)	16 to 24	1,420 L (375 gal)
1032	6.3 × 1.7 × 0.3 m (21 × 5.6 × 1 ft)	920 kg (2,030 lb)	22 to 32	1,890 L (500 gal)
1940	7.8 × 1.7 × 0.3 m (26 × 5.6 × 1 ft)	1,150 kg (2,535 lb)	27 to 40	2,350 L (620 gal)

Operational

Pallets have two-way forklift access, forklift guards, a drain, static ground connections, and grounding lugs.

Transportability

Can be handled by forklift; no lifting lugs.

Manufacturer

Shields Environmental Corporation
19800 MacArthur Blvd.
Suite 510A
Irvine, CA 92715

Telephone (714) 260-3880
Toll-free (800) 588-8439
Facsimile (714) 260-2200

OTHER DATA

Shields also manufactures prefabricated, weatherproofed, and fire-rated buildings for storing hazardous materials; constructed of 12-gauge, noncommercial grade A-570 steel, and featuring palletized grids for efficient use of space.

REFERENCES

(1) Manufacturer's Literature

SYNTECHNICS
TRACK COLLECTOR PAN SYSTEM

No. 27

DESCRIPTION

A fluid-collection system for railroad beds to contain spillage of corrosive materials at chemical loading/unloading and railroad refueling areas. It consists of a centre collector pan and two lateral side collector pans, prefabricated of fibreglass-reinforced plastic (FRP).

OPERATING PRINCIPLE

The centre pan fits between the rails and side pans are held in place with backfill. Collector pans catch any spillage from tank cars and empty into the crossdrains placed perpendicularly between pan sections and designed to interface with a drainage system.

STATUS OF DEVELOPMENT AND USAGE

Commercially available product in general use.

TECHNICAL SPECIFICATIONS

Physical

Six types of systems range from 6 to 12 m (20 to 39 ft) long and 3, 3.6, and 4.3 m (10, 12, and 14 ft) wide and fit rail sizes 80# to 135#. They can fit curved track up to 21 degrees with standard tooling. Collector pans are manufactured of chemical-resistant, fibreglass-reinforced plastic.

Operational

These systems are meant to be permanently installed at a facility to prevent spills. If a spill occurs elsewhere, the components could be used to construct an emergency containment system. The FRP collector pans are suitable for service from -46 to 54°C (-50 to 130°F). The FRP material, of which the systems are 35% by weight, has a tensile strength, compressive strength, flexural strength, and shear strength of 10,340 kPa (15,000 psi) and a Young's modulus of elasticity rating of 6.9×10^6 kPa (1.0×10^6 psi). It takes about six hours to install a 6 m (20 ft) centre pan with two side pans, including excavation, installing drainage piping, backfill, and mounting hardware on the cross ties.

Transportability

While these systems are meant for permanent installation at a facility, the fibreglass material is lightweight and they can be transported to a spill site and installed by a small crew without heavy handling equipment. The length of the collector pans, from 6 to 12 m (20 to 39 ft), is the prime consideration when choosing a means of transportation.

PERFORMANCE

Since 1974, more than 69,000 m (227,000 ft) of Track Collector Pan Systems have been manufactured and shipped to over 800 installation sites in the United States and Canada. Approximately 25% of this has been shipped to Class I Railroads, with the remainder going to Short Line Regional Railroads and Industrial locations.

Manufacturer

Syntechnics Incorporated
700 Terrace Lane
Paducah, KY 42003
U.S.A.

Telephone (502) 898-7303
Facsimile (502) 898-7306

REFERENCES

(1) Manufacturer's Literature.

TROUBLE SHOOTER
PORTABLE CONTAINMENT BERMS

No. 28

DESCRIPTION

Portable containment berms designed for quick deployment to contain leaking tanks, drums, pails, other storage vessels, and valves or fittings. **Series A** are also well suited to capturing discharges from decontamination operations or acting as a secondary containment system for bulk transfers and for isolating hazardous waste. **Series F** berms are smaller berms suited for use as containment berms or drip pads, depending on the volume required to be contained. They can also be used as decontamination pads for washdown of personnel or equipment.

OPERATING PRINCIPLE

Series A - A one-piece membrane unit with semi-rigid sides supported by 5-cm (2-in) wide aluminum L-brackets that forms a berm capable of containing liquids or solids.

Series F - A one-piece membrane unit with rigid sides supported by closed cell foam cylinders [10 or 15 cm (4 or 6 in) in diameter] that forms a berm capable of containing liquids or solids.

STATUS OF DEVELOPMENT AND USAGE

Commercially available products in general use.

TECHNICAL SPECIFICATIONS

Physical

Both Trouble Shooter Series A and Series B berms are available in seven standard sizes as well as custom sizes. Depending on the application, the unit can be fabricated from a variety of heavy-duty and chemical-resistant materials. Standard material is Bio Fab 240. On the Series A berms, support brackets are made of aluminum and are designed so as not to come into contact with the material being contained within the berm. The following are detailed specifications for the standard sizes of Series A and Series F berms.

Series A

Model	Size	Draft	Maximum Capacity
A 600	2.4 × 2.4 m (8 × 8 ft)	46 cm (18 in)	2,270 L (600 gal)
A 800	2.4 × 3.0 m (8 × 10 ft)	46 cm (18 in)	3,030 L (800 gal)
A 1000	3.0 × 3.0 m (10 × 10 ft)	46 cm (18 in)	3,800 L (1,000 gal)
A 1200	3.0 × 3.7 m (10 × 12 ft)	46 cm (18 in)	4,540 L (1,200 gal)
A 1500	3.7 × 3.7 m (12 × 12 ft)	46 cm (18 in)	5,680 L (1,500 gal)
A 6000	3.7 × 11 m (12 × 35 ft)	61 cm (24 in)	22,700 L (6,000 gal)
A 10000	4.9 × 14 m (16 × 45 ft)	61 cm (24 in)	37,850 L (10,000 gal)

1.2 Containment - On Land

Series F

Model	Size	Draft	Maximum Capacity
F55	1.2 × 1.2 m (4 × 4 ft)	15 cm (6 in)	210 L (55 gal)
F90	1.5 × 1.5 m (5 × 5 ft)	15 cm (6 in)	320 L (84 gal)
F125	1.8 × 1.8 m (6 × 6 ft)	15 cm (6 in)	455 L (120 gal)
F225	2.4 × 2.4 m (8 × 8 ft)	15 cm (6 in)	815 L (215 gal)
F350	3.0 × 3.0 m (10 × 10 ft)	15 cm (6 in)	1,170 L (335 gal)
F-4-35	1.2 × 1.2 m (4 × 4 ft)	10 cm (4 in)	135 L (35 gal)
F-4-55	1.5 × 1.5 m (5 × 5 ft)	10 cm (4 in)	210 L (55 gal)
F-4-75	1.8 × 1.8 m (6 × 6 ft)	10 cm (4 in)	285 L (75 gal)
F-4-140	2.4 × 2.4 m (8 × 8 ft)	10 cm (4 in)	530 L (140 gal)
F-4-220	3.0 × 3.0 m (10 × 10 ft)	10 cm (4 in)	835 L (220 gal)

Operational

Trouble Shooter Series A and Series F berms can be set up on most level, relatively smooth surfaces, including grass, asphalt, concrete, sand, and soil. However, the surface must remain stable and intact while supporting the anticipated weight. Bio Fab 240 liner material will not crack at temperatures as low as -34.4°C (-30 °F) and has a blocking (sticking of material) resistance of up to 82.2°C (180°F). The following products have been tested by the manufacturer and found to be chemically compatible with the Trouble Shooter Berm.

A = little or no effect *B = minor-to-moderate effect* *C = severe effect*
T = No data. Likely to be acceptable *X = No data. Not likely to be acceptable*

Acetic acid (5%)	B	Jet A	A
Acetic acid (50%)	C	JP-4 Jet fuel	A
Ammonium phosphate	T	JP-5 Jet fuel	A
Ammonium sulphate	T	JP-8 Jet fuel	A
Antifreeze (ethylene glycol)	A	Kerosene	A
Animal oils	A	Magnesium chloride	T
Aqua regia	X	Magnesium hydroxide	T
ASTM Fuel A (100% Iso-octane)	A	Methanol	A
ASTM Oil #2 [(Flash point 240°C(464°F)]	A	Methyl alcohol	A
ASTM Oil #3	A	Methyl ethyl ketone	X
Benzene	X	Mineral spirits	A
Calcium chloride solution	T	Naphtha	A
Calcium hydroxide	T	Nitric acid (5%)	B
Chlorine solution (20%)	A	Nitric acid (50%)	C
Concentrated ammonium hydroxide	A	Perchloroethylene	C
Corn oil	A	Phenol	X
Crude oil	A	Phenol formaldehyde	B
Diesel fuel	A	Phosphoric acid (50%)	A
Ethanol	A	Phthalate plasticizer	C
Ethyl acetate	C	Potassium chloride	T
Ethyl alcohol	A	Potassium sulphate	T
Fertilizer solution	A	Raw linseed oil	A
Fuel Oil #2	A	SAE-30 oil	A
Fuel Oil #3	A	Salt water (25%)	B
Furfural	X	Sea water	A
Gasoline	B	Sodium acetate solution	T
Glycerine	A	Sodium hydroxide (60%)	A
Hydraulic fluid	A	Sodium phosphate	T
Hydrochloric acid (50%)	A	Sulphuric acid (50%)	A
Hydrofluoric acid (5%)	A	Tanic acid (50%)	A
Hydrofluoric acid (50%)	A	Toluene	C
Hydrofluorosilicic acid (30%)	A	Transformer oil	A
Isopropyl alcohol	T	Turpentine	A

Urea formaldehyde	A	Water (93°C, 200°F)	A
UAN	A	Xylene	X
Vegetable oil	A		

Spills collected within a Trouble Shooter berm should be neutralized or transferred to permanent containers as soon as possible. The berm is a temporary holding medium intended to provide hours or a few days of containment. Chemical resistance data are based on an exposure limit of 28 days. Material samples are available to customers for immersion testing. Punctures can be repaired on site.

Power/Fuel Requirements

None

Transportability

The Trouble Shooter Series berms are easy to transport.

PERFORMANCE

Although documentation was not received on the application of Trouble Shooter berms to spills of hazardous materials, the manufacturer has thoroughly tested the liner materials for chemical resistance, chemical permeability, strength, water absorption, wicking, high temperature blocking (sticking), dead load, abrasion resistance, and ease of repair in the field. Documentation of this testing is available from the manufacturer.

CONTACT INFORMATION

Manufacturer

ThermaFab Incorporated
200 Rich Lex Drive
Lexington, SC 29072
U.S.A.

Telephone (803) 794-2543
Facsimile   (803) 796-0999

REFERENCES

(1)   Manufacturer's Literature.

Section 1.3

Containment on Water

SPILL CONTAINMENT/DEFLECTION BARRIERS - GENERAL LISTING No. 29

DESCRIPTION

Boom devices that use mechanical, sorbent, or pneumatic means or high-pressure water jets to contain and concentrate spilled products for recovery or *in-situ* burning, or to deflect or exclude spilled products from areas that are particularly threatened, e.g., water intakes and shorelines. Booms can also be used to deflect a moving slick into a backwater or convenient/accessible collection point along a river or shoreline.

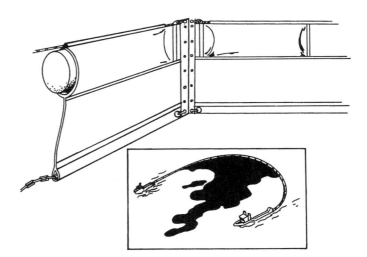

It should be noted that deflection barriers are suitable only to contain or deflect chemical spills that float in water, i.e., maintain a specific gravity <1, and preferably, for chemicals compatible with the boom materials.

OPERATING PRINCIPLE

There are four types of spill containment/deflection barriers.

Mechanical Booms - Fence-type devices designed for use in inshore, harbour, river, and offshore environments, and relying on a vertical curtain extending above and below the surface of the water. Mechanical booms are horizontally supported by air- or foam-filled flotation and vertically supported by some form of ballast, typically lead weights, heavy chains, or water chambers. Upper, middle, and/or lower tension members are commonly used to enhance a boom's load distribution and sea-keeping characteristics.

Sorbent Booms - Comparatively simple devices consisting of a porous stocking stuffed with sorbent material designed for use in confined, slow-moving waters. Due to their natural buoyancy, the need for special flotation or ballast is minimized and contaminants can be contained and recovered just above and below the water surface.

Pneumatic Barriers - Specifically suited for permanent, zero, or low-current applications, such as harbours, canals, and ship channels, pneumatic barriers use submerged perforated piping to release a curtain of compressed air bubbles, creating lateral currents at the water surface that deflect or deter the spread of floating contaminants.

High Pressure Water Barriers - Water jets sprayed from a series of nozzles control and oppose the advancing spill.

STATUS OF DEVELOPMENT AND USAGE

In their application to oil spill response, mechanical, sorbent, and to a lesser extent, pneumatic barriers have been well tested, and are commercially available and in wide use. Documentation of the use of these barriers in chemical spill response is comparatively limited, restricted to tank or laboratory testing of a few hazardous substances. In one case in 1994, both mechanical and sorbent booms were reportedly used to contain and later recover part of a 2-tonne sytrene monomer spill in Port Moody harbour in British Columbia. Although fireproof booms are being used more for *in-situ* burning in oil spill response, the application of fireproof booms, pneumatic barriers, and high-water pressure barriers to chemical spill response is generally less explored.

TECHNICAL SPECIFICATIONS

Physical

Mechanical Booms - These can be characterized according to the location (inshore, harbour, river, and offshore) and the nature of their application, e.g., *in-situ* burning, long-term harbour installation. Booms are usually selected for specific applications according to their size and reserve buoyance (the buoyancy in excess of that required to keep them floating). The following are some typical physical specifications. See also the specifications provided in the table.

Dimensions
- Overall height (cm)
- Draft (cm)
- Freeboard (cm)
- Section length (m)
- Storage volume (m^3/m)

Weight
- Per unit length (kg/m)
- Per section (kg)
- Ballast (kg/m)

Construction
- *Fabric* -Type (a critical factor in chemical spill response)
  - Colour
  - Tensile and tear strength (kg)
  - Cold crack temperature (°C)
- *Flotation* - Type (closed-cell foam, sealed air chamber, inflatable)
  - Location (built-in, attached, outriggered)
  - Shape (cylindrical, rectangular, spherical)
  - Reserve buoyancy (kg/m) and weight ratio
- *Tension Members* - Type (wire cable, chain)
  - Number and location (upper, middle, lower, vertical)
  - Tensile strength (kg)
- *Other* - Type of ballast (lead weights, heavy chain, water chambers)
  - End connectors (ASTM, Universal Slide, Hinged Plate, etc.)
  - Anchor points and handholds

Typical Boom Specifications

	Inshore	Harbor	River	Offshore
Freeboard (m)	0.15 to 0.3	0.3 to 0.5	0.15 to 0.18	0.6 to 1.0
Draught (m)	0.3 to 0.6	0.6 to 1.0	0.19 to 0.31	1.2 to 1.8
Max. Wave Height (m)	0.3 to 0.6	0.6 to 1.2	0.3 to 1.0	1.2 to 2.4
Weight (kg/m)	2 to 8	5 to 45	2.8 to 3.4	6 to 75
Storage (m^3/m)	0.02 to 0.05	0.3 to 0.4	0.04 to 0.05	0.09 to 1.9
Reserve Buoyancy (kg/m)	0.9 to 30	0.4 to 50	4.4 to 4.6	4.2 to 44

Sorbent Booms - Physical characterization is similar to other forms of sorbent and should include the following:
- Boom dimensions: Length - typically 3 m
  - Diameter - typically 20 to 25 cm
- Sorption capacity (kg/cm^3, L/m)
- Dry and saturation weight (kg/m, kg/section)
- Material type(s), reusability, storage and disposal requirements
- Chemical-attracting and water-repelling properties

Pneumatic Barriers - Physical specifications vary greatly depending on the size and application of the system installed. Typical factors include:
- Pipe dimensions (length, diameter, weight)
- Pipe durability (corrosion resistance, physical impact)
- Pipe perforation (holes/square area)
- Air compressor capacity
- Various installation and maintenance specifications

High Pressure Water Barriers - As with pneumatic barriers, physical specifications vary with the type of system. Capacity of the water pump and frequency/number of jet sprays are the most obvious factors to be considered.

Operational

Mechanical Booms - Although probably the most available form of barrier and the quickest to deploy, mechanical booms may not be an effective means of containment/deflection at a remote site or in rapidly flowing water. Chemical spills can be contained in currents of up to 1.5 km/h (1.0 knots), and deflection is possible as long as the perpendicular vector of the current on the boom does not exceed its containment limit (approximately 1.5 km/h).

Boom failure may occur in the following situations.

Entrainment	Strong currents force chemicals to be drawn under the boom.
Drainage	The thickness of the spill exceeds the depth of the skirt.
Splashover	Wind or waves carry chemicals over the freeboard.
Boom Submergence	Forces render the boom's flotation inadequate.
Boom Planing	Strong wind and current velocities move in opposite directions.
Structural Breakdown	Weak or poorly constructed joints or tension members cause the structure to break down.
Fabric Breakdown	Gross failure caused by the attack of highly solvent chemicals; this is unlikely during short-term deployment.
Improper Deployment	Anchors do not hold boom in high currents or tidal cycles.

Sorbent Booms - Most sorbent booms are fitted with carabineers (fireman's hooks) so they can be connected end-to-end and overlapped. They are most effective in calm or slow-moving water bodies with a relatively small amount of spilled material.

Pneumatic Barriers - Pneumatic barriers have been tried on oil spills in flowing watercourses, but their use is limited. Their costly, time-consuming deployment generally restricts the use of air-curtain barriers for emergency spill response.

High Pressure Water Barriers - Basically developed for oil spill control, water spray systems are poorly tested with chemical spills and would probably increase the likelihood of mixing and dissolution of contaminants into an affected water body.

Power/Fuel Requirements

Power or fuel may be required to inflate some mechanical booms and to operate pneumatic and water barriers fitted with pumping devices.

Transportability

Transportation and handling requirements vary depending on the type and size of barrier, with sorbent booms requiring one to two persons to transport and pneumatic barriers requiring a team of specialized personnel, as well as mechanical lifting capability and significant transport capacity.

PERFORMANCE

Mechanical Booms provide containment and deflection potential for spills of hazardous materials if they can be deployed in a timely and effective manner.

Sorbent Booms have a restricted application to chemical spill response, i.e., for smaller spill volumes, with compatible substances, and in calmer waters.

Pneumatic Barriers are not yet a viable spill control alternative unless applied to a long-term, shallow water, low current situation that is expected to result in an ongoing, continuous discharge of floating chemicals.

High Water Pressure Barriers have not yet proven to have significant containment or deflection capability for chemical spills. Their cost, weight, and power requirements may limit their use to only very large spills of hazardous materials.

REFERENCES

(1) Schulze, Robert (ed.), "World Catalog of Oil Spill Response Products", Fifth Edition, World Catalog Joint Venture (JV), City Press, Baltimore, MD (1995).

Section 1.4

Containment In/Under Water

PORTADAM BARRIER SYSTEM No. 30

DESCRIPTION

A patented coffer dam capable of retaining contaminated water in rivers and streams until hazardous materials are removed. It consists of a braced metal structure that supports a reinforced fabric membrane.

OPERATING PRINCIPLE

The frame structure is set up at an angle to direct hydro-static pressure down onto the ground. The frame is braced against lateral forces and correctly spaced to fit the membrane. The membrane is pleated and attached to the frame by high-strength loops, extended, and clamped to the ground by water pressure.

STATUS OF DEVELOPMENT AND USAGE

Developed to the commercial stage in Canada by Conenco International Limited of Markham, Ontario, under Environment Canada sponsorship. The company no longer operates and the system is now considered a concept.

TECHNICAL SPECIFICATIONS

Physical

A modular system consisting of a frame constructed of galvanized steel or aluminum tubular sections, with a 1.5 to 2.4 m (5 to 8 ft) draught, braces consisting of chains or steel plates, and a membrane constructed of a series of pleats, tailored from high-strength, durable, coated fabric and with a standard length of 9 and 18 m (30 and 60 ft).

Operational

The system can be used in water depths to 3 m (10 ft). The basic 1.5 m-(5 ft-) high frame can be extended up to 2.4 m (8 ft) by means of telescopic extension. Spacing between the frames is 38 cm (15 in), centre to centre. Depending on soil conditions, bracing chains or X-frames are placed at 7.6 m (25 ft) centres. Special pleating in the membrane ensures correct distribution over the support structure. Coated fabrics are available to resist oil, acids, etc., as well as to perform at low temperatures.

Transportability

For emergency purposes, the dams should be supplied on pallets that can be loaded by a forklift onto trucks or trailers. Each pallet should contain enough frames and membranes for a 50 m (160 ft) length of dam, and a kit containing other equipment such as shovels, ropes, knives, sandbags, cartridge-operated fastening gun, sledge hammers, spare ties, chain-edged sealing sheet for maneuvering over leaks or holes in flood walls, and a membrane repair kit.

Canadian Contact

Emergencies Science Division
Environmental Technology Centre
Environment Canada
3431 River Road
Ottawa, Ontario
K1A 0H3

Telephone (613) 998-9622
Facsimile (613) 991-9485

REFERENCES

(1) Conenco International Limited, "Development of Portadam Oil Barrier for Rivers and Streams - Final Report" (April, 1979).

Section 1.5

Containment - Vapour Control

1.5 Containment - Vapour Control

CHEMICALLY ACTIVE COVERS - GENERAL LISTING

No. 31

DESCRIPTION

A chemically active material or agent that when applied to cover a chemical spill reacts with the spilled material to reduce or eliminate the release of vapour, thereby decreasing the hazardous nature of the spill. See also the following specific entries in Section 5.1, Spill Treating Agents.

> Ansul Spill Treatment Kits and Applicators (No. 111)
> CAPSUR, MEXTRACT, and PENTAGONE Spill Agents (No. 112)
> Cartier Spill Response Kits (No. 113)
> Epoleon Odor-Neutralizing Agents (No. 114)
> Foul-Up Hazardous Wet-Waste Solidifying Agent (No. 115)
> Mercon Spill-treating Kits (No. 116)
> Nochar Solidifying Agents (No. 117)

OPERATING PRINCIPLE

A covering material is considered to be chemically active if it readily reacts with the spilled compound to neutralize or otherwise decrease its inherent toxicity. The active mechanisms of such materials, which include sorption, ion exchange, oxidation, and neutralization, act to reduce the release of vapour from the spilled chemical.

STATUS OF DEVELOPMENT AND USAGE

Some chemically active covers are simply scrap metals or earthen materials that are readily available, while others are compounds specifically formulated for spill response and commercially marketed as spill-treating or control agents.

TECHNICAL SPECIFICATIONS

Physical

Materials that can be used as chemically active covers include scrap iron, sulphide ores, pyrite, clays, diatomaceous earth, manganese dioxide, proteinaceous wastes such as wool, carbon compounds such as activated carbon, bonechar, calcium carbonates such as lime, limestone, and chalk, gypsum, sulphur, potassium permanganate, alum, alumina, ferric sulphate, and certain commercial ion exchange resins. Specifically formulated spill control agents and sorbents are also commercially available.

Operational

When using an active cover, each spilled chemical must be considered individually as there are no universally active covering materials. Each spill should therefore be evaluated by a specialist familiar with the spilled material and its chemistry. As chemically active covers rely on chemical reactions, variables such as temperature, concentrations of both the spilled compound and the cover itself, and application rate will affect the cover's efficiency. Covering material must be selected carefully as some combinations of cover and spilled chemical can react vigorously or even violently. One such field experience in the early 1970s involved a misguided attempt to neutralize a rail tank car spill of acid by adding caustic soda, instead of limestone.

Commercial agents that neutralize, sorb, solidify, or gel, are typically developed for liquid immobilization rather than vapour hazard control. These products, however, usually have some influence on the evapouration rate and hence vapour concentrations. As sorbents decrease the vapour hazard only when they are unsaturated, they must be continuously replenished in order to be effective. If they become saturated, the vapour hazard increases due to the large increase in surface area. Gelling agents achieve a gel formation through the interaction of a high-molecular-weight molecule (macromolecule) and a liquid. This form of liquid-phase modification has been extensively developed for liquid immobilization rather than for vapour hazard control. The gel structure is a combination of physical and chemical interaction that generally results in the formation of a two or three dimensional network of macro-molecular cages, entrapping the liquid phase. The formation of a gel generally has some influence on the evapouration rate and hence on the vapour concentrations from a spilled chemical. The evapouration rate is reduced by forming a continuous cover of gelled material to encapsulate the more volatile spilled liquid.

Transportability

Transportability varies with the type and amount of cover to be used.

CONTACT INFORMATION

Activated Carbon

A C Carbone Canada Incorporated
300 Brosseau Street
St. Jean sur Richelieu, Quebec
J3B 2E9

Telephone (514) 348-1807
Facsimile (514) 348-3311

Air Filter Sales and Service Limited
101 Courtland Avenue
Concord, Ontario
L4K 3T5

Telephone (905) 669-5470
Toll-free   (800) 387-8820 (Canada)
Facsimile (905) 669-5463

Airguard Industries, Division of Delcon Filtration
2700 Steeles Avenue West
Concord, Ontario
L4K 3C8

Telephone (905) 660-0688
Facsimile (905) 660-7852

Anderson Water Systems Limited
44 Head Street
Dundas, Ontario
L9H 3H3

Telephone (905) 627-9233
Facsimile (905) 628-6623

Calgon Carbon Corporation
P.O. Box 717
Pittsburgh, PA 15230-0717
U.S.A.

Telephone (412) 787-6700
Toll-free   (800) 422-7266 (North America)
Facsimile   (412) 787-4523

Canadian Supplier for Calgon Carbon Corporation -
STANCHEM, Inc.
43 Jutland Road
Etobicoke, Ontario
M8Z 2G6

Telephone (416) 259-8231
Toll-free   (800) 268-0358 (Ont./Que.)
Facsimile (416) 259-6175

Cameron-Yakima, Inc.
P.O. Box 1554
Yakima, WA 98907
U.S.A.

Telephone (509) 452-6605
Facsimile (509) 453-9912

Carbtrol Corporation
51 Riverside Avenue
Westport, CT 06880
U.S.A.

Telephone (203) 226-5642
Toll-free   (800) 242-1150 (North America)
Facsimile (203) 226-5322

FARR Incorporated
2785 Francis Hughes Avenue
Laval, Quebec
H7L 3J6

Telephone (514) 629-3030
Facsimile (514) 662-6035

General Filtration Division of Lee Chemicals Ltd.
1119 Yonge Street
Toronto, Ontario
M4W 2L7

Telephone (416) 924-0721
Facsimile (416) 960-8750

Hoyt Corporation - Canada Branch
210 Ronald Drive
Montreal, Quebec
H4X 1M8

Telephone (514) 481-9409
Facsimile (514) 481-5880

J. V. Manufacturing Company, Incorporated
963 Ashwaubenon Street
Green Bay, WI 54304
U.S.A.

Telephone (414) 337-4944
Toll-free   (800) 334-9092
Facsimile (414) 337-6282

1.5 Containment - Vapour Control

National General Filter Ltd.
7351 Chouinard
Lasalle, Quebec
H8N 2L6

Telephone (514) 364-0341
Toll-free   (800) 363-2135 (Canada & parts of U.S.)
Facsimile (514) 366-2110

Q-Air Environmental Controls
43 Teal Avenue, Unit 4
Stoney Creek, Ontario
L8E 3B1

Telephone (905) 662-6831
Facsimile   (905) 662-8983

Zedex Products Incorporated
300 Steelcase Road West, Suite 23
Markham, Ontario
L3R 2W2

Telephone (905) 475-3753
Facsimile   (905) 475-2381

Commercial Sorbents - For detailed information on commercially available sorbents, refer to Section 2.5, Sorbents.

Commercial Spill Treating Agents - For further information on commercially available spill control agents, refer to Section 5.1, Spill Treating Agents.

Earthen Materials/Minerals/Chemicals - For information on local availability of earthen materials, minerals, and chemicals, consult trade indexes, such as *Fraser's Canadian Trade Directory* and the *Thomas Register,* or telephone directories under headings such as Aggregates, Minerals, Sand and Gravel, or Chemicals.

Scrap Metals - For information on local availability of scrap metals, consult local telephone directories under the heading "Scrap Metal".

REFERENCES

(1) Bennet, G.F., F.S. Feates, and I. Wilder, *Hazardous Materials Spills Handbook,* McGraw-Hill Book Company, Toronto, Ont. (1982).
(2) Holmes, J. M. and C.H. Byers, *Countermeasures to Airborne Hazardous Chemicals*, Noyes Publications, Park Ridge, NJ (1990).
(3) Manufacturer's Literature.
(4) Noyes, R., *Handbook of Pollution Control Processes*, Noyes Publications, Park Ridge, NJ (1991).

INDUCED AIR MOVEMENT - GENERAL LISTING   No. 32

DESCRIPTION

Air movement, usually mechanically induced, is used to dilute hazardous chemical vapours to below their threshold limit value or lower explosive limit, thereby providing a safe working environment in which to clean up the spill. See also Coppus Portable Ventilators (No. 33) in this section.

OPERATING PRINCIPLE

The dilution technique involves transporting and mixing uncontaminated air with the vapours released from a chemical spill. The volume of uncontaminated air must be large enough to maintain the concentration of hazardous chemical vapours below their threshold limit value or lower explosive limit.

STATUS OF DEVELOPMENT AND USAGE

Mechanical ventilators are commercially available and are designed specifically for ventilating confined spaces, fume removal, and tanker de-gassing. Very large air-moving equipment using surplus jet engines can be produced for removing snow from railroad tracks and for removing fog at airports.

TECHNICAL SPECIFICATIONS

Physical

The mechanical ventilators designed for ventilating confined spaces, fume removal, and tanker/vessel de-gassing are available with air delivery capacities from 2 to 30,000 m^3/h (70 to 11×10^5 ft^3/h), and can be driven electrically, by gasoline, compressed air, or steam/air turbine drive, water turbine drive, or hydraulic motor drive. The jet-driven air movers can generate a 1,050 km/h (650 mph) air blast, which delivers approximately 2.8×10^6 m^3/h (10^7 ft^3/h) of air for up to six hours of independent operation.

Operational

Operational specifications can be calculated from data generated with natural dispersion models. These models predict evapouration rates for spills of volatile hazardous chemicals to range from 1 to 3.5 m^3 (35 to 4,400 ft^3) of vapour released per hour per square metre of spill surface. If an average threshold limit value of 10 ppm is assumed, 1.0×10^5 to 3.5×10^6 m^3 of uncontaminated air per square metre of spill surface must be provided hourly to keep the concentration of the hazardous chemical at this limit. If an average lower explosive limit of one percent by volume is assumed, 1.0×10^2 to 3.4×10^5 m^3 of uncontaminated air must be delivered hourly to keep the vapour concentration at this lower limit. These estimates indicate the need for very high capacity air-moving equipment or ventilators. One problem that must be considered is the increase in evapouration rates that may occur. If the release is a vapourizing liquid, the fan may significantly increase the total quantity of toxic or flammable vapour generated.

Power/Fuel Requirements

Power or fuel requirements vary depending on the type of air mover used. Power and fuel sources can include gasoline, jet fuel, diesel, compressed air, high-pressure water, electricity, or hydraulic pressure from existing hydraulic lines.

Transportability

Transportability varies with the type of air mover used. Air movers vary from lightweight, hand-held units to self-propelled, rail car assemblies.

PERFORMANCE

Specific performance data is provided for Coppus Portable Ventilators in No. 33.

1.5 Containment - Vapour Control

CONTACT INFORMATION

The following is contact information for all devices discussed including those listed as separate entries in this section.

Coppus Portable Ventilators

Tuthill Corporation/Coppus Portable Ventilation Division
P.O. Box 8000
Millbury Industrial Park
Millbury, MA 01527-8000
U.S.A.

Telephone   (508) 756-8391
Facsimile   (508) 756-8375

Affiliated-Dynesco, Division of Wajax Industries Limited
2720 Steeles Avenue West
Concord, Ontario
L4K 4N5

Telephone   (416) 660-3600
Facsimile   (416) 660-9574

Quickdraft Industrial Exhausters

Quickdraft, Division of C.A. Litzler Company Incorporated
1525 Perry Drive S.W., P.O. Box 80659
Canton, OH 44710
U.S.A.

Telephone   (216) 477-4574
Facsimile   (216) 477-3314

REFERENCES

(1)   Bennet, G.F., F.S. Feates, and I. Wilder, *Hazardous Materials Spills Handbook,* McGraw-Hill Book Company, Toronto, Ont. (1982).
(2)   Holmes, J. M. and C.H. Byers, *Countermeasures to Airborne Hazardous Chemicals*, Noyes Publications, Park Ridge, NJ (1990).
(3)   Manufacturer's Literature.
(4)   Noyes, R., *Handbook of Pollution Control Processes*, Noyes Publications, Park Ridge, NJ (1991).

COPPUS PORTABLE VENTILATORS No. 33

DESCRIPTION

Air, electric, steam, water, or gasoline-driven mechanical ventilators used to dilute hazardous chemical vapours to below their threshold limit value or lower explosive limit, thereby providing a safe working environment in which to mitigate the spill.

OPERATING PRINCIPLE

The dilution technique involves transporting and mixing uncontaminated air with the vapours released from a chemical spill. The volume of uncontaminated air must be sufficiently large to maintain the concentration of hazardous chemical vapours below their threshold limit value or lower explosive limit.

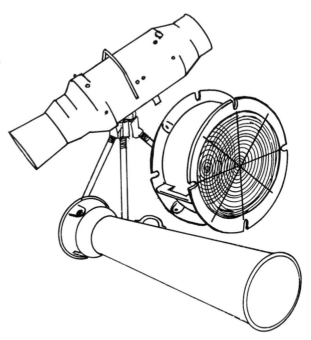

STATUS OF DEVELOPMENT AND USAGE

Mechanical ventilators are commercially available that are designed specifically for ventilating confined spaces, fume removal, and tanker de-gassing.

TECHNICAL SPECIFICATIONS

Physical

Coppus Portable Ventilators are available in a wide variety of models. The following are the general specifications for eight Coppus ventilators.

Product	Models	Fan Type	Drive	Air Capacity Range	Weight
Vano	2	Vaneaxial	Electric Motor	2,600 to 5,100 m^3/h	27 to 44 kg (60 to 97 lb)
TA	1	Axial	Electric Motor	9,400 m^3/h	39 kg (86 lb)
Portavent	4	Centrifugal	Electric Motor	950 to 2,700 m^3/h	26 to 52 kg (57 to 115 lb)
Ventair	7	Centrifugal	Electric Motor and Gasoline Engine	2,800 to 18,000 m^3/h	50 to 300 kg (110 to 660 lb)
Jectair	5	N/A	Air/Steam Venturi	2,800 to 15,000 m^3/h	2.7 to 19 kg (6 to 42 lb)
FR	6	Axial	Compressed Air Reaction Fan	2,500 to 29,000 m^3/h	18 to 64 kg (40 to 140 lb)
CP-20	1	Axial	Air/Steam Turbine	19,000 m^3/h	55 kg (120 lb)
Marine Ventilators	4	Axial	Air/Steam and Water Turbine	7,800 to 14,000 m^3/h	30 to 55 kg (66 to 120 lb)

Coppus fan casings are manufactured from either high-grade aluminum castings or heavy-gauge steel, coated with epoxy powder for corrosion resistance. Fans are directly connected, which eliminates maintenance of drive belts, chains, and pulleys.

Operational

Vano - This small, hand-held vaneaxial ventilator moves large volumes of air through flexible ducts. The electric motor drive is available totally enclosed or explosion-proofed. Folding tripod and transport cart are available. Suggested uses include: welding and toxic fume exhaust, spot cooling of product and equipment, and ventilation of tanks, tank cars, manholes, vats, vessels, and airplane compartments.

TA - This skid-mounted, multi-purpose, steel-cased, tubeaxial blower/exhauster moves air in high volume. The electric motor drive is available totally enclosed or explosion-proofed. The TA can be used with or without ducts.

Portavent - The powerful, non-overloading centrifugal blade design is ideal for a variety of fume and dust removal applications. Four models are available with totally enclosed or explosion-proof electric motors from 0.4 to 1.5 kW (0.5 to 2 hp). Due to its high pressure capability, it can be used with long runs of small-diameter duct. A unique "milk stool" base allows for selective positioning.

1.5 Containment - Vapour Control

Ventair - This heavy-duty, high-volume ventilator is available in five models with totally enclosed or explosion-proof electric motors and in two models with gasoline engines. The Ventair can be used with or without ducts.

Jectair - This lightweight, high-volume compressed air (or steam) air mover is suited to sites where lightweight, adjustable volume flow is required. It is available in five sizes, has no moving parts, and can be used with or without ducts.

FR Series - These compressed, air-driven reaction fan type ventilators are available in two models and are designed to fit directly onto tank openings for removal of fumes and dust. Flanges mate with 5 and 61 cm (20 and 24 in) tank openings. Carrying handles are cast into the housing to allow ease of transportation to the site. The unit can also be rolled on its flanges. The RF ventilators have permanently lubricated, sealed, heavy-duty bearings. Identical flanges on both sides allow the RF ventilators to be used as supply or exhaust units.

CP-20 - Similar to the FR Series, but turbine-driven. The turbine can be powered by either steam or compressed air.

Marine Ventilators - Coppus marine ventilators are tubeaxial units that operate on steam, compressed air, or water. They are suited to a wide range of on-board applications including gas-freeing, drying, and ventilating operations on oil tankers and similar vessels. All marine ventilators have flanged designs allowing them to be directly mounted to tank openings. Adaptors are available to allow use of flexible duct and to fit other openings, such as standard "Butterworth" deck openings.

Power/Fuel Requirements

Compressed air, pressurized water, electricity, steam, or gasoline.

Transportability

Coppus Ventilators are small, lightweight units, most of which are considered "tools" that can be taken from job to job. Depending on the model, they weigh between 2.7 and 300 kg (6 to 670 lb).

PERFORMANCE

Coppus Ventilators have been used at refineries, petrochemical and chemical processors, construction sites, airports, railroads, shipyards, steel mills, and manufacturing plants to disperse fumes from tanks, towers, vessels, compartments, holds, manholes, crawl areas, and confined spaces. No detailed documentation is available on their use at hazardous chemical spill sites.

CONTACT INFORMATION

Manufacturer

Tuthill Corporation/
Coppus Portable Ventilation Division
P.O. Box 8000
Millbury Industrial Park
Millbury, MA 01527-8000
U.S.A.

Telephone (508) 756-8391
Facsimile (508) 756-8375

Canadian Representative

Affiliated-Dynesco, Division of Wajax Industries Limited
2720 Steeles Avenue West
Concord, Ontario
L4K 4N5

Telephone (416) 660-3600
Facsimile (416) 660-9574

REFERENCES

(1) Manufacturer's Literature.

INERT GAS BLANKETS - GENERAL LISTING   No. 34

DESCRIPTION

Inert gases such as carbon dioxide (CO_2) and nitrogen (N_2) can be used as a vapour control countermeasure. The gases provide a gas blanket that isolates the spilled chemical from the air and inhibits evapouration of the spilled material.

OPERATING PRINCIPLE

Heavier than air, inert gases such as CO_2 will settle like a blanket over the top of a spilled material contained within a depression such as a berm or a trench, and inhibit the material's evapouration into the air. Lighter-than-air gases such as N_2 can be used to prevent hazardous vapours from escaping from damaged tanks or vessels. They can also contain vapours from surface spills if used in conjunction with a mechanical cover to keep the lighter-than-air gas blanket in contact with the spilled material.

STATUS OF DEVELOPMENT AND USAGE

Nitrogen and carbon dioxide are readily available and inexpensive gases. Both have been used in spill situations to blanket products from the air and reduce vapour concentrations. Both gases have also been used to purge tanks and pipelines, and transfer products from vessel to vessel.

TECHNICAL SPECIFICATIONS

Physical

Carbon Dioxide

Chemical formula	CO_2
Molecular weight	44
Specific gravity @ 21°C and 1 atm	1.522 (air = 1)
Specific gravity @ 0°C and 1 atm	1.524 (air = 1)

Carbon dioxide is a compound of carbon and oxygen, in proportions by weight of about 27% carbon to 73% oxygen. A gas at normal atmospheric temperatures and pressures, carbon dioxide is colourless, odourless, and about 1.5 times as heavy as air. Carbon dioxide gas is relatively nonreactive and nontoxic. It will not burn and it will not support combustion or human life. When dissolved in water, carbonic acid (H_2CO_3) is formed. The pH of carbonic acid varies from 3.7 at atmospheric pressure to 3.2 at 23.4 atm.

Nitrogen

Chemical formula	N_2
Molecular weight	28
Specific gravity @ 21°C and 1 atm (air = 1)	0.967

Nitrogen makes up the major portion of the atmosphere (78% by volume, 76% by weight). It is a colourless, odorless, tasteless, nontoxic, and almost totally inert gas. It is colourless as a liquid, nonflammable, will not support combustion, and is not life-supporting. It is only slightly soluble in water and most other liquids and is a poor conductor of heat and electricity.

Operational

Nitrogen and carbon dioxide are often used because of their low cost, availability, and inertness. The gas used must be compatible with the product to be blanketed. For example, carbon dioxide reacts with ammonia while nitrogen is inert to almost all compounds at ambient temperature. The amount of gas required depends on the size of the spill and the material to be blanketed. To eliminate the risk of fire or explosion with a combustible product, the concentration of oxygen (O_2) in the vapour space above the product must be well below the explosive range of the product's vapour/oxygen ratio.

1.5 Containment - Vapour Control

The N_2/air and CO_2/air concentrations as well as the maximum recommended oxygen concentrations for several explosive materials are listed in the following table. These levels are suitable when an ignition source is unlikely.

Flammable Vapour or Gas	Nitrogen/Air		Carbon dioxide/Air	
	Minimum O_2 for Combustion %	Maximum Recommended O_2 %	Minimum O_2 for Combustion %	Maximum Recommended O_2 %
Acetone	13.5	11.0	15.5	12.5
Benzene	11.0	9.0	14.0	11.0
Carbon Monoxide	5.5	4.5	6.0	5.0
Ethanol	10.5	8.5	13.0	10.5
Ethylene	10.0	8.0	11.5	9.0
Hydrogen	5.0	4.0	6.0	5.0
Hydrogen Sulphide	7.5	6.0	11.5	9.0
Methane	12.0	9.5	14.5	11.5
Methanol	10.0	8.0	13.5	11.0
Propylene	11.5	9.0	14.0	11.0

Inert gas covers are often an additional benefit that results from the use of either solid CO_2 (dry ice) or liquid nitrogen for cryogenic cooling of a spill. Carbon dioxide fire extinguishers should not be used as they are designed to discharge a liquid rather than a vapour.

Power/Fuel Requirements

None

Transportability

Carbon dioxide is contained, shipped, and stored in either liquified or solid form. Liquified CO_2 can be supplied in compressed gas cylinders, tank cars, cargo tanks, or portable tanks and ISO containers. Solid CO_2 (dry ice) has a temperature of -79 °C (-109°F) and must be protected with thermal insulation during storage to minimize loss through sublimation. Blocks of dry ice weighing from 23 to 27 kg (50 to 60 lb) are usually wrapped in heavy kraft paper bags and stored and shipped in insulated containers and storage boxes of varying sizes. Extruded dry ice pellets are usually stored and shipped in bulk or bags packed in insulated containers. Nitrogen is shipped as a non-liquified gas at pressures of 14,000 kPa (2,000 psi) or above, and also as a cryogenic fluid at pressures below 1,400 kPa (200 psi). Nitrogen gas is authorized for shipment in cylinders, tube tank cars, and tube trailers. Liquid nitrogen is shipped as a cryogenic fluid in vacuum-insulated cylinders, and in insulated portable tanks, tank trucks, and tank cars.

PERFORMANCE

No documentation was obtained for spills of hazardous materials.

CONTACT INFORMATION

Carbon Dioxide - Gas and Solid

Union Carbide Canada Inc.
1210 Sheppard Avenue East
Box 38
Willowdale, Ontario
M2K 1E3

Telephone (416) 490-0052
Facsimile (416) 490-0051

Liquid Carbonic Incorporated
140 Allstate Parkway
Markham, Ontario
L3R 5Y8

Telephone (905) 477-4141
Facsimile (905) 477-6760

Nitrogen

Norwood Carbonation CO$_2$ Supply Company
118 Arrow Road
North York, Ontario
M9M 2M1

Telephone (416) 233-5321
Facsimile (416) 745-2664

Air Products Canada Limited
2090 Steeles Avenue East
Brampton, Ontario
L6T 1A7

Telephone (905) 791-2530
Facsimile (905) 791-6808

Cominco Limited
220 Burrard Street
Suite 400
Vancouver, British Columbia
V6C 3L7

Telephone (604) 682-0611
Facsimile (604) 685-3041

Union Carbide Canada Inc.
1210 Sheppard Avenue East
Box 38
Willowdale, Ontario
M2K 1E3

Telephone (416) 490-0052
Facsimile   (416) 490-0051

Liquid Carbonic Incorporated
140 Allstate Parkway
Markham, Ontario
L3R 5Y8

Telephone (905) 477-4141
Facsimile   (905) 477-6760

REFERENCES

(1) Bennet, G.F., F.S. Feates, and I. Wilder, *Hazardous Materials Spills Handbook,* McGraw-Hill Book Company, Toronto, Ont. (1982).
(2) Blakey, P. and G. Orland, "Using Inert Gases for Purging, Blanketing, and Transfer", in: *Chemical Engineering,* (May 28, 1984).
(3) Compressed Gas Association Inc., *Handbook of Compressed Gases,* Third Edition, Van Nostrand Reinhold, New York, NY (1990).
(4) Holmes, J. M. and C.H. Byers, *Countermeasures to Airborne Hazardous Chemicals,* Noyes Publications, Park Ridge, NJ (1990).
(5) Noyes, R., *Handbook of Pollution Control Processes,* Noyes Publications, Park Ridge, NJ (1991).

1.5 Containment - Vapour Control

MECHANICAL COVERS - GENERAL LISTING — No. 35

DESCRIPTION

A material used to create a physical cover, or "lid", over a spilled chemical to prevent the release of vapours. Common mechanical covers include large geotextile sheets, freestanding structures or tents, spray-on covers such as urethane, and buoyant particles. See also the following entries in Section 1.2 and in this section.
   Froth-Pak Portable Foam System (No. 22)
   Airotech EcoSpheres (No. 36)
   Sprung Instant Structures (No. 37)

OPERATING PRINCIPLE

A cover is placed over the spilled chemical creating a physical barrier to the release of hazardous vapours. Mechanical covers contain the vapours so that they can be collected and removed or treated in situ.

STATUS OF DEVELOPMENT AND USAGE

Synthetic membrane covers serve as linings for oxidation ponds, evapouration ponds, tanks, aeration ponds, and waste treatment ponds. These materials are also used to cover landfills to control odors and blowing debris. Floating covers have been used to contain vapours over lagoons. Freestanding structures have been used in soil remediation projects to contain vapours released by contaminated soils. Sprayed covers have been investigated as a means of vapour control, although foams seem to offer more potential in this regard. Buoyant particles are commonly used in open storage tanks, ponds, and reaction vessels to restrict evapourative losses.

TECHNICAL SPECIFICATIONS

Physical

Synthetic Membrane Covers - Synthetic membranes or geotextiles are available in varying thicknesses of a number of materials such as butyl rubber, EPDM, Neoprene, Hypalon, polyethylene, vinyl chloride, and polyurethane.

Freestanding Structures - Freestanding structures are available in unlimited sizes with free-span construction.

Sprayed Covers - Urethane has been the material of choice with sprayed covers.

Buoyant Particle Covers - Buoyant particles are available in polypropylene, high-density polyethylene, and low-density polyethylene.

Operational

Synthetic Membrane Covers - When selecting covering membranes and methods of deployment, the following factors must be considered: probable spill size; compatibility of the membrane with the spilled material; weight and cost of the membrane; portability, availability, and reuse requirements; and interfacing with the vapour and liquid removal operations. Sheets can be provided with reinforcing fabrics for added strength. Long lengths are feasible, but larger sections are usually made by an appropriate joining method, which includes mechanical systems such as zippers, Velcro, or welding. In all but small spills, deployment may be difficult.

Freestanding Structures - These can be constructed off site and lifted over the site with a crane or other lifting device. The structures are modular, portable, easy to relocate, and designed to withstand severe winds and snowloading. Negative pressures can be maintained within the structure, aiding in vapour recovery from ground spills.

Sprayed Covers - These are applicable primarily to spills on land. In all but small spills, deployment may be difficult.

Buoyant Particle Covers - Buoyant particles can cover up to 91% of the surface area of a spill, reducing evapourative losses and insulating the spilled material against heat, thereby reducing concentrations of vapours above the spill.

Power/Fuel Requirements

None

Transportability

Transportability varies with the type and amount of cover to be used.

PERFORMANCE

Any available performance data are presented in Entries No. 22, 36, and 37.

CONTACT INFORMATION

The following is contact information for all covers discussed.

Synthetic Membrane Covers

Fluid Systems Incorporated
1245 Corporate Blvd., Suite 300
Aurora, IL 60504
U.S.A.

Telephone (708) 898-1161
Toll-free (800) 346-9107 (North America)
Facsimile (708) 898-1179

For detailed information on other geotextiles, refer to Section 3.3, Liners.

Freestanding Structures

Sprung Instant Structures Limited
1001 - 10th Avenue, S.W.
Calgary, Alberta
T2R 0B7

Telephone (403) 245-3371
Toll-free (800) 661-1163
Facsimile (403) 229-1980

Sprayed Covers

Insta-Foam Products, Inc.
1500 Cedarwood Drive
Joliet, IL 60435-3187
U.S.A.

Telephone (815) 741-6800
Toll-free (800) 800-3626 (North America)
Facsimile (815) 741-6822

Insta-Foam Products, Inc.
P.O. Box 21
Etobicoke, Ontario
M9C 4V2

Telephone (416) 622-6844/45
Facsimile (416) 626-2818

Local distributors throughout Canada and the United States retail the company's products.

Buoyant Particles

Airotech Incorporated
Boyle Centre, 120 East Ninth Avenue
Homestead, PA 15120-1600
U.S.A.

Telephone (412) 462-4404
Facsimile (412) 462-8816

OTHER DATA

For specifications, refer to specific entries for mechanical covers (No. 22, 36, and 37). For further information on geotextiles, refer to Section 3.3, Liners.

REFERENCES

(1) Bennet, G.F., F.S. Feates, and I. Wilder, *Hazardous Materials Spills Handbook,* McGraw-Hill Book Company, Toronto, Ont. (1982).
(2) Holmes, J.M. and C.H. Byers, *Countermeasures to Airborne Hazardous Chemicals*, Noyes Publications, Park Ridge, NJ (1990).
(3) Manufacturer's Literature.
(4) Noyes, R., *Handbook of Pollution Control Processes,* Noyes Publications, Park Ridge, NJ (1991).

AIROTECH ECOSPHERES No. 36

DESCRIPTION

Polypropylene and high-density and low-density polyethylene spheres used as a mechanical cover over chemical spills.

OPERATING PRINCIPLE

The floating spheres can cover up to 91% of the surface area of a spill, reduce evaporative losses, and insulate the spilled material against heat to further reduce evapourative losses, thus reducing the concentrations of vapours above the spill.

STATUS OF DEVELOPMENT AND USAGE

Commercially available product used as floating covers for controlling evapouration loss, heat loss, oxygen absorption, algae growth, foam production, and for fire protection. Also marketed as packing for wet scrubbers.

TECHNICAL SPECIFICATIONS

Physical

Material	Polypropylene, high-density polyethylene (HDPE), or low-density polyethylene (LDPE)
Sizes	Available in diameters of 19, 25, 38, 51, and 76 mm (0.8, 1.0, 1.5, 2.0, and 3.0 in)
Colour	Natural white. LDPE spheres are available by special order in a wide range of colours.

Operational

EcoSpheres will cover 91% of the surface area regardless of the sphere diameter, reducing the evapourative losses and minimizing the formation of vapour clouds. Depending on the size of sphere used, 3 to 3.7 kg/m^2 (10 to 12 oz/ft^2) of EcoSpheres are required to cover a spill with one layer. Slows the spread of flames and assists fire extinguishing agents to extinguish fire. EcoSpheres can lower the ice formation point by 68°C (20°F) for aqueous liquids.

Power/Fuel Requirements

None

Transportability

EcoSpheres are lightweight and should be easy to transport. The amount required to cover the spill must be considered.

PERFORMANCE

No documentation was available on the ability of EcoSpheres to reduce evapouration or vapour concentrations for a spill of hazardous chemicals. The following are the results of a study conducted by the manufacturer on the insulative capabilities of the spheres.

Approximate Heat Loss in kW/m^2 at 40% Relative Humidity

Water Temperature (°C)	Without EcoSpheres	With One Layer of EcoSpheres	Heat Savings (%)
40	0.9	0.7	25
50	1.7	1	44
60	2.9	1.2	59
70	4.8	1.7	64
80	7.5	2.1	72
90	12	3.1	73

Manufacturer

Airotech Incorporated
Boyle Centre, 120 East Ninth Avenue
Homestead, PA 15120-1600

Telephone (412) 462-4404
Facsimile (412) 462-8816

REFERENCES

(1) Manufacturer's Literature.

SPRUNG INSTANT STRUCTURES

No. 37

DESCRIPTION

Modular structures, which can be quickly erected, dismantled, or relocated, designed to provide a sealed enclosure that maintains a positive or negative pressure environment for capturing and containing emissions of hazardous vapours for removal or in-situ treatment.

OPERATING PRINCIPLE

Designed and engineered according to the Membrane Stress Theory, the structure is constructed of lightweight aluminum arches integrally connected to an all-weather outer membrane. The structure is freestanding and can be assembled directly over the spill site or assembled off-site and crane-lifted or moved on a track wheel system to the spill site.

STATUS OF DEVELOPMENT AND USAGE

Commercially available on a lease or full-purchase basis, Sprung Instant Structures are used worldwide for transuranic waste treatment and storage, radioactive vault storage, soil remediation, asbestos abatement, cleanup of hazardous waste sites, and for other uses such as warehousing, construction, and exhibitions.

TECHNICAL SPECIFICATIONS

Physical

Sprung Instant Structures are free-span design, constructed from extruded aluminum arches, integrally connected to an all-weather membrane of PVC-coated polyester scrim, treated with inhibitors to prevent degeneration from ultraviolet rays. The membrane is fire-retardant, i.e., self-extinquishing. The modular design allows built-up areas of unlimited length, easy extension of existing structures, and quick dismantling and relocation. The following are detailed specifications for the standard structure modules.

Width	Module Length	Structure Length	Height	Weight
9 m (30 ft)	3 m (10 ft)	unlimited	4.8 m (16 ft)	7.3 kg/m^2 (1.5 lb/ft^2)
12 m (40 ft)	3 m (10 ft)	unlimited	5.6 m (18 ft)	7.3 kg/m^2 (1.5 lb/ft^2)
15 m (50 ft)	3 m (10 ft)	unlimited	7.2 m (24 ft)	7.3 kg/m^2 (1.5 lb/ft^2)
18 m (60 ft)	3 m (10 ft)	unlimited	8 m (26 ft)	7.3 kg/m^2 (1.5 lb/ft^2)
27 m (89 ft)	3 m (10 ft)	unlimited	11 m (36 ft)	12 kg/m^2 (2.5 lb/ft^2)
31 to 40 (100 to 130 ft)	4.6 m (15 ft)	unlimited	11 to 14 m (35 to 46 ft)	18 kg/m^2 (3.6 lb/ft^2)

Operational

With little or no ground preparation, structures from 9 to 40 m (30 to 130 ft) wide and of unlimited length can be erected in one to six days. Erection with a crew of six to ten unskilled labourers can proceed at a rate of 110 to 140 m^2 (1,200 to 1,500 ft^2) per eight-hour day, or faster with a larger work force and longer work day. They can be erected with minimal or no foundation on any reasonably flat, firm surface and provide a sealed enclosure that can be customized to maintain a positive or negative pressure. A negative pressure environment can actually "vacuum" the vapours from porous surfaces such as soil and prevent vapour from escaping by pulling fresh air into the structure through any leaks, rather than allowing vapour to escape. Ventilation and in-situ vapour treatment systems are easily installed within the flexible membrane.

1.5 Containment - Vapour Control

Designs are also available for double-skinned structures or structures within structures to create double-wall containment. A personnel door is provided with each structure. Additional doors, double personnel doors, single or double personnel doors with panic hardware, vehicle doors, and oversize vehicle doors are also available as options. The structures can be designed with double doors to create a vestibule to minimize vapour release. A vehicle can then enter through the first door, which closes behind it before the second door is opened. The structures can be designed to withstand windloads of more than 209 km/h (130 mph) and to shed snow. A crane or cherry picker is required to erect structures more than 15 m (50 ft) wide.

Power/Fuel Requirements

Sprung Instant Structures are freestanding and require no power supply. Environmental control options, such as lighting, ventilation, air conditioning, or vapour treatment systems, will require appropriate power sources.

Transportability

Each module of the Instant Structures can be further broken down for transportation. The aluminum arches come apart and the membrane covering can be removed and folded into a compact bundle. Any size of structure can be transported on a flatbed truck.

PERFORMANCE

Sprung Instant Structures have been used successfully by the U.S. EPA as components in remediation at Superfund Sites. One example is a trial excavation of waste material at the McColl Superfund Site in Fullerton, California in 1990. The site is an inactive hazardous waste disposal facility used to dispose of acidic refinery sludge from the production of aviation fuel during the Second World War and oil exploration in the 1950s. In the 1980s, residents of nearby housing developments complained of odours from the site. The U.S. EPA and the California Department of Health Services subsequently found extensive contamination at the site. The trial was to determine whether waste could be excavated with conventional equipment without releasing significant amounts of volatile organic compounds (VOCs) and sulphur dioxide (SO_2) to the surrounding community. To safeguard against this, a Sprung Instant Structure [18 m wide, 49 m long, and 8 m high (60 × 160 × 26 ft)] was erected over the excavation site. The waste was excavated within the structure with a diesel-powered backhoe. During excavation of a tar layer, levels of SO_2 and total hydrocarbons increased dramatically within the enclosure and reached five-minute average values of 1,000 ppm for SO_2 and 492 ppm for total hydrocarbon content. The exhaust treatment system within the enclosure removed up to 99.9% of the SO_2 and up to 90.7% of the total hydrocarbon content. For additional information on this project, see U.S. EPA (1990).

CONTACT INFORMATION

Manufacturer

Sprung Instant Structures Limited
1001 - 10th Avenue S.W.
Calgary, Alberta
T2R 0B7

Telephone (403) 245-3371 or (800) 661-1163
Facsimile   (403) 229-1980

REFERENCES

(1) Manufacturer's Literature.
(2) Personal communication, Peter Bos, Sprung Instant Structures, Calgary, Alta. (November, 1995).
(3) U.S. EPA (United States Environmental Protection Agency), "SITE Program Demonstration of a Trial Excavation at the McColl Superfund Site", Superfund Innovative Technology Evaluation, Fullerton, CA (November, 1990).

VAPOUR SUPPRESSING FOAMS - GENERAL LISTING — No. 38

DESCRIPTION

A range of foams that control the vapour hazard from certain classes of hazardous materials through the formation of a continuous layer of bubbles or a thin film. A **finished foam** is the homogeneous blanket obtained by mixing a foam concentrate, water, and air to create an aggregate of air-filled bubbles. A **foam solution** is the homogeneous mixture of water and the foam concentrate. **Foam generators** are mechanical devices that agitate the foam solution to produce a bubble structure which makes up the foam and discharges it onto the spill.

OPERATING PRINCIPLE

The foam blanket can temporarily reduce the concentration of vapour above the spill surface, decrease the evapouration rate, provide a barrier to thermal solar radiation, and in some cases inhibit ignition or flame propagation if ignition does occur. Low expansion foams mitigate a spill by forming a barrier to vapourization. High expansion foams appear to engulf the vapours, thereby reducing the vapour concentration directly above the foam layer.

STATUS OF DEVELOPMENT AND USAGE

Commercially available products formulated for and widely used to extinguish fires resulting from the ignition of flammable spills and to mitigate vapour release from spills of hazardous materials.

TECHNICAL SPECIFICATIONS

Physical - The following are the three **categories of foam**.

Low Expansion Foams - These foams have an expansion ratio of between 2:1 and 20:1 and have proven in actual use to be an effective means of controlling, extinguishing, and securing most flammable liquid (Class B) fires.

Medium Expansion Foams - These foams are primarily used to suppress the release of hazardous chemical vapours. They have an expansion ratio of between 20:1 and 200:1. Foams with expansion ratios between 45:1 and 55:1 produce the optimal foam blanket for vapour mitigation of highly water-reactive chemicals and low boiling organics.

High Expansion Foams - High expansion foam liquid concentrates are synthetic, detergent-type foaming agents designed for firefighting in confined spaces. This concentrate is mixed at 1 to 1/2% with water and then expanded with air to form high expansion foam. When used with a High Expansion Foam Generator, it makes a foam with an average expansion ratio of 500:1. It can also be used with generators producing foams from 200:1 to 1,000:1.

The following are the six **types of foam** used today, all of which are either protein-derived material or surfactant-based materials. Each is used for a specific purpose and set of circumstances.

Protein - These are the first type of mechanical foam to be marketed and used extensively. These foams are produced by hydrolysis of waste protein material such as hoof and horn. Stabilizing additives and inhibitors are included to prevent corrosion, resist bacterial decomposition, and control viscosity. The "alcohol foams" designed for use with polar compounds are normally protein-based, although polar AFFF foams [Alcohol Resistant Aqueous Film Forming Foam (AR-AFFF)] are now available.

Fluoroprotein Foam (FP) - These foams are formed by adding special fluorochemical surfactants to protein foam.

Film Forming Fluoroprotein Foam (FFFP) - These are a mixture of fluorochemical surfactants and protein foam.

Aqueous Film Forming Foam (AFFF) - These foams are a mixture of fluorochemical surfactants and synthetic foaming agents that creates an aqueous film. This film is a thin layer of foam solution with unique surface energy characteristics that spreads rapidly across the surface of a hydrocarbon fuel causing dramatic fire knock-down.

Alcohol Resistant Aqueous Film Forming Foam (AR-AFFF) - These foams are produced by mixing synthetic stabilizers, foaming agents, fluorochemicals, and alcohol-resistant, membrane-forming additives. Polar solvents and water-miscible fuels such as alcohols are destructive to non-alcohol resistant types of foam. Alcohol mixes aggressively with the water in the foam and destroys the foam blanket.

1.5 Containment - Vapour Control

Synthetic Detergent Foam (Medium and High Expansion) - This is a mixture of synthetic foaming agents with additional stabilizers. Medium expansion synthetic detergent foam is used to suppress hazardous vapours. Specific foams are required for specific chemicals. Acid, alkaline, and neutral products, such as hydrocarbons and polar solvents, all require a specific foam. These foams are designed to produce stable blankets, resist product pickup, and to have slow drainage rates, characteristics that are essential for controlling the emission of vapours from hazardous materials. Concentrate supplied in pails or drums is mixed with water and air through an eductor to produce the finished foam.

Operational - Low-expansion foams provide the longest time delay before vapour breaks through and are less influenced by wind, rain, or high temperature than other foams. However, they require larger volumes of water than high-expansion systems to be effective. This water can spread the spill or overflow impoundments as it drains from the foam structure. High-expansion foams require about half the water than the equivalent cover of low-expansion foams.

The decision to use high expansion foams will be influenced by wind conditions, the containment of the spill, and the nature of the spilled material. With winds less than 16 km/h (10 mph), it should be possible to maintain an adequate high-expansion blanket. At greater wind speeds, the foam mass will have to be contained with fencing or some other structure.

Foam is not recommended for use on materials that can be stored as liquids but are vapour at ambient conditions, such as propane, butadiene, and vinyl chloride. Firefighting foam is not recommended for use on materials that react with water. The capabilities of the different types of foams to suppress or otherwise minimize the release of vapours from various chemicals, based on experience and U.S. EPA reporting, are shown in the following table (Holmes and Byers, 1990).

Product	Recommended	Satisfactory	Not Recommended
Polar Solvents	Alcohol-resistant		
Silicon tetrachloride vapours	High expansion surfactant		
Octanol	Alcohol-resistant	AFFF	
		Fluoroprotein	
		Protein	
LNG	Surfactant		AFFF
Vinyl acetate	Alcohol-resistant, fluoroprotein	AFFF	
Ethane	Surfactant	Alcohol-resistant	
		Fluoroprotein	
		Protein	
Ethylene	Surfactant	Alcohol-resistant	
		Fluoroprotein	
		Protein	
Hepatane	Alcohol-resistant	AFFF	
Octane	Alcohol-resistant	AFFF	
		Fluoroprotein	
		Protein	
		Surfactant	
Benzene	Alcohol-resistant	Fluoroprotein	
		Protein	
		Surfactant	
Ethyl Benzene	AFFF	Surfactant	
	Fluoroprotein		
	Alcohol-resistant		
	Surfactant		
Toluene	Alcohol-resistant	AFFF	
	Protein	Surfactant	
	Fluoroprotein		
Cyclohexane	Alcohol-resistant	AFFF	
	Protein	Surfactant	
	Fluoroprotein		
Gasoline	Alcohol-resistant	AFFF	
	Fluoroprotein		
	Surfactant		

Product	Recommended	Satisfactory	Not Recommended
Kerosene	Alcohol-resistant Fluoroprotein Surfactant	AFFF	
Naphtha	Alcohol-resistant Fluoroprotein Surfactant	AFFF	
Paint thinner	Alcohol-resistant Protein Fluoroprotein	AFFF Surfactant	
Ammonia	Surfactant	AFFF Alcohol Protein Fluoroprotein	
Triethylamine	Alcohol-resistant	Fluoroprotein	
Ethyl ether	Alcohol-resistant		
n-butyl acetate	Alcohol-resistant	Fluoroprotein AFFF Protein Surfactant	
Methyl acrylate	Alcohol-resistant		
Octane	Alcohol-resistant	Protein Fluoroprotein Surfactant AFFF	
Benzene	Alcohol-resistant	Protein Fluoroprotein Surfactant AFFF	
Cyclohexane	Protein Alcohol resistant Fluoroprotein	Surfactant	
Ethylene oxide	Alcohol-resistant Polar liquid		
Vinyl chloride	Alcohol-resistant		

Power/Fuel Requirements - Foam application equipment, including pumps, hose, and a foam generator or eductor, is required to apply foam to a spill. Eductors typically need water supplied at a rate of 38 m^3/s (60, 95, 125, or 250 gpm) and pressures of between 1,200 to 1,400 kPa (180 to 200 psi).

Transportability - Foam concentrates are typically supplied in readily-transportable pails or drums.

PERFORMANCE

Experimental results using several foams on different chemicals are shown in the following table.

Total Vapour Suppression Time [50 cm (20 in) of foam]

Chemical	Type of Foam	Total Suppression Time (minutes)
Toluene	Emulsiflame 2%	11
Cyclohexane	Emulsiflame 2%	20
Ethylbenzene	Emulsiflame 2%	11
Cyclohexane	Lorcon Full-Ex 2%	22
Benzene	Lorcon Full-Ex 2%	25
Toluene	Lorcon Full-Ex 2%	40
Ethylbenzene	Lorcon Full-Ex 2%	40

1.5 Containment - Vapour Control

CONTACT INFORMATION

The following are several representative manufacturers, with comments on the products they manufacture or supply.

Manufacturer	Product(s)	Comments
3M Canada Incorporated P.O. Box 5757 London, Ontario N6A 4T1 Telephone (519) 451-2500 Toll-free (800) 410-6880 Facsimile (519) 452-6262	3M Light Water ATC Stabilizer FS-7000	Light Water ATC can extinguish flammable liquid hazards and suppress toxic and corrosive vapours from the following classes of chemical compounds: alcohols, liquid hydrocarbons, ketones, esters, ethers, chlorinated hydrocarbons, amines, acrylates, organic acids, isocyanates.
Ansul Fire Protection International Dept. One Stanton Street Marinette, WI 54143-2542 U.S.A. Telephone (715) 732-7411 Facsimile (715) 735-3471		Ansul foams, formerly Lorcon foams, are manufactured and distributed by Wormald U.S., Inc. in the United States. Ansul Products are available throughout Canada from Levitt Safety Ltd.
Chubb National Foam 150 Gordon Drive, P.O. Box 270 Exton, PA 19341-1350 U.S.A Telephone (610) 363-1400 Facsimile (610) 524-9239	Aer-O-Foam (Protein) and Aer-O-Foam (Fluoroprotein) Aer-O-Water (AFFF) Universal Plus and Gold (AR-AFFF) Vapour Shield (Alkali) Vapour Shield (Acid) Vapour Shield (Organic)	Suitable for Class B fires and hydrocarbon vapour suppression. Suitable for Class A and B fires and hydrocarbon vapour suppression. Suitable for Class A, B, and C fires and hydrocarbon and polar solvent vapour suppression. Suitable for alkaline vapour suppression. Suitable for acid vapour suppression. Suitable for organics vapour suppression. Chubb National Foam was formerly National Foam Systems Incorporated.

Common Terminology Associated with Foam

The following are some common terminology and acronyms associated with the use of foams, adapted from the United Nations Environment Programme report "Theory and Practice of Foams in Chemical Response".

Application Rate - The rate at which foam solution is applied. Usually expressed in litres of solution per square metre of fire surface area per minute $[L/(m^2 \cdot min)]$.
AFFF - Aqueous Film Forming Foam Compound
AR-AFFF - Alcohol-resistant Aqueous Foam
Base Injection - The introduction of foam underneath the surface of a liquid.
Burn Back Resistance - The ability of a foam blanket to resist direct flame impingement.
Bund Area - Dyke Area
Concentration - The amount of foam compound contained in a given volume of foam solution. The amount of foam compound actually used determines the percentage of concentration.
Critical Application Rate - The lowest rate at which foam solution can be applied to a fire.
Dissolved Solids - Solid matter left after evaporating all water from a foam compound.
Drainage Rate - The rate at which water drains from a finished foam.
Expansion Ratio - The ratio of the volume of total finished foam to the volume of foam solution from which it is made.
FFFP- Film-Forming Fluoroprotein Foam
Flashbacks - Reignition of flammable liquid caused by exposure of vapours to a source of ignition such as hot metal surfaces or a spark.

Fluoroprotein Foam Compound - A foam compound based on hydrolyzed protein with surface-active fluorocarbon surfactants added.
Finished Foam -The homogeneous blanket obtained by mixing compound, water, and air and consisting of an aggregate of air-filled bubbles.
Foam Solution - The homogeneous mixture of water and concentrate (compound).
FP - Fluoroprotein Foam
Gentle Application - Application of foam to the surface of a liquid fuel via a backboard, tank wall, or other surface.
High Expansion Foam - Foam with an expansion ratio from 200 to 1,000:1.
Inductor - A device, usually in a water line, for introducing foam compound to water to form a solution.
Induction Rate - The percentage of foam compound mixed or induced into the water supply to achieve a solution.
Low Expansion Foam - Foam with an expansion ratio of between 2 and 20:1.
Medium Expansion Foam - Foam with an expansion ratio of between 20 and 200:1.
Non-aspirated Foam - Foam with a typical expansion ratio from 0 to 2:1. Only Film Forming Compounds are suitable.
Polar Solvents - Generally water-miscible solvents that require special foam compounds, e.g., alcohols and ketones.
Pour Point - The lowest temperature at which a foam compound is fluid enough to pour - generally a few degrees above freezing.
Pre-mix Solution - A mixture of water and foam compound in correct proportions.
Sediment - Insoluble particles in the foam concentrate.
Stability - A term used with foam concentrates to determine the permanence and security of a foam blanket.
Sub-surface Injection - Base injection.
Solution Transit Time - The time taken for the foam solution to pass from the point at which foam compound is introduced to the water supply to when aeration takes place.

For further information on commercially available, specifically formulated spill control agents and sorbents, refer to the specific entries in Section 5.1, Spill Treating Agents.

Foam application equipment is generally available from foam manufacturers.

REFERENCES

(1) Bennet, G.F., F.S. Feates, and I. Wilder, *Hazardous Materials Spills Handbook,* McGraw-Hill Book Company, Toronto, Ont. (1982).
(2) Holmes, J. M. and C.H. Byers, *Countermeasures to Airborne Hazardous Chemicals*, Noyes Publications, Park Ridge, NJ (1990).
(3) Manufacturer's Literature.
(4) Noyes, R., *Handbook of Pollution Control Processes,* Noyes Publications, Park Ridge, NJ (1991).

Section 2.1

Removal - From Land

ENVIROPAC EMERGENCY SPILL CLEANUP KITS No. 39

DESCRIPTION

Prepackaged spill cleanup kits for quick response to chemical, herbicide/pesticide, oil/petrochemical, or polychlorinated biphenyl (PCB) spills.

OPERATING PRINCIPLE

The kits can be readily transported to a spill site for quick cleanup. Kits include drums, containment materials, protective clothing, respiratory equipment, cleanup materials, and instructions. Kits are packaged in either steel or polyethylene Enviropac Vaults. Recovered materials can be stored in the vault pending disposal.

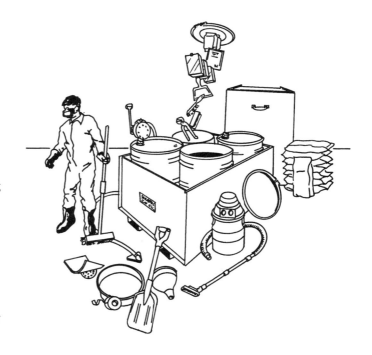

STATUS OF DEVELOPMENT AND USAGE

Commercially available product in general use. The PCB Spill Cleanup Kits (model SEP-455) are presently in place at Armed Forces bases across Canada and several hundred Enviropac Vaults were used when PCBs were removed from the Distant Early Warning (DEW) line sites.

TECHNICAL SPECIFICATIONS

Physical

Seven kits are available with configurations based on their intended application and with equipment for two or four people. The four-person kits are contained in steel or polyethylene Enviropac Vaults with the following specifications.

	Polyethylene Vault	Steel Vault
Material	Molded polyethylene with UV inhibitor	16 gauge carbon steel with polyurethane coating
Weight	72 kg (160 lb) vault only	145 kg (320 lb) vault only
Dimensions	1.4 × 1.4 × 1.2 m (4.5 × 4.5 × 3.8 ft)	1.1 × 1.5 × 1 m (3.8 × 5 × 3.5 ft)
Capacity	Four 208-L (55-gal) drums	Four 208-L (55-gal) drums
Sump Capacity	760 L (200 gal)	2,000 L (315 gal)
Colour	Safety green	Safety green

Kit contents vary depending on the application. Standard kits are General Purpose for general chemical spills, Herbicide/Pesticide, Oil/Petrochemical, and Polychlorinated Biphenyl (PCB). Custom designed kits are also available for special applications. The kits may contain the following:

Containers and Access Equipment
208-L (55-gal) DOT-17E closed top drum
208-L (55-gal) DOT-17E open top drum
PCB label
19-L (5-gal) DOT-17C/H steel pail and lid
19-L (5-gal) polyethylene pail and lid
208-L (55-gal) drum liner, 1.7 mm (65 mil) poly
0.05-mm (2-mil) gusseted bags,
    30 × 20 × 53 cm (12 × 8 × 21 in)
Grounding cable, 3.7 m (12 ft), #8
Bung wrench
Pliers

Testing, Neutralizing and Containment
pH paper
Acid neutralizer, 0.95 L (1 qt)
Alkaline neutralizer, 0.95 L (1 qt)
Neutralizer spray pump
Flexible drain cover, 46 × 46 cm (18 × 18 in)
Absorbent sheets, 46 × 46 cm (18 × 18 in)
Safestep sorbent, 11-kg (25-lb) bag
Absorbent chips, floating
Absorbent boom
Barricade tape
"Danger-Chemical Spill" sign

Cleanup Material
Scoop shovel, non-sparking, poly
Scoop shovel, steel
Push broom
Brush and squeegee, driveway type
Bench brush
Deck brush
Funnel, non-sparking, poly
Drip pan
Shop rags
Hand pump, 2.4-m (8-ft) hose, bung adaptor
Pump suction strainer
Modified wet/dry vacuum
15 m (50 ft) of 1.6-cm (5/8-in) vinyl hose, with shutoff
1,1,1-trichloroethane, 3.8 L (1 gal)
Mineral spirits solvent, 3.8 L (1 gal)

Protective Clothing/Equipment
Saranex/Tyvek coveralls
Saranex/Tyvek shoe covers
Neoprene boots
Neoprene gloves
Nitrile gloves
Vinyl glove liners
Chemical splash goggles
Duct tape, 1 roll
Toxic dust and mist respirators
Half mask respirator and filter retainers
Respirator cartridge
Respirator filter

Operational

Enviropac Vaults satisfactorily exceed all sump capacity regulations in the United States. The steel vaults can be stacked to a maximum of three high.

Power/Fuel Requirements

The cleanup equipment has no power or fuel requirements. A drum-lifting sling and a forklift or other lifting device are required to load full, 208-L (55-gal) drums into the storage vault after the cleanup is completed.

Transportability

The kits contained in the Enviropac Vaults must be moved with a forklift. Kits are available on wheels for moving on smooth surfaces.

PERFORMANCE

Performance of the kits has not been documented. All components are standard, off-the-shelf items that should adequately remove the material for which the kits are designed.

CONTACT INFORMATION

Manufacturer

Environmental Container Corporation
31700 S.W. Riedweg Road
Cornelius, OR 97113-9609
U.S.A.

Telephone (503) 693-0774
Toll-free   (800) 729-7137
Facsimile (503) 648-3136

OTHER DATA

Available accessories include drain valve kits, drum-lifting slings, hinge kits, kits on wheels, and barrel-dispensing racks.

REFERENCES

(1) Manufacturer's Literature

Section 2.2

Removal - From Surface Water

MECHANICAL SKIMMERS - GENERAL LISTING   No. 40

DESCRIPTION

Stationary or mobile systems that mechanically recover floating contaminants from water.

OPERATING PRINCIPLE

Skimmers are classified according to the following general collection principles.

Weir Skimmers - A fixed or self-leveling lip at the spill-water interface allows contaminants to overflow into a collection sump, which is cleared by an onboard or remotely-operated pump.

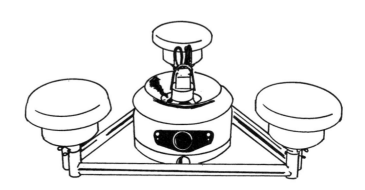

Suction Devices - Spilled products are drawn through a skimming head. Recovered fluids are then pumped to temporary storage facilities or into a vacuum truck.

Sorbent-surface Skimmers - Contaminants are collected with a rotating chemical-attracting surface, such as a drum, disc, brush, belt, or rope mop, which is then scraped or squeezed clean of collected material.

Hydrodynamic and Other Devices - Progressive cavities, hydrocyclone chambers, rotating paddle belts, submersion planes, and boom/skimmer combinations are used to redirect, submerge, or concentrate contaminants for recovery.

STATUS OF DEVELOPMENT AND USAGE

Most mechanical skimmers have been well tested for oil spill response and are commercially available and widely used. Their use for chemical spill response has not been well documented, however, and is restricted to tank or laboratory testing of a few hazardous substances.

TECHNICAL SPECIFICATIONS

Physical

Type	Simple vs auger weir, suction, drum, disc, rope mop, submersion plane, etc.
	Associated dimensions and materials
Dimensions	Overall height and width; length; skimming width; draft; discharge port diameter
Construction	
Materials	Type (a critical factor in long-term chemical spill response)
	Ruggedness and durability
Flotation	Type (closed-cell foam, sealed air chamber)
	Location (built-in flotation chambers, outriggered floats)
Other	Debris handling
	Lifting and tethering points and handholds
Pump	Type and manufacturer
	Length, height, and width
	Intake and discharge port diameters
	Weight
	Engine power/fuel requirements
	Self-priming/ability to run dry/debris tolerance
	Capacity
	Construction (liners, connectors, pump housing, and prime mover materials)

Operational - Skimmers designed for use in ditches, manholes, lakes, rivers, harbours, coastal, and offshore environments are commercially available, but recovery of chemical spills will probably only be attempted in smaller water courses with the aid of lightweight, portable skimmer devices, perhaps in conjunction with booming operations. Typical operational specifications for mechanical skimmers are listed according to collection principle. **Note: Recovery efficiency** is the percentage of spilled product contained in the total quantity of liquid collected, and **recovery rate** is the volume of spilled product collected per unit of time.

Weir Skimmers - Simple, portable design allows for inexpensive recovery of relatively non-viscous fluids. Although weir skimmers are known to recover large amounts of water, efficiency further diminishes during recovery of viscous fluids and thin slicks, and in conditions of rough water and high debris content.

Suction Devices - Simple operation and high mobility allow suction devices to be used in confined, shallow, or varied conditions. Collection of large amounts of water should be expected, as should diminished efficiency in thin slicks and rough water conditions. Vacuum trucks are effective in recovering viscous, weathered products and some forms of debris.

Sorbent-surface Skimmers - Varied designs can achieve high contaminant-to-water recovery rates for a broad range of chemical products, even in moderately rough water. Cost and debris-handling vary with size and sophistication of skimmer.

Hydrodynamic and Other Devices - It has been found that these devices do not have significant application to chemical spill response. Little information is available on any recent advancements in chemical spill recovery on surface water by mechanical devices.

Power/Fuel Requirements - Power or fuel (electric, petroleum) are required to operate pumps and hydraulic power packs, although exact power and fuel requirements vary according to skimmer type and sophistication.

Transportability - Transportation and handling requirements vary according to the type of skimmer. A small, lightweight electric weir skimmer can easily be carried and deployed by one or two persons, while a rugged, high-capacity hydrodynamic skimmer designed for offshore use may require equipment to transport and handle several tons of cargo and with overhead lifting capability.

PERFORMANCE

The performance of mechanical skimmers varies greatly even among the same generic class. Skimming capacities advertised by manufacturers are often overstated when compared to performance data collected during testing or *in-situ* operation. Most performance data for mechanical skimmers involve the recovery of oil, not chemicals. Ideally, the performance of each device should be individually assessed with the products likely to be recovered, in order to accurately assess the skimmer's actual capability. The following are general comments on the performance of the various types of mechanical skimmers.

Weir Skimmers provide a high recovery capacity if fitted with self-leveling weirs and/or high throughput pumps. Recovery efficiencies and rates improve dramatically when a thick layer of contaminant is to be recovered and when water conditions are relatively calm and debris-free.

Suction Devices have performance characteristics similar to weir skimmers, but they also recover viscous materials and small pieces of debris. They are well suited to inland and nearshore watercourses where vacuum trucks, temporary storage, and/or fluid separation facilities are available.

Sorbent-surface Skimmers typically offer good recovery rates with high recovery efficiency, particularly in calm water conditions. Although they vary widely in performance, sorbent-surface skimmers can provide very high levels of spill recovery of some chemical products even in relatively thin slick conditions.

Hydrodynamic and Other Devices have limited application to chemical spill response due to their generally large size and complex deployment and operation. Recently developed small skimming devices specifically suited to recovery of hazardous materials may offer new alternatives in chemical spill response.

CONTACT INFORMATION

The "World Catalog of Oil Spill Response Products" (Schulze, 1995) provides general specifications and contact information for commercially produced mechanical skimmers.

REFERENCES

(1) Schulze, Robert (ed.), "World Catalog of Oil Spill Response Products", Fifth Edition, World Catalog Joint Venture (JV), Port City Press, Baltimore, MD (1995).

Section 2.3

Removal - From Groundwater

CEE PRODUCT RECOVERY SYSTEMS No. 41

DESCRIPTION

Automatic hydrocarbon separation and recovery systems designed to remove contaminants from water underground.

OPERATING PRINCIPLE

A surface-mounted, air-driven, double diaphragm pump draws product through a skimmer and discharges it to a product recovery tank on the surface. The skimmer floats on the water in the well, allowing it to automatically follow the fluctuations of the groundwater and positioning the skimmer intake at the hydrocarbon/water interface.

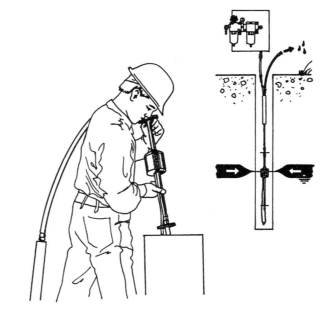

STATUS OF DEVELOPMENT AND USAGE

Commercially available product in general use.

TECHNICAL SPECIFICATIONS

Physical

The following are specifications for the standard configurations of Shallow and Deep Well Systems.

Pumping system	**Shallow Well System** - surface-mounted, air-driven, double diaphragm pump, constructed of aluminum, Viton, and brass. Also available in stainless steel and polypropylene with elastomers of urethane. **Deep Well System** - 61 cm (24 in) or 122 cm (48 in) submersible, air-driven, bladder pump, constructed of stainless steel, brass, Teflon, and Viton
Skimmers	5 cm (2 in) and 10 cm (4 in) diameter **Selective Skimmers** and **Specific Gravity Skimmers**
Control panels	First Response Systems have basic control panels with programmable pneumatic timers, 5-µm (0.19-mil) filter, air pressure regulator, and 0.01-µm (0.000 4-mil) air filter. An indicator inside the control panel turns pink if oil passes through the main filter. Standard Recovery systems have similar controls housed in a steel, double-locked box. The control panels on the standard systems also indicate if the controls are turned on, the fuel tank is full, and if there is a high water condition in the well. The standard systems also come with high-water and tank-full shutoffs.
Hoses	Systems are supplied with industrial grade hoses. Fuel hoses are urethane-covered, steel-reinforced, nylon-core hose, with a 125,000 kPa (18,000 psi) burst pressure and a static ground running the length of the hose.

Operational

The following are maximum capacities for the standard configurations.

	Selective Skimmers		Specific Gravity Skimmers	
	5 cm (2 in)	**10 cm (4 in)**	**5 cm (2 in)**	**10 cm (4 in)**
Shallow Well Double Diaphragm Pump	1,370 L/day (360 gpd)	3,800 L/day (1,000 gpd)	2,660 L/day (700 gpd)	8,200 L/day (2,160 gpd)
Deep Well Bladder Pump 61 cm (24 in)	380 L/day (100 gpd)	608 L/day (160 gpd)	608 L/day (160 gpd)	608 L/day (160 gpd)
Deep Well Bladder Pump 122 cm (48 in)	760 L/day (200 gpd)	1,210 L/day (320 gpd)	1,210 L/day (320 gpd)	1,210 L/day (320 gpd)

The **Selective Skimmer** removes all free-floating hydrocarbon on groundwater to a sheen of about 500 µm (0.02 in) without removing any water. The skimmer float slides up and down on a guide tube that centres the device in the well and also carries oil from the skimmer up out of the well. The skimmer float follows the fluctuation of water level in the well and maintains its designed inlet level at the oil/water interface. A semipermeable screen inside the skimmer float allows oil to pass into the skimmer, but repels water, unless the skimmer is forced 5 cm (2 in) beneath the water. There are no electrical switches or sensors in or around the skimmer. When the oil has been removed to a sheen, the skimmer merely passes air to the pumps, which are designed to run dry without damage.

The **Specific Gravity Skimmer** requires a minimum of 2.5 cm (≥ 1 in) of fuel and removes oil alone or oil and water separately from underground spill sites, using separate pumps. The skimmer system maintains the oil inlet above the oil/water interface by floating freely on groundwater. The float is guided and centred in the well so that the float cannot hang up in the well casing. Water is drawn up from beneath the skimmer through a stainless steel tube which also positions the skimmer in the well. Oil is drawn through a second tube and the fluid level in the well is sensed in a third.

The double diaphragm pump, which can be used to pump recovered product, water, or both, operates outside the well casing, drawing fluids from the well and pushing them to a treatment system or holding tank. It draws fluids from a depth of 6.7 m (22 ft) and pushes them against a 690 kPa (100 psi) head. Sustained pumping rates of 0 to 30 L/min (0 to 8 gpm) with surges of up to 57 L/min (15 gpm) are possible. The bladder pump operates inside the well and draws oil up, out of the skimmer, and to the surface. It operates at depths of 76 m (250 ft) and in wells as small as 5 cm (2 in) in diameter. It is designed to meet stringent air quality standards. As the hydrocarbon never comes into contact with the air inside the pump, the exhausted air is very clean. The 61 cm (24 in) bladder pump transfers 608 L/day (160 gpd) with a cycle rate of four times per minute. The 122 cm (48 in) bladder pump removes 1,210 L/day (320 gpd) with a cycle rate of four times per minute. Both pumps and skimmers can be steam-cleaned without damage.

Power/Fuel Requirements

A supply of compressed air is required.

Transportability

While each individual component is person-portable, a pickup truck is required to transport entire systems.

PERFORMANCE

Clean Environment Equipment (CEE) is a leader in quality equipment for groundwater remediation and leachate extraction. CEE has been supplying equipment to oil refineries, gasoline stations, terminals, pipelines, airports, industrial manufacturing plants, transportation yards, landfills, and public and private companies for 15 years.

CONTACT INFORMATION

Manufacturer

Clean Environment Equipment
1133 Seventh Street
Oakland, CA 94607
U.S.A.

Telephone (510) 891-0880
Toll-free (800) 537-1767
Facsimile (510) 444-6789

OTHER DATA

Clean Environment Equipment manufactures a line of skimmers and pumps that remove floating hydrocarbons to a sheen and are applicable across a broad range of applications to depths of 91 m (300 ft) in 5 cm (2 in) diameter wells or larger. Water drawdown is available to create a cone of depression and special **AutoPumps** are also available for groundwater recovery and total fluids removal. These pumps do not require surface controls and operate effectively without high quality clean air and use air "on demand", thus being very air efficient. CEE also manufactures completely secured portable systems that can easily be moved between sites. These include single axle and larger trailer units for both product recovery systems and water table depression systems. Custom options are also available.

REFERENCES

(1) Manufacturer's Literature

DYNAMIC PETRO-BELT HYDROCARBON SKIMMER

No. 42

DESCRIPTION

A belt skimmer that removes floating hydrocarbons from groundwater and deposits the recovered product and minimal water into a waste product barrel.

OPERATING PRINCIPLE

An oleophilic, hydrophobic belt, which circulates through the length of the well bore, attracts and retains hydrocarbons while repelling water. At the surface, the hydrocarbons are scraped off both sides of the belt and deposited into a waste product barrel. The system will clean fluctuating water surfaces to a sheen.

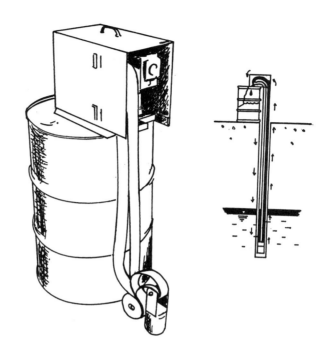

STATUS OF DEVELOPMENT AND USAGE

Commercially available product in general use.

TECHNICAL SPECIFICATIONS

Physical - The skimmer is 48 cm (19 in) long, 36 cm (14 in) wide, 38 cm (15 in) high, and weighs 39 kg (85 lb). The polyurethane belt is available in submergence depth sizes of 3, 4.5, 6, 7.5, 9, and 11 m (10, 15, 20, 25, 30, 35 ft).

Operational - The Petro-Belt removes floating product down to a surface sheen and can recover floating hydrocarbons directly from 5 or 10 cm (2 or 4 in) and larger monitor wells. The unit removes up to 0.004 L/s (2 gph) of diesel fuel, a depths down to 18 m (60 ft). The belt has a smooth, thermal welded seam so that oil can be scraped off both sides. The estimated service life of the belt is two years.

Power/Fuel Requirements and Transportability - Depending on the model, 115 or 220 VAC, 60 Hz electric power is required. The system is person-portable and equipped with a carrying handle.

PERFORMANCE

The units will remove up to 8 L (2 gal) per hour of diesel fuel with lower recovery rates for lighter hydrocarbons and higher for heavier hydrocarbons. Skimming depths of 18 m (60 ft) have been successfully tested by the manufacturer.

CONTACT INFORMATION

Manufacturer

Dynamic Process Industries
1900 West Northwest Highway
Dallas, TX 75220
U.S.A.

Telephone (214) 556-0010
Facsimile (214) 556-9149

OTHER DATA

The Petro-Belt Skimmer is also available in an **enclosed model** with a vapour-tight cabinet which virtually eliminates the escape of hydrocarbon vapours to the atmosphere during recovery operations. A **sub-grade model**, designed for sites where aboveground installation is prohibited or impractical, is intended for use with a Transfer Package or an 11-L (3-gal) tank that fits inside the vault.

REFERENCES

(1) Manufacturer's Literature

ESI MOBILE RECOVERY TRAILER No. 43

DESCRIPTION

A complete groundwater pump and treat system on wheels, to which an optional soil vapour extraction system can be added for treating contaminated soils.

OPERATING PRINCIPLE

This self-contained, pre-plumbed, and pre-wired unit is capable of recovering contaminated groundwater and free phase hydrocarbon from multiple wells. An optional soil vapour extraction system is available to remove volatile vapours from soil. Treatment options include an oil-water separator, an air stripper, and liquid and vapour phase carbon adsorption canisters.

STATUS OF DEVELOPMENT AND USAGE

Commercially available product in general use.

TECHNICAL SPECIFICATIONS

Physical

The system is available in a 95 L/min (25 gpm) size and a 38 L/min (10 gpm) size. The following outlines the standard equipment.

Pumps - The Mobile Recovery Trailer system uses ejector pumps which are available in diameters from 3.8 to 20 cm (1.5 to 8 in), and in lengths from 0.5 to 1.8 m (1.5 to 6 ft). The pump's flow rates range from 1.9 L/min (0.5 gpm) for the 3.8 cm diameter, 0.5 m long (1.5 in dia., 1.5 ft long) pump to 220 L/min (58 gpm) for the 20 cm diameter, 1.8 m long (8 in dia, 6 ft long) pump. Air consumption for the respective pumps is 0.0005 m^3/s (1.0 cfm) and 0.0160 m^3/s (34 cfm).

Electrical - All electrical components including fittings, wiring, starters, lights, thermostats, and motors meet the requirements for Class I, Division II, Group D.

Piping - System piping is constructed of schedule 40 steel pipe, connected with Victaulic type couplings or welding. All piping within the container is coated with rust-inhibiting paint.

Compressed Air System - This is supplied to operate the pumping systems. The 95 L/min (25 gpm) container uses two units and the 38 L/min (10 gpm) container uses one compressor. All compressors are explosion-proof and use a 450 L (120 gal) ASME Code horizontal air receiver with automatic water drain. To reduce operational speed and noise, and to increase system longevity, 5.6 kW (7.5 hp) cast-iron block and heads are used on the compressors.

Oil-Water Separation System - A coalescing oil-water separator, manufactured by Ejector Systems, Inc. (ESI), is used to pre-treat the recovered liquid.

Air Stripping System - A Cascade low profile air stripper, manufactured by ESI, is used to treat the recovered liquids.

Soil Vapour Extraction System (optional) - Soil vapour extraction systems can be incorporated into the container. Since such systems are site-specific, Ejector Systems, Inc. must be consulted to determine feasibility and selection.

Liquid and Vapour Phase Carbon Adsorption Systems - A TIGG Corporation N-1000 Radial flow unit is used in both the 38 and 95 L/min (10 and 25 gpm) systems for vapour treatment. The canister measures 1.1 m long × 1.1 m wide × 1.8 m high (3.7 × 3.7 × 5.8 ft). For liquid treatment, two Barnebey & Sutcliffe Corporation L170 units in series are used in the 38 L/min (10 gpm) system. The canisters measure 0.6 m diameter × 0.8 m high (1.9 × 2.9 ft). The 95 L/min (25 gpm) system uses two Barnebey & Sutcliffe Corporation L500 units in series for liquid treatment. The canisters measure 0.9 m diameter × 1.2 m high (2.9 × 4 ft).

Trailer/Container - All systems are housed in a trailer-mounted container. The container's inside dimensions are 6.1 m long × 2.3 m wide × 2.6 m high (20 × 7.5 × 8.5 ft). The container has a 2.5 m (99 in) double opening, curbside door for side entry to the unit and a full-swing double opening rear door for rear entry. It is insulated with an R-rating of 11.2 and can be equipped with an optional explosion-proof electric heater, with thermostat, capable of a 12,652 kJ (12,000 BTU) output. The floor is sealed and covered with a petroleum-resistant Nitrile matting. Three explosion-proof lights are mounted on the ceiling. The container is ventilated with an explosion-proof ventilation fan.

Operational

Ejector System's pumps are entirely pneumatic and have only two moving parts which reduces maintenance. They can be used to pump floaters such as gasoline and fuel oils, sinkers such as creosote or chlorinated solvents, or dissolved hydrocarbons only. These containerized systems feature redundancy in monitoring high-tank levels. In all alarm situations, the groundwater pumping system will be interlocked to shut down and cannot be started again until all sensors are within their normal operating range and a reset button is manually pushed. The following equipment is monitored: oil-water separator - product and water side; air stripper, high and low pressure; air stripper, effluent level.

Power/Fuel Requirements

The system uses 220 VAC, single phase power. Maximum amperage draw will be 100 amps for the 95 L/min (25 gpm) system and 80 amps for the 38 L/min (10 gpm) system.

Transportability

The systems are built on trailers for mobility.

CONTACT INFORMATION

Manufacturer

Ejector Systems, Inc.
910 National Avenue
Addison, IL 60101
U.S.A.

Telephone (708) 543-2214
Facsimile (708) 543-2014

REFERENCES

(1) Manufacturer's Literature

KECK PRODUCT RECOVERY SYSTEM   No. 44

DESCRIPTION

A system that separates and retrieves floating hydrocarbons from wells 5 cm (2 in) in diameter or larger and automatically pumps the recovered product to a collection tank.

OPERATING PRINCIPLE

Pump, skimmer, and controller are incorporated into one package for use in wells 5 cm (2 in) or larger. A small diameter bladder pump, housed inside the unit, pumps the hydrocarbons to a collection tank at the surface. A floating buoy travels along the length of the pump to accommodate fluctuations in the water table.

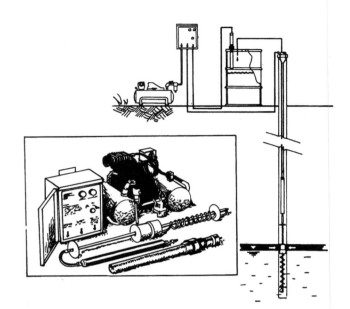

STATUS OF DEVELOPMENT AND USAGE

Commercially available product in general use.

TECHNICAL SPECIFICATIONS

Physical

Pump
Diameter	3.8 cm (1.5 in) O.D.
Length	47 cm (19 in)
Weight	2 kg (4.5 lb)
Materials	303 and 304 stainless steel, polyvinyl chloride (PVC), and brass

Skimmer
Diameter	9.5 or 4.1 cm (3.8 or 1.6 in) O.D.
Length	102 cm (40 in)
Weight	1.8 kg (4 lb)
Materials	304 stainless steel, polyethylene, polypropylene, brass

Controls
Height	30 cm (12 in)
Width	25 cm (10 in)
Depth	15 cm (6 in)
Weight	8.2 kg (18 lb)

Air Compressor
Height	46 cm (18 in)
Length	61 cm (24 in)
Width	31 cm (12 in)
Power	0.56 kW (0.75 hp)
Tank capacity	14 L (3.8 gal)

Operational

The Keck Product Recovery System fits all wells 5 cm (2 in) in diameter and larger, and can accommodate groundwater fluctuations in excess of 61 cm (24 in) (longer travel attachments are available). The skimmer can remove floating contaminant down to less than 3 mm (0.12 in) over the effective travel range of the skimmer. The system can achieve recovery rates of greater than 530 L/day (140 gpd). Air consumption is 0.000 7 m^3/s @ 620 kPa (1.5 cfm @ 90 psi). The operating temperature range for the whole system is 0 to 38°C (32 to 100°F).

Power/Fuel Requirements

110 VAC electric power is required to run the air compressor that powers the unit.

2.3 Removal - From Groundwater

Transportability

Person-portable.

CONTACT INFORMATION

Manufacturer

Keck Instruments Incorporated
P.O. Box 345, 1099 West Grand River
Williamston, MI 48895
U.S.A.

Telephone (517) 655-5616
Facsimile (517) 655-1157

Canadian Distributors

Geneq Incorporated
223 Signet Drive
Weston, Ontario
M9L 1V1

Telephone (416) 747-9889
Facsimile (416) 747-7570

Geneq Incorporated
8047 Jarry East
Montreal, Quebec
H1J 1H6

Telephone (514) 354-2411
Facsimile (514) 354-6948

OTHER DATA

The **Keck Canister** is also available. It is a lightweight, passive floating hydrocarbon recovery device for wells 5 or 10 cm (2 or 4 in) in diameter. The canister can be easily transported to any site, set up in minutes, and requires no power source. The passive, no-pump system collects and stores floating hydrocarbons from slow recharging wells in a self- contained canister that can be drained when brought to the surface.

REFERENCES

(1) Manufacturer's Literature

NEPCCO PETROPURGE PUMP SYSTEMS

No. 45

DESCRIPTION

Complete electric or pneumatic hydrocarbon recovery systems with automatic controls for below ground oil/water separation. Floating or sinking product versions are available.

OPERATING PRINCIPLE

The PetroPurge Pump and HydroPurge Groundwater Drawdown unit form the dedicated dual pump recovery system. The HydroPurge unit pumps water from the recovery well, lowering the groundwater level and creating a cone of depression. Separate phase hydrocarbons then float from all directions along the hydraulic gradient and accumulate in the recovery well. When the layer of oil reaches a predetermined thickness, the PetroPurge Pump is activated to extract the separate phase hydrocarbons from groundwater and pump them both to a recovery tank. A sinking layer at the bottom of the water column can be recovered with screen attachments available for any pump model.

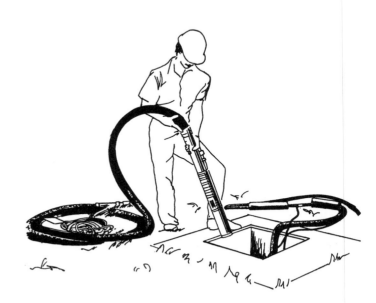

STATUS OF DEVELOPMENT AND USAGE

Commercially available products in general use.

TECHNICAL SPECIFICATIONS

Physical

Diameter	13 cm (5 in)
Length	109 cm (43 in)
Weight	16 kg (36 lb)

The PetroPurge pump is available with a stainless steel or aluminum housing, a PVC inlet valve, Celcon seals, and diffusers to reduce vibration and wear. Included with the system are 11 m (35 ft) of power cable, a special inlet, starting capacitor, and 4.5 m (15 ft) of 3.8 cm (1.5 in) I.D. pump hose.

Operational

Pumping Rates	0 to 57 L/min (0 to 15 gpm)
Operating Head	4.6 to 16 m (15 to 52 ft)

The pump inlet includes a screen to prevent damage from debris.

Power/Fuel Requirements

220 VAC, 60 Hz, single-phase electric power is required.

Transportability

Person-portable.

PERFORMANCE

The PetroPurge Pump system is part of a complete line of proven and dependable, standard and custom NEPCCO groundwater treatment equipment. NEPCCO products are designed to be compatible and include soil vapour treatment systems (see No. 57, NEPCCO SoilPurge System), catalytic converters, and air strippers.

CONTACT INFORMATION

Manufacturer

NEPCCO
2140 Northeast 36th Avenue
Ocala, FL 34470
U.S.A.

Telephone (904) 867-7482
Facsimile (904) 867-1320

OTHER DATA

When several recovery wells are operating at a site, a **Multiwell Control Panel** is available which allows simultaneous control of up to four HydroPurge/PetroPurge pump pairs or up to eight HydroPurge units. Panels are available in weathertight or explosion-proof designs. Other available options are **solid-state or pneumatic control panels, multi-interface sensing probes**, the **ChemPurge System** for recovering viscous or corrosive liquids, and the **Total Fluids Option** for applications requiring an above-ground oil/water separator.

REFERENCES

(1) Manufacturer's Literature.

ORS SCAVENGER HYDROCARBON RECOVERY SYSTEMS

No. 46

DESCRIPTION

Probe Scavenger and **Filter Scavenger** systems use membrane and sensor technology to separate product from water in-situ, eliminating the need for an oil/water separator at the surface. They can be used to recover floating hydrocarbons from groundwater as well as dense non-aqueous phase liquids (DNAPLs) that have sunk below the groundwater surface.

OPERATING PRINCIPLE

The **Probe Scavenger** uses a submersible, self-priming pump. A sensor differentiates between hydrocarbon contaminants and water, so that water-free oil can be retrieved. In the **DNAPL Probe Scavenger**, a conductivity probe distinguishes DNAPL from water, allowing water-free product recovery. The pump operates only when the intake is entirely in the DNAPL layer, recovering 100% water-free chemicals. The recovered contaminants are then automatically pumped from the well into a recovery tank. When the recovery tank is full, a sensor shuts down the pump to prevent overflow.

The **Filter Scavenger** uses a patented intake cartridge to separate and retrieve a wide range of hydrocarbons from water. The cartridge consists of a buoy with a specially manufactured oleophilic/hydrophobic mesh screen that floats precisely at the oil/water interface and allows hydrocarbons to pass through into a reservoir while repelling water.

STATUS OF DEVELOPMENT AND USAGE

Commercially available products used in thousands of projects worldwide since 1975.

TECHNICAL SPECIFICATIONS

Physical

Scavenger Systems are constructed of Teflon and stainless steel to protect against corrosive halogenated compounds. Pump and motor assembly are supplied with petroleum-resistant power cord and sensor has petroleum-resistant cable and weatherproof connectors. Two filter cartridges are available for the **Filter Scavenger** - 100-mesh cartridges for light hydrocarbons such as gasoline or diesel and 60-mesh cartridges for heavy hydrocarbons such as hydraulic oils. All systems are modular in design so that components can be interchanged in the field. These systems are approved for use in Class 1, Division 1 locations. The following are specifications.

	Probe Scavenger	DNAPL Probe Scavenger	Large Diameter Filter Scavenger	Small Diameter Filter Scavenger
Applications	5 cm (2 in) or larger recovery wells	Recovery wells with DNAPLs	Trenches, small bodies of water, 62 cm (24 in) or larger wells with fluctuating water levels	10 cm (4 in) or larger recovery wells with fluctuating water levels
Recovery Rate	Up to 76 L/min (20 gpm)	Up to 15 L/min (4 gpm)	Up to 15 L/min (4 gpm)	Up to 1.9 L/min (.5 gpm)
Maximum Operating Depth	46 m (150 ft)	46 m (150 ft)	3.7 m (12 ft)	46 m (150 ft)

	Probe Scavenger	DNAPL Probe Scavenger	Large Diameter Filter Scavenger	Small Diameter Filter Scavenger
Down Well Module Dimensions/ Weight	9 cm dia. x 175 cm length (3.5 x 69 in)/ 20 kg (50 lb)	9 cm dia. x 122 cm length (3.5 x 48 in)/ 20 kg (50 lb)	48 cm dia. x 26 cm high x 13 cm draft (19 x 10 x 5 in)/ 5 kg (11 lb)	9 cm dia. x 175 cm length (3.5 x 69 in)/ 20 kg (50 lb)
Oil/Water Separation	Density sensor distinguishes hydrocarbons from water	Conductivity sensor distinguishes hydrocarbons from water	Oleophilic/hydrophobic membrane	Oleophilic/hydrophobic membrane
Options	Lifting winch, cable, SitePro controls, explosion-proof (NEMA 4/7 controls), in-line water pump, surface-mounted pump, and tank-full shutoff	Lifting winch, cable, NEMA 7 controls, surface-mounted pump, and tank-full shutoff	60-mesh membrane for viscous hydrocarbons, salt water applications and aromatic hydrocarbons, and deep-well modifications available	Lifting winch, cable, SitePro controls, NEMA 7 controls, in-line water pumps, tank-full shutoff, and wellhead electrical junction boxes.

Operational

The **Scavenger Product Recovery Systems** are designed for use in small diameter wells [as small as 5 cm (2 in) in diameter] or in large diameter wells. Different operating modes allow for variable rates of recovery, differing viscosities and densities, and a variety of soil conditions. Three position controls can be set for hand, off, or auto, providing automatic, 24-hour, continuous operation. The system controls are explosion-proof and the pump has a thermal overload and permanent split capacitor. The winch, cable, and stand are spark-proof. Recovery rates are claimed to be up to 76 L/min (20 gpm). The **DNAPL Probe Scavenger** can be used in wells as small as 10 cm (4 in) in diameter. The Shallow Well model can be used with maximum efficiency in wells up to 6 m (20 ft) deep. The Deep Well models can be used in wells up to 31 m (100 ft) deep.

Power/Fuel Requirements

115 or 230 VAC, electric power is required.

Transportability

Person-portable

PERFORMANCE

According to the manufacturer, the **DNAPL Probe Scavenger** can recover the following common heavier-than-water hydrocarbons.

1,1 dichloroethane
1,1 dichloroethylene
1,1,1 trichloroethane
1,1,2 trichloroethane
1,2 dichloroethane
1,2 transdichloroethylene
Carbon tetrachloride
Chlorethane
Chlorobenzene

Chloroform
Methyl chloride
Methylene chloride
Tetrachloroethane
Tetrachloroethylene
Trichloroethylene
Trichlorofluoromethane
Vinyl chloride

CONTACT INFORMATION

Manufacturer

ORS Environmental Systems, A Division of Sippican, Inc.
32 Mill Street
Greenville, NH 03048
U.S.A.

Telephone (603) 878-2500
North America Toll Free (800) 228-2310
Facsimile (603) 878-3866

OTHER DATA

Similar to the Filter Scavenger, the **Filter Bucket** is a passive system using the same membrane technology and with the same separation capabilities but requiring no electricity, making it ideal for remote and hazardous locations. It can be used in any area 25 cm (10 in) or larger in diameter. With a capacity of 2 L (.5 gal), it is particularly well suited for retrieving small quantities of hydrocarbons.

To accelerate recovery, all Probe Scavengers and Filter Scavengers can be used with **ORS Groundwater Extraction Pumps** which create a "cone of depression" in the water table, accelerating the flow of contaminants towards the recovery well. The available **SITEPRO 2000 control panel** automatically monitors and controls the pumps and has telemonitoring capability for remote access to the site. These pumps are available in several models, including **submersible pumps** designed for wells up to 61 m (200 ft) in depth. Their small diameter, multi-stage design can be used in wells as small as 10 cm (4 in) in diameter and where high output pressure is required. **Surface-mounted and jet pumps** can be used in narrow gauge wells that do not exceed 6.1 m (20 ft) in depth. These centrifugal pumps have Vitron seals and are available in a number of flow ranges.

REFERENCES

(1) Manufacturer's Literature

PETROTRAP AND SKIMRITE FREE PRODUCT SKIMMERS

No. 47

DESCRIPTION

PetroTrap - A passive skimmer system that separates light oil products and a wide range of other hydrocarbons from groundwater in wells 5 or 10 cm (2 or 4 in) in diameter.

SkimRite - A system that separates and retrieves light oil products and a wide range of other hydrocarbons from wells 5 cm (2 in) in diameter or larger and automatically pumps the recovered product to a collection tank.

OPERATING PRINCIPLE

PetroTrap - The skimmer is lowered into the well and suspended using the lanyard/vent tube. The unit begins recovering product as soon as product is available. The canister is raised to the surface and emptied manually through the drain valve at the bottom.

SkimRite - Pump, skimmer, and controller are incorporated into one package for use in wells 5 cm (2 in) in diameter or larger. A small diameter bladder pump housed inside the unit pumps the hydrocarbons to a collection tank at the surface. A floating buoy travels along the length of the pump to accommodate fluctuations in the water table.

STATUS OF DEVELOPMENT AND USAGE

Commercially available products in general use.

TECHNICAL SPECIFICATIONS

Physical

PetroTrap - This system is available for wells 5 and 10 cm (2 and 4 in) in diameter. It is made of polyvinyl chloride (PVC), stainless steel (303 Series), UHMW polyethylene, and brass. The following are specifications for both sizes.

	10 cm (4 in) PetroTrap	5 cm (2 in) PetroTrap
Diameter	8.9 cm (3.5 in)	4.5 cm (1.8 in)
Length	1.5 m (5 ft)	2 m (6.4 ft)
Weight	8.2 kg (18 lb)	2.8 kg (6.3 lb)
Capacity	2 L (0.5 gal)	0.7 L (0.2 gal)

SkimRite - This system is 4.5 cm (1.8 in) in diameter, 1.6 m (5.2 ft) long, and weighs 4.8 kg (11 lb). Every SkimRite system includes a choice of 5, 10 or 15 cm (2, 4, or 6 in) well cap, a 1.1 kW (1.5 hp) air compressor with air filtration, product tank-overfill protection, and 31 m (100 ft) of 9 mm (3/8 in) polyethylene discharge line, 8 mm (5/16 in) nylon air line, and 9 mm (3/8 in) nylon vent line.

Operational

PetroTrap - This system fits wells 5 and 10 cm (2 and 4 in) in diameter. The standard system comes with a 7.6 m (25 ft) lanyard/vent tube and can accommodate groundwater fluctuations as great as 61 cm (24 in). The unit removes free hydrocarbons to a sheen. The canister must be raised to the surface and emptied manually through the drain valve at the bottom. It is ideal for use in wells in which a very low yield of free product is expected and for use as a monitoring device to delineate migrating plumes.

SkimRite - This system fits all wells 5.1 cm (2 in) in diameter or larger. The standard system can operate in wells to depths of 31 m (100 ft), or with additional hoses, can operate to depths of 76 m (250 ft). The system achieves recovery rates of greater than 380 L/day (100 gpd), depending on the layer thickness of the product. Hydrocarbons are removed to the point where they are no longer visible (i.e., no sheen). The unit can accommodate groundwater fluctuations of up to 61 cm (24 in). The pneumatic control unit is factory-adjusted so no adjustments are necessary or possible. Air consumption is 0.40 cfm.

Power/Fuel Requirements

The **PetroTrap** does not require power and the **SkimRite** requires 115 VAC electric power to run the air compressor.

Transportability

Person-portable

PERFORMANCE

Both these systems are used by hundreds of consulting firms and end users throughout Canada, the United States, and many other countries. Major oil companies use the **PetroTrap** at service stations. The **SkimRite** is used by one state highways department as a mobile unit to clean up diesel fuel spills. Another U.S. firm is using the **SkimRite** with the outdoor enclosure and heat option to recover product at a site more than 480 km (300 miles) from their location.

CONTACT INFORMATION

Manufacturer

Enviro Products Incorporated
1431 Rensen Street, Suite A
Lansing, MI 48910
U.S.A.

Telephone (517) 887-1222
Canada Toll Free (800) ENVIRO 4
Facsimile   (517) 887-8374

OTHER DATA

Options available for the **SkimRite** system include multiple pump systems, outdoor enclosures for the compressor, custom well head configurations to accommodate depression pumps, and sensors and custom hose lengths.
Enviro Products Incorporated produces a full line of underground spill sampling, recovery, and treatment equipment, in addition to well supplies.

REFERENCES

(1) Manufacturer's Literature

PROTEC RECIP PUMP AND ROTARY PUMP

No. 48

DESCRIPTION

Surface-driven, down-hole reciprocating and rotary pumps for installing in groundwater extraction wells and trench sumps at groundwater recovery and leachate collection projects. The Rotary Pump is effective for difficult pumping applications.

OPERATING PRINCIPLE

The down-hole pump assembly of the **Recip Pump** is anchored by a string of stationary hold-down pipe. The pump piston is attached to the bottom of a string of reciprocating tubing that runs inside the hold-down pipe. The surface cam-drive alternately lifts and lowers the tubing, giving the down-hole pump its reciprocating motion. The positive displacement pump lifts fluid to the surface. The down-hole pump barrel of the **Rotary Pump** is connected to the bottom of a string of production tubing running inside the well casing. The down-hole pump rotor is connected to the drivehead drive shaft by a string of drive rods that run inside the production tubing. An electric motor mounted on the drive head drives the system.

STATUS OF DEVELOPMENT AND USAGE

Commercially available products in general use.

TECHNICAL SPECIFICATIONS

Physical

Recip Pump - Pump rates range from fractional to 61 L/min (16 gpm). The surface drive assembly is 81 cm (32 in) high, 35 cm (14 in) wide, and 41 cm (16 in) deep. The down-hole pump is made of stainless steel and Teflon. The following are specifications for the four models available.

	RP1	RP2	RP3	RP5
Pump outside diameter	4.8 cm (1.9 in)	6 cm (2.4 in)	7.3 cm (2.9 in)	11 cm (4.5 in)
Standard motor	0.75 kW (1 hp)	1.1 kW (1.5 hp)	1.5 kW (2 hp)	2.2 kW (3 hp)

Rotary Pump - The surface drive assembly is 23 cm (9 in) high, 20 cm (8 in) wide, and 51 cm (20 in) long. The down-hole pump is made of materials suitable for pumping highly corrosive fluids. Pump rates range from fractional to 160 L/min (42 gpm). The following are specifications for the ten models available.

Shallow Well	RX1A-SW	RX 1-SW	RX3-SW	RX4-SW	RX5-SW
Pump outside diameter	4 cm (1.6 in)	7.6 cm (3 in)	9 cm (3.5 in)	10 cm (4 in)	10 cm (4 in)
Standard motor	0.4 kW (0.5 hp)	0.6 kW (0.8 hp)	1.5 kW (2 hp)	2.2 kW (3 hp)	5/6 kW (7.5 hp)
Deep Well	RX1A	RX1	RX3	RX4	RX5
Pump outside diameter	4.2 cm (1.7 in)	7.6 cm (3 in)	9 cm (3.5 in)	10 cm (4 in)	10 cm (4 in)
Standard motor	0.6 kW (0.8 hp)	0.8 kW (1 hp)	2.2 kW (3 hp)	3.7 kW (5 hp)	7.5 kW (10 hp)

Operational

Both these pumps are positive displacement pumps. Pump operation is easily interfaced with down-hole level controls, tank level controls, safety shutdowns, timers, and other controls. The manufacturer will provide pump curves for each model of the Recip and Rotary Pump. The **Recip Pump** has a pumping rate directly proportional to strokes per minute and stroke length. Stroke length is adjusted manually by changing the position of the drive rod connection to the cam. Strokes per minute depend on the motor rpm selected (1140, 1750, or 3450 rpm). An electronic variable speed drive is also available. The Recip Pump can run dry without damaging the pump and is available in four sizes.

	RP1	RP2	RP3	RP5
Pump-setting depth	0.3 to 61 m (1 to 200 ft)	0.3 to 61 m (1 to 200 ft)	0.3 to 61 m (1 to 200 ft)	0.3 to 61 m (1 to 200 ft)
Pumping rate - Fixed-drive speed				
1140 rpm motor	3 to 8 L/min (0.8 to 2.1 gpm)	5.3 to 12 L/min (1.4 to 3.2 gpm)	0.5 to 19 L/min (2.2 to 4.9 gpm)	19 to 38 L/min (5 to 10 gpm)
1750 rpm motor	4.6 to 12 L/min (1.2 to 3.1 gpm)	8 to 18 L/min (2.1 to 4.8 gpm)	12 to 28 L/min (3.2 to 7.3 gpm)	30 to 61 L/min (8 to 16 gpm)
3450 rpm motor	9.5 to 24 L/min (2.5 to 6.2 gpm)	16 to 32 L/min (4.2 to 8.5 gpm)	25 to 49 L/min (6.5 to 13 gpm)	N/A N/A
Pumping rate - Variable-drive speed	.38 to 24 L/min (0.1 to 6.2 gpm)	5.3 to 32 L/min (1.4 to 8.5 gpm)	8.4 to 49 L/min (2.2 to 13 gpm)	11 to 61 L/min (3 to 16 gpm)

The **Rotary Pump** has no valves or other moving parts, except the rotor. The basic pumping unit has a fixed operating speed and variable speed capability can be added. This pump is an effective "problem solver" for difficult pumping applications such as high specific gravity sinkers, e.g., TCE, EDC, chlorinated solvents and organics, creosote, and/or high viscosity sinkers and floaters, e.g., creosote and heavy hydrocarbons such as bunker oil. The smooth, low-speed rotational pumping action does not emulsify the recovered product. The surface drive equipment is preassembled, connects directly to the well head, and is small enough to be installed in limited space, i.e., below-grade well head cellar.

Shallow Well	RX1A-SW	RX 1-SW	RX3-SW	RX4-SW	RX5-SW
Pump-setting depth	≤18 m (≤60 ft)	≤18 m (≤60 ft)	≤18 m (≤60 ft)	≤18 m (≤60 ft)	≤18 m (≤60 ft)
Pumping rate - Fixed-drive speed	2.7 L/min (0.7 gpm)	10 L/min (2.7 gpm)	42 L/min (11 gpm)	80 L/min (21 gpm)	160 L/min (42 gpm)
Pumping rate - Variable-drive speed	0.38 to 2.7 L/min (0.1 to 0.7 gpm)	1.1 to 10 L/min (0.3 to 2.7 gpm)	3.8 to 42 L/min (1 to 11 gpm)	7.6 to 80 L/min (2 to 21 gpm)	15 to 160 L/min (4 to 42 gpm)
Deep Well	RX1A	RX 1	RX 3	RX 4	RX5
Pump-setting depth	18 to 61 m (60 to 200 ft)	18 to 61 m (60 to 200 ft)	18 to 61 m (60 to 200 ft)	18 to 61 m (60 to 200 ft)	18 to 61 m (60 to 200 ft)
Pumping rate - Fixed-drive speed	2.7 L/s (0.7 gpm)	10 L/min (2.7 gpm)	42 L/min (11 gpm)	80 L/min (21 gpm)	160 L/min (42 gpm)
Pumping rate - Variable-drive speed	0.38 to 2.7 L/min (0.1 to 0.7 gpm)	1.1 to 10 L/min (0.3 to 2.7 gpm)	3.8 to 42 L/min (1 to 11 gpm)	7.6 to 80 L/min (2 to 21 gpm)	15 to 160 L/s (4 to 42 gpm)

Power/Fuel Requirements

An electric or pneumatic power supply is required.

Transportability

Person-portable.

Manufacturer

Protec, A Division of Patton Enterprises Incorporated
7136 South Yale, Suite 200
Tulsa, OK 74136, U.S.A.

Telephone (918) 493-6101
Facsimile (918) 493-2054

REFERENCES

(1) Manufacturer's Literature

TIMCO ISOMEGA RECOVERY PUMP

No. 49

DESCRIPTION

Separates and retrieves floating products from wells 5 cm (2 in) in diameter or larger and automatically pumps the recovered product to a collection tank.

OPERATING PRINCIPLE

Pump, skimmer, and controller are incorporated into one package for use in wells 5 cm (2 in) in diameter or larger. A flexible Teflon bladder, a 51 cm (20 in) chamber, and weir permit skimming of floating product at the water surface in the well.

STATUS OF DEVELOPMENT AND USAGE

Commercially available product in general use.

TECHNICAL SPECIFICATIONS

Physical

Diameter   4.2 or 6 cm (1.7 or 2.4 in)
Length   0.8, 1.3, or 1.8 m (2.5, 4.3, or 6 ft)
Inlet screen slot   0.51 mm (0.02 in).

Operational - This pump fits all wells 5 cm (2 in) in diameter and larger. The standard system can operate in wells up to 31 m (100 ft) in depth. Flow rate depends on controller type, depth to water surface, pump submergence, internal diameters of pressure and recovery lines, capacity of pump body, flow rate of compressed air or gas, adjustment of the cycling, and the presence of particulate in the well.

Power/Fuel Requirements - A gas-powered generator is required to run the air compressor and controller.

Transportability - Person-portable

CONTACT INFORMATION

Manufacturer

Timco Manufacturing Inc.
851 15th Street, P.O. Box 8
Prairie du Sac, WI 53578
U.S.A.

Telephone (608) 643- 8534
Facsimile (608) 643-4275

Canadian Representatives

Christensen Mining Products
6119 Center Street, S.W.
Calgary, Alberta
T2A 0C5

Telephone (403) 243-3340
Facsimile   (403) 243-2710

R & R Drilling
699 Malenfant Boulevard, Dieppe Industrial Park
Dieppe, New Brunswick
E1A 6E9

Telephone (506) 859-8680
Facsimile   (506) 859-7086

REFERENCES

(1) Manufacturer's Literature

Section 2.4

Removal - From Subsurface Soils

CARBTROL REGENERATIVE AND MULTI-PHASE EXTRACTION SYSTEMS

No. 50

DESCRIPTION

These vapour extraction systems are suited for removing organic hydrocarbons from subsurface soils and aboveground soil piles. The soil vent system should be used in conjunction with an off-gas or vapour treatment system.

OPERATING PRINCIPLE

This technology is applicable to highly volatile contaminants in porous soils. Wells are drilled into the contaminated soil and cased with slotted vent pipe. A turbine type vacuum exhauster creates areas of low pressure in the soil around the well casings which vaporizes the contaminant and carries it to the surface.

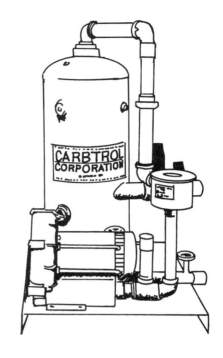

STATUS OF DEVELOPMENT AND USAGE

Commercially available products in general use.

TECHNICAL SPECIFICATIONS

Physical - The **Regenerative Soil Vapour Extractor** is available in a range of capacities up to 7.2 m^3/min (250 cfm) air flow and a maximum vacuum of 20 kPa (6 in Hg). The system includes a regenerative, low maintenance blower, with an explosion-proof motor, and a 28 L (7.5 gal) moisture trap for removing entrained water. The system is supplied with an air dilution valve, temperature and vacuum gauges, an inlet filter and vacuum relief valve to prevent blower damage, and comes factory-piped, wired, and mounted on a steel skid. The **Multi-phase Extraction System** is available in four models ranging in capacity from 1.8 to 9.6 m^3/min (75 to 400 cfm) air flow, from 28 to 150 L/min (7.5 to 40 gpm) water flow, and vacuums up to 83 kPa (25 in Hg). It uses a water-sealed liquid ring pump which can often be directly coupled to the vacuum wells or trenches. Carbtrol air/water, oil/water, and adsorption separation technologies are then used to treat the extracted components.

Operational - The **Regenerative Soil Vapour Extractor** features a compact and low maintenance design, which requires minimum space and allows for easy handling. It is designed to be interfaced with Carbtrol activated carbon vapour treatment systems. The **Multi-phase Extraction System** combines free product, groundwater, and soil vapour extraction capabilities in a single skid-mounted extraction/treatment package. Vacuums of up to 83 kPa (25 in Hg) produced by the fluid vac system greatly increase the removal rate of groundwater in low permeability soils. When the extracted fluids contain heavy phase-free product, a pre-separation tank is used to limit emulsification. Removal efficiencies for both systems vary with contaminant, soil, and hydrological properties.

Power/Fuel Requirements - An electric power supply is required.

Transportability - The systems are skid-mounted and have forklift fittings.

CONTACT INFORMATION

Manufacturer

Carbtrol Corporation
51 Riverside Avenue
Westport, CT 06880
U.S.A.

Telephone (203) 226-5642
Facsimile   (203) 226-5322

REFERENCES

(1) Manufacturer's Literature

EG&G ROTRON REGENERATIVE BLOWERS FOR ENVIRONMENTAL REMEDIATION

No. 51

DESCRIPTION

A complete line of environmental blowers designed for use in soil vapour extraction systems. The regenerative type of blowers simultaneously produce both vacuum and pressure, which can be used to strip contaminants from soils and to push the vapours into treatment systems.

OPERATING PRINCIPLE

Wells are drilled into the contaminated soil and are cased with slotted vent pipe. The regenerative blower, as part of a complete soil vapour extraction system, creates areas of low pressure in the soil around the well casings. The induced vacuum in the soil vaporizes the contaminant and carries it to the surface where it can be destroyed using an off-gas or vapour treatment system. This technology is applicable to highly volatile contaminants in porous and permeable soils.

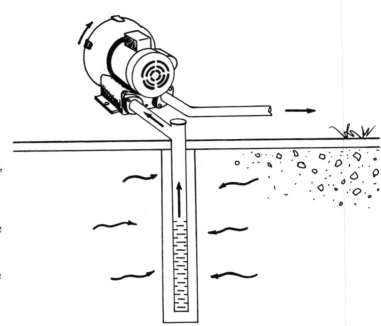

STATUS OF DEVELOPMENT AND USAGE

Commercially available product marketed for many environmental applications including soil remediation, landfill degassing, air stripping, radon removal, and bio-remediation.

TECHNICAL SPECIFICATIONS

Physical

EG&G Rotron produces a complete line of environmental blowers from fractional through 45 kW (60 hp) models. Standard motors include 115/230 VAC, 230/460 VAC, and 575 VAC, all at 50/60 Hz. Explosion-proof motors and custom-built blowers are available to handle a variety of flammable, toxic, and /or caustic gases. The following are dimensions for the standard blowers.

Model Number	Weight	Depth	Width	Height
DR202Y9	15 kg (32 lb)	28 cm (11 in)	23 cm (9 in)	23 cm (9 in)
DR303AE9	16 kg (36 lb)	32 cm (13 in)	23 cm (9 in)	26 cm (10 in)
DR353BR9	25 kg (56 lb)	28 cm (11 in)	23 cm (9 in)	28 cm (11 in)
DR404AL72	28 kg (61 lb)	36 cm (14 in)	30 cm (12 in)	30 cm (12 in)
DR454R72	33 kg (73 lb)	36 cm (14 in)	32 cm (13 in)	34 cm (14 in)
DR505AS72	37 kg (82 lb)	36 cm (14 in)	36 cm (14 in)	37 cm (15 in)
DR513BR72	40 kg (87 lb)	32 cm (13 in)	32 cm (13 in)	40 cm (16 in)
DR523K72	54 kg (118 lb)	37 cm (15 in)	32 cm (13 in)	40 cm (16 in)
DR606CK72	49 kg (108 lb)	40 cm (16 in)	37 cm (15 in)	40 cm (16 in)
DRS5K72E	37 kg (82 lb)	46 cm (18 in)	40 cm (16 in)	46 cm (18 in)
DR6D89	60 kg (133 lb)	48 cm (19 in)	42 cm (17 in)	42 cm (17 in)
DR707D89X	77 kg (169 lb)	46 cm (18 in)	42 cm (17 in)	42 cm (17 in)
DRS7X72	84 kg (185 lb)	58 cm (23 in)	42 cm (17 in)	42 cm (17 in)
DRS75BC72	95 kg (210 lb)	58 cm (23 in)	40 cm (16 in)	46 cm (18 in)
DR8BB7S	117 kg (258 lb)	61 cm (24 in)	48 cm (19 in)	51 cm (20 in)
DRS85BM72	N/A	78 cm (31 in)	48 cm (19 in)	52 cm (21 in)
DRS9BM72	190 kg (420 lb)	78 cm (31 in)	48 cm (19 in)	51 cm (20 in)

Model Number	Weight	Depth	Width	Height
DRP9BM72C	190 kg (420 lb)	78 cm (31 in)	48 cm (19 in)	51 cm (20 in)
DR10BE72W	260 kg (572 lb)	66 cm (26 in)	52 cm (21 in)	58 cm (23 in)
DRS11BP72	474 kg (1,044 lb)	84 cm (33 in)	52 cm (21 in)	58 cm (23 in)
DR12BE72VW	193 kg (426 lb)	61 cm (24 in)	51 cm (20 in)	58 cm (23 in)
DRP13BP72C	312 kg (687 lb)	91 cm (36 in)	51 cm (20 in)	58 cm (23 in)
DRS13BP72	311 kg (686 lb)	91 cm (36 in)	51 cm (20 in)	58 cm (23 in)
DR14DW72W	348 kg (767 lb)	69 cm (27 in)	58 cm (23 in)	69 cm (27 in)
DRP15EE72C	465 kg (1,025 lb)	90 cm (35 in)	58 cm (23 in)	71 cm (28 in)
DRS15EE72	465 kg (1,025 lb)	90 cm (35 in)	58 cm (23 in)	71 cm (28 in)

Operational

Rotron blowers are available with maximum air flow capacities ranging from 8.4 to 51 m^3/min (29 to 1,800 cfm), maximum pressures up to 34 kPa, and maximum vacuum ratings of 22 kPa. Blowers are available in direct or remote drive models. Remote driven blowers can be mounted in any position. The drive motors are capable of carrying full-rated load continuously with a temperature rise not exceeding 100°C (212°F) above an ambient temperature of 40°C (104°F). These blowers adapt to changing conditions, e.g., if soil impedance increases, the blowers pull harder, and vice versa. Rotron blowers meet all safety and acoustical noise standards.

Power/Fuel Requirements

Depending on the model, 115/230, 230/460, or 575 VAC, 50/60 Hz, electric power is required.

Transportability

Smaller blowers are person-portable, while larger blowers have built-in lifting rings and require lifting equipment and a truck for transportation

PERFORMANCE

Removal efficiencies vary with contaminant, soil, and hydrological properties. Rotron Regenerative Blowers have been used successfully in more than 6,000 soil vapour extraction applications over the past 10 years.

CONTACT INFORMATION

Manufacturer

EG&G Rotron, Industrial Division
North Street
Saugerties, NY 12477
U.S.A.

Telephone (914) 246-3401
Facsimile   (914) 246-3802   EG&G is represented internationally by over 100 sales representatives.

OTHER DATA

EG&G Rotron produces a **moisture separator** for use in conjunction with the Rotron blower to make vapour extraction safer and more efficient. By separating and containing entrained liquids extracted during the removal process, the moisture separator helps protect the blower from corrosion damage caused by excess moisture and protects the end treatment system from further contamination. The moisture separator is positioned between the blower and the extraction well and contains a vacuum relief valve that protects the blower from overheating by detecting blockage in the line, and a float system that protects the blower from flooding when the moisture separator is full. No routine maintenance is required other than draining the accumulated liquid. Rotron provides the services of **Environmental Application Engineers** to assist in designing a complete soil vapour extraction system using their system components.

REFERENCES

(1)   Manufacturer's Literature

GLOBAL VACUUM EXTRACTION AND VAPOUR LIQUID SEPARATOR MODULES

No. 52

DESCRIPTION

The **Vacuum Extraction Module** removes volatile organic chemicals from soils and the **Vapour Liquid Separator Module** protects downstream vacuum extraction and vapour treatment equipment from entrained water. Both network with other modules to form a comprehensive subsurface remediation system.

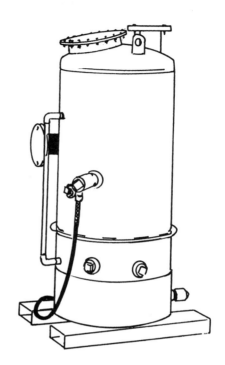

OPERATING PRINCIPLE

Vacuum Extraction Module - Wells are drilled into the contaminated soil and are cased with slotted vent pipe. A vacuum system creates areas of low pressure in the soil around the well casings which vaporizes the contaminant and carries it to the surface where it can be destroyed using an off-gas or vapour treatment system. This technology is applicable to highly volatile contaminants in porous soils.

Vapour Liquid Separator Module - By separating and containing entrained liquids, which are extracted during the removal process from the gas flow stream, the liquid separator protects the blower from corrosion damage caused by excess moisture and the end treatment system from further contamination. The liquid separator is positioned between the blower and the extraction well. Entrained water is drawn through the Separator Module and centrifugally separated. Before exiting the vessel, the air passes through a mist eliminator to remove small water droplets. Separated water is contained within a holding tank.

STATUS OF DEVELOPMENT AND USAGE

Commercially available products in general use.

TECHNICAL SPECIFICATIONS

Physical

Vacuum Extraction Module - Global offers a full range of blower types, including positive displacement, regenerative, and liquid ring for vacuums up to 95 kPa (28 in Hg) and flow rates up to 34 m^3/min (1,200 cfm). The standard Module includes a vacuum blower, explosion-proof motor and motor controls, flanged inlet and outlet, inlet filter, outlet silencer, vacuum relief valve, temperature and vacuum gauge, sample ports, and manual dilution air valve. The system components are skid-mounted and housed in a sound-attenuating steel cabinet with lockable doors.

Vapour Liquid Separator Module - Includes an epoxy-coated steel vessel with internal baffle plates and de-mister, explosion-proof, high-level shutdown switch, liquid level indicator, flanged inlet and outlet, drain port, and additional ports for bayonet heater and automatic level controls. The following are specifications for the two standard models.

	VSM 5	VSM 10
Weight	272 kg (600 lb)	408 kg (900 lb)
Height	2 m (6.5 ft)	2.3 m (7.5 ft)
Diameter	0.8 m (2.5 ft)	0.9 m (3 ft)

Operational

Vacuum Extraction Module - This module is housed in a durable, weather-resistant steel cabinet, with lockable double-hinged doors that provide security and easy access to all components. The cabinet and the discharge silencer maintain total sound levels at below 80 dB. Flanged inlet and outlet and single point electrical connection allow for easy hook-up and transportability. This module is selected based on the airflow and vacuum required for subsurface conditions. The following table shows the range of flow and vacuum capacities for the different types of blowers.

2.4 Removal - From Subsurface Soils

	Regenerative Blower	Turbotron Blower	Positive Displacement Blower	Liquid Ring Blower
Well Head Vacuum	24 kPa (7 in Hg)	41 kPa (12 in Hg)	58 kPa (17 in Hg)	95 kPa (28 in Hg)
Air Flow	0 to 10 m^3/min (0 to 350 cfm)	0 to 31 m^3/min (0 to 1,110 cfm)	0 to 34 m^3/min (0 to 1,200 cfm)	0 to 20 m^3/min (0 to 700 cfm)
Power	2.2 to 3.7 kW (3 to 5 hp)	7.5 to 37 kW (10 to 50 hp)	7.5 to 37 kW (10 to 50 hp)	15 to 45 kW (20 to 60 hp)

Vapour Liquid Separator Module - These modules have large liquid holding capacities to prevent frequent shutdowns due to high liquid level. An optional pump and automatic liquid level drain controls, available from Global, are recommended when high levels of water are present in the air stream, or dual extraction of vapours and groundwater is expected. The following are operating specifications.

	VSM 5	VSM 10
Air flow	14 m^3/min (500 cfm)	28 m^3/min (1,000 cfm)
Liquid capacity	280 L (75 gal)	570 L (150 gal)

Power/Fuel Requirements

An electric power supply is required to power the Vacuum Extraction Module, which draws the air through the Vapour Liquid Separator Module. The Vapour Liquid Separator Module itself requires no power source. These systems are designed to be powered and controlled through Global's Power Distribution Module and are part of Global's complete line of soil and groundwater remediation products which are designed to interface easily and network to form a comprehensive subsurface remediation system.

Transportability

The **Vacuum Extraction Module** is housed in a steel cabinet and the **Vapour Liquid Separator Module** is skid-mounted. Both have forklift channels and lifting lugs for transportation. Trailer-mounted systems are also available.

CONTACT INFORMATION

Manufacturer

Global Technologies, Incorporated
4927 North Lydell Avenue
Milwaukee, WI 53217
U.S.A.

Telephone (414) 332-5987
Facsimile (414) 332-4375

Canadian Distributor

Mapleleaf Environmental Equipment
36 Geneva Court
Brockville, Ontario
K6V 1N1

Telephone (613) 498-1876
Facsimile (613) 345-7633

Mapleleaf Environmental Equipment
5919 5 Street, S.E.
Calgary, Alberta
T2H 1L5

Telephone (403) 255-9083
Facsimile (403) 252-8712

OTHER DATA

These modules are part of Global's line of soil and groundwater remediation products, which also includes **Vapour Treatment Modules** (see No. 156, **Chloro-Cat Catalytic Oxidizer** in Section 5.5), and **Power Distribution Modules**. Each module interfaces easily, and when networked, the Global modules form a comprehensive subsurface remediation system that works continuously and automatically.

REFERENCES

(1) Manufacturer's Literature.

HORIZONTAL DIRECTIONAL DRILLING FOR TRENCHLESS REMEDIATION

No. 53

DESCRIPTION

A drilling technology that enables horizontal soil vent piping to be installed without excavation in environmentally sensitive areas, under urban congestion, and in hazardous conditions.

OPERATING PRINCIPLE

A specialized drill rig is positioned at the remote drill site, at a slight angle off the horizontal plane. A pilot hole is drilled along a pre-surveyed path at the desired depth and under any obstacles. The hole is enlarged with a pull-back reamer and the well or vent casing is pulled through the reamed hole. State-of-the-art drilling equipment is used, including proven directional survey tools that give continuous, accurate data in real time during pilot hole drilling.

STATUS OF DEVELOPMENT AND USAGE

A specialized drilling service offered worldwide and commonly used to install utility pipelines under river crossings or other sensitive areas.

TECHNICAL SPECIFICATIONS

Physical - Constructed in 1988, this drilling rig incorporates the latest technology for long distance controlled drilling operations. Directional survey equipment includes Sharewell's Tru Tracker System which accurately determines the location of the drill bit during the drilling operation, and Sharewell's Magnetic Guidance System which guides the bit.

Operational - Services are offered either on a turnkey or a day-rate basis.

Power/Fuel Requirements - Diesel fuel.

Transportability - The rig is transported in four truckloads. It is mounted on trailer bases specifically designed for ease of access to difficult sites with minimal environmental disturbance.

PERFORMANCE

Documented performance data for applications in chemical spill response were not received. The technology is a proven and acceptable method of crossing under rivers and other areas, especially where environmental protection is a concern. It has been widely used in South America, Europe, Asia, Africa, as well as in North America as an alternative to the traditional open-cut practice. Since 1984, R.H. Woods Ltd. has been involved in drilling nearly 15,240 m (50,000 ft) of horizontally drilled crossings in North America, ranging from 91 m (300 ft) to over 1,240 m (4,068 ft).

CONTACT INFORMATION

Manufacturer and Contractor

R.H. Woods Ltd.
R.R. 3
Watford, Ontario
N0M 2S0

Telephone (519) 849-5440
Facsimile (519) 849-5444

REFERENCES

(1) CHMM Inc., "Product and Service Releases", in: *Hazardous Materials Management* (June, 1992).
(2) Manufacturer's Literature.

HRUBOUT® IN-SITU SOIL PROCESSOR   No. 54

DESCRIPTION

A transportable, in-situ soil remediation technology that vaporizes and oxidizes volatile, less volatile, and involatile petroleum and certain chemical contaminants, leaving the soil in a sterilized state.

OPERATING PRINCIPLE

The Hrubout® process involves injecting heated air at temperatures of up to 650°C (1,200°F) into and below the zone of contamination via vertical or horizontal perforated piping. The heated air vaporizes the soil moisture, thereby steam-distilling the volatile and semi-volatile contaminants. As water in the soil evaporates, the soil pores become less constricted, soil permeability is increased, and greater air flows carry higher air temperatures. While most volatiles are removed with the soil water, heavier constituents of the hydrocarbon chain remain. These contaminants are removed by slow oxidation within the soil at a temperature of approximately 427°C (800°F). At the surface, which is covered with an impermeable barrier, the vapours are drawn off by a vacuum blower and directed into an incinerator.

STATUS OF DEVELOPMENT AND USAGE

Commercially available product in general use.

TECHNICAL SPECIFICATIONS

Physical

A Hrubout® system consists of a burner/blower, a perforated piping system, an impermeable surface barrier, a vacuum blower, and an incinerator. The entire unit is trailer-mounted, 12 m (38 ft) long, 2.4 m (8 ft) wide, has a stack height of 4 m (13 ft), and weighs 9.1 t (10 tons). The burner/blower supplies air flows from 121 to 650°C (250 to 1,200°F) and creates an underground pressure gradient of 138 to 255 kPa (20 to 37 psi). The perforated pipe delivers the superheated air to the soil. The surface is sealed with heavy-gauge aluminum foil and is maintained under a vacuum. The vacuum blower unit simultaneously produces a vacuum under the barrier to strip contaminants from the soils and assist in capturing vent gas, and creates a positive pressure for pushing the vapours into the incinerator.

The incinerators are built to California State specifications. Vapour is destroyed at 816°C (1,500°F), with a 0.6 second retention time for a 99.5% VOC destruction rate. An auxiliary air blower, rated at 12 m^3/min (420 cfm), automatically dilutes incoming vapours if they approach the LEL. Other safety features include automatic shutdown for high and low gas pressure, flame failure, and high temperature. Heated air is provided by an adiabatic burner with a rating of 3.2 mm Btu. Air is compressed by twin air blowers powered by a 112 kW (150 hp) electric motor delivering 52 m^3/min (1,840 cfm). Burners for both incineration and injection are fueled by either propane or natural gas. Smaller units using a heat exchanger are also available, mounted on 4.9 m (16 ft) trailers and weighing approximately 4.5 t (5 tons). They operate quietly and can be run unmanned.

Operational

With the Hrubout® process, soil can be remediated with minimal disturbance to the site. This can be significant if refilling and replacing existing structures is a consideration. The systems operate 24 hours a day and can be operated unmanned depending on the site. The process features a quick rig-up and take-down time. It is compact and relatively fast when compared to techniques such as bioremediation. The site is sterilized after the process so nutrients and moisture must be added back to the soil if revegetation is required.

Power/Fuel Requirements

The manufacturer stresses that versatility can be offered through various options for fuel and power. Burners for both incineration and injection are fueled by either propane or natural gas. A source of electric power is required for the blowers and the control systems. The specifications for electric power supply vary with specific site applications.

Transportability

The larger unit is mounted on a 12 m (38 ft) long trailer, is 2.4 m (8 ft) wide, has a stack height of 4 m (13 ft), and weighs 9.1 t (10 tons). The smaller unit is mounted on a 4.9 m (16 ft) trailer and weighs approximately 4.5 t (5 tons).

PERFORMANCE

The recently patented in-situ process was developed after three years of research. A steam recovery process was considered using both steam injection and vapour recovery wells. The company eventually turned to using heated air and steam distillation by using the residual water and interstitial moisture in the vadose zone. With hot air as a source and a new surface collection system, vapour recovery holes were eliminated, as well as the steam boiler with its water preparation system. The condenser/liquid collection was also eliminated. Before final design of the heated air unit, considerable field testing was done in both the vadose and water table zones to determine the transmission of air flow through various soil media. These tests provided the final data for the unit design. While documented performance data were not received, according to the manufacturer, the process assures safe, economical, and complete destruction of contaminants based on sound scientific principles. The process is approved by the Texas Water Commission.

CONTACT INFORMATION

Manufacturer

Hrubetz Environmental Services Incorporated
5949 Sherry Lane
Suite 525
Dallas, TX 75225
U.S.A.

Telephone (214) 363-7833
Facsimile   (214) 691-8545

REFERENCES

(1) Hrubetz, M., "An Attractive New Alternative in Soil Remediation Technology", in: *Environmental Waste Management Magazine* (September-October, 1991).
(2) Manufacturer's Literature.
(3) U.S. EPA, "Superfund Innovative Technology Evaluation Program", 5th, 6th, and 7th Editions.

LAMSON CENTRIFUGAL EXHAUSTERS FOR ENVIRONMENTAL REMEDIATION

No. 55

DESCRIPTION

A complete line of centrifugal blowers and exhausters used to pressurize and evacuate air in soil contaminated with volatile organic compounds.

OPERATING PRINCIPLE

Wells are drilled into the contaminated soil and are cased with slotted vent pipe. The exhauster, as part of a complete soil vapour extraction system, creates areas of low pressure in the soil around the well casings which vaporizes the contaminant and carries it to the surface where it can be destroyed using an off-gas or vapour treatment system. Applicable to highly volatile contaminants in porous and permeable soils.

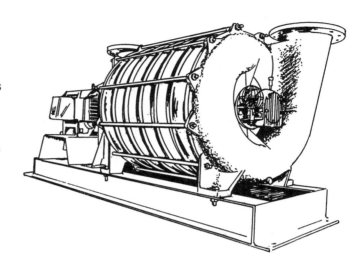

STATUS OF DEVELOPMENT AND USAGE

Commercially available product marketed for many environmental applications including soil remediation, landfill degassing, air stripping, radon removal, and bioremediation.

TECHNICAL SPECIFICATIONS

Physical - Constructed of cast-iron housings with cast aluminum, phenolic-coated, Teflon-coated, or stainless steel impellers, stainless steel or Monel shafts (Monel is a Trademark of Huntington Alloys), and labyrinth, carbon ring, or positive rotating shaft seals.

Operational - Available in models with maximum air flow capacities of between 5.4 to 72 m^3/min (200 to 2,400 cfm), maximum pressures of up to 310 kPa (45 psi), and maximum vacuum ratings of 54 kPa (16 in Hg). All but the smallest units are normally direct-driven through a flexible coupling by 3,600 rpm, two-pole, 60 Hz electric motors. The machine and driver are mounted on a common heavy-duty base and rest upon vibration isolators. When speed requirements dictate, power can be transmitted through gear increasers or V-belt drives. All exhausters are self-governing and power consumption varies with the load. Data on analysis of sound level and vibration are available from the manufacturer.

Power/Fuel Requirements - While 60 Hz, electric power is standard for the Lamson centrifugal exhausters, other drivers such as steam turbines or internal combustion engines may be used.

Transportability - Smaller blowers are person-portable. Lifting equipment and a truck are needed for larger ones.

CONTACT INFORMATION

Manufacturer

Lamson Corporation
P.O. Box 4857
Syracuse, NY 13221, U.S.A.

Telephone (315) 433-5500
Facsimile (315) 433-5451

OTHER DATA

Lamson Corporation also offers a line of compressors and exhausters called **Turbotron**, which is a reliable alternative to positive displacement blowers, featuring pressures up to 207 kPa (30 psi), vacuums to 54 kPa (16 in Hg), quiet, pulse-free operation with low vibrations, requires no special foundation, and low maintenance with only one moving part and no timing gears. Lamson exhausters have proven themselves in over 50 years of service.

REFERENCES

(1) Manufacturer's Literature.

M-D ROTARY POSITIVE DISPLACEMENT BLOWERS

No. 56

DESCRIPTION

Rotary positive blowers which can be used for soil vapour extraction and vapour recovery.

STATUS OF DEVELOPMENT AND USAGE

Commercially available product in general use.

TECHNICAL SPECIFICATIONS

Physical - A complete line of rotary positive displacement blowers for both pressure and vacuum applications. The standard line features blower models up to 150 kW (200 hp), with flows up to 114 m^3/min (4,000 cfm) with either vertical or horizontal flow. For custom packages, blower design, materials of construction, seals, and all auxiliary components can be selected to ensure performance and compatibility with the type of product and operating conditions expected. M-D can supply custom packages with two-stage blowers, high-vacuum boosters, special lubrication systems, or liquid injection systems, with flows from 5.4 to 230 m^3/min (200 to 8,000 cfm) and pressures up to 790 kPa (115 psi).

Power/Fuel Requirements - An electric power supply is required.

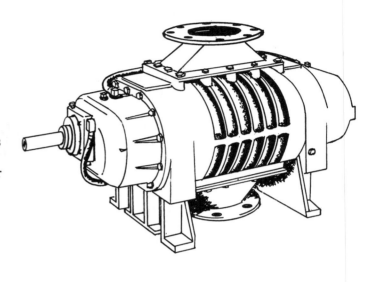

PERFORMANCE AND CONTACT INFORMATION

M-D Pneumatics have been a major manufacturer of rotary positive displacement blowers since 1953. Every M-D Blower is subjected to a mechanical test run before leaving the factory. M-D is ISO 9001 registered, which is the most comprehensive standard issued by the International Organization for Standards applicable to quality for engineering, production, and service. Removal efficiencies vary with contaminant, soil, and hydrological properties.

Manufacturer	*Canadian Representatives*	*Canadian Representatives*
M-D Pneumatics Division Tuthill Corporation 4840 West Kearney St., Box 2877 Springfield, MO 65801-0877 U.S.A. Telephone (417) 865-8715 Facsimile (417) 865-2950	Franklin Supply Company 1900, 300 - 5th Avenue, S.W. Calgary, Alberta T2P 3C4 Telephone (403) 263-7290 Facsimile (403) 234-7698 (IMCO) Industrial Machine Co. 5004 Timberlea Blvd., Unit 38 Mississauga, Ontario L4W 5C5 Telephone (905) 206-1617 Facsimile (905) 206-1663	IMCO Industriel Inc. 4000 Cote Vertu St-Laurent, Quebec H4R 1V4 Telephone (514) 956-7760 Facsimile (514) 956-7766 Marine, Industrial & Technical 306 King William Road St. John, New Brunswick E2L 4E3 Telephone (506) 635-8914 Facsimile (506) 635-8152

REFERENCES

(1) Manufacturer's Literature

NEPCCO SOILPURGE SOIL VAPOUR TREATMENT SYSTEM

No. 57

DESCRIPTION

Skid-mounted vapour extraction system for removing volatile materials from soils. It should be used with an off-gas or vapour treatment system.

OPERATING PRINCIPLE

Wells are drilled into the contaminated soil and are cased with slotted vent pipe. A blower creates areas of low pressure in the soil around the well casings which vaporizes the contaminant and carries it to the surface where it can be destroyed using an off-gas or vapour treatment system.

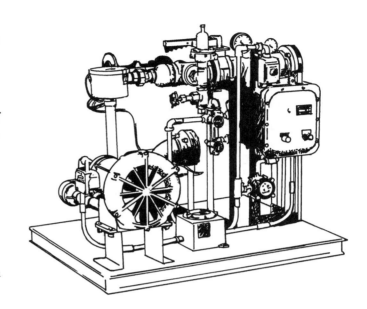

STATUS OF DEVELOPMENT AND USAGE

Commercially available product in general use.

TECHNICAL SPECIFICATIONS

Physical - Systems include a regenerative blower with explosion-proof motor, starter, overload protection and connector, in addition to inlet/outlet, vacuum/pressure gauges and temperature indicators, inlet bleed valve, self-actuating vacuum relief valve, and 10-μm inlet particle filter. All components are mounted on a skid, complete with carbon steel piping and electrical connections. Five standard models are available. All are 1.2 m (4 ft) wide, 1.2 m (4 ft) high, from 0.9 m (3 ft) to 1.2 m (4 ft) long, and weigh from 200 kg (450 lb) to 450 kg (1,000 lb).

Operational - These systems are completely packaged, pre-wired, and fully mobile on skids or trailers. The motors on all five models are 230 volt, 60 cycles, and all are 1-phase except for model 050 which is 3-phase.

	015	**020**	**030**	**050**	**050S**
Power	1.2 kW (1.5 hp)	1.5 kW (2 hp)	2.2 kW (3 hp)	3.7 kW (5 hp)	3.7 kW (5 hp)
Minimum	15 kPa (59 in H_2O)	12 kPa (50 in H_2O)	19 kPa (75 in H_2O)	20 kPa (82 in H_2O)	42 kPa (170 in H_2O)
Maximum Flow	59 L/s (125 cfm)	75 L/s (160 cfm)	94 L/s (200 cfm)	132 L/s (280 cfm)	94 L/s (200 cfm)
Minimum Flow	0 L/s (0 cfm)	47 L/s (100 cfm)	50 L/s (105 cfm)	70 L/s (150 cfm)	0 L/s (0 cfm)

Power/Fuel Requirements and Transportability - A supply of 230 VAC, 1 or 3 phase, 60 Hz electricity is required to power the system. The systems are available skid-mounted or trailer-mounted.

PERFORMANCE AND CONTACT INFORMATION

This system is part of a complete line of proven and dependable, standard and custom NEPCCO groundwater treatment equipment. NEPCCO products are designed to be compatible and include **hydrocarbon recovery systems** (see No. 45, **NEPCCO PetroPurge Pump Systems**), **catalytic converters**, and **air strippers**.

Manufacturer

NEPCCO
2140 Northeast 36th Avenue
Ocala, FL 34470, U.S.A.

Telephone (904) 867-7482
Facsimile (904) 867-1320

REFERENCES

(1) Manufacturer's Literature.

PEGO SOIL VAPOUR EXTRACTION UNITS

No. 58

DESCRIPTION

This system is suited for removing organic hydrocarbons from subsurface soils and above-ground soil piles. It should be used in conjunction with off-gas or vapour treatment systems.

OPERATING PRINCIPLE

Wells are drilled into the contaminated soil and are cased with slotted vent pipe. The extraction unit creates areas of low pressure in the soil around the well casings which vaporizes the contaminant and carries it to the surface where it can be destroyed using an off-gas or vapour treatment system. Applicable to highly volatile contaminants in porous soils.

STATUS OF DEVELOPMENT AND USAGE

Commercially available product in general use.

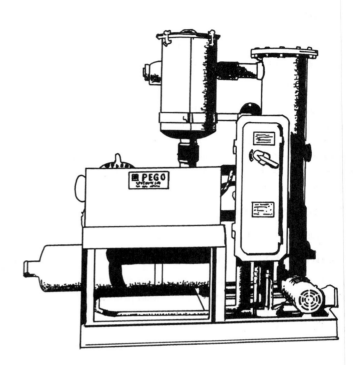

TECHNICAL SPECIFICATIONS

Physical - Units range from 3 m^3/min (100 cfm) to 144 m^3/min (5,000 cfm) in capacity. Pego offers self-contained, liquid-ring vacuum pump systems for severe environments and rotary positive blowers for normal conditions. Systems can be built to any customer-specified configuration, although standard units are available. Specifications are available from the manufacturer. Units include an entrainment separator, which incorporates a de-mister to capture entrained liquid for treatment and disposal or reuse. Special self-contained internal combustion engine units, which use the soil vapour as fuel, are available. Systems are factory-piped, wired, and mounted on a steel skid.

Operational - All systems incorporate a control package that allows automatic operation, liquid separation, and delivery of gas to an off-gas treatment system. Other automatic controls include flame ionization detection analyzer, and data logging of hydrocarbon levels, gas flow, and inlet temperature. Pego also supplies the **Therm-Cat Catalytic Oxidizer** for simultaneous destruction of vapours, start-up support for every system installed, and on-stream follow-up.

Power/Fuel Requirements and Transportability - An electric power supply is required on standard units. Units with self-contained internal combustion engine use soil vapour as fuel, supplemented with propane. Systems are skid- or trailer-mounted.

PERFORMANCE AND CONTACT INFORMATION

The vaporization system has proven effective in cleaning soil spills of light hydrocarbons, without digging up the soil. Testing the air pulled from the soil shows that the vapour concentration diminishes over time. The well is allowed to rest without vacuum and then re-tested to prove that concentration does not increase. Pego has assembled packages featuring virtually every type and size of air- and gas-handling process equipment for more than 100 customers.

Manufacturer

Pego Systems Incorporated
P.O. Box 90038
1196 E. Willow Street
Long Beach, CA 90809, U.S.A.

Telephone (310) 426-1321
Facsimile (310) 490-0633

REFERENCES

(1) Manufacturer's Literature.

RETOX
(REGENERATIVE THERMAL OXIDIZER)

No. 59

DESCRIPTION

A 98% thermally efficient regenerative oxidizer with a built-in vapour extraction system that can be used for vapour extraction and treatment from soils, and treatment of emissions from groundwater air stripping.

OPERATING PRINCIPLE

The RETOX oxidizer consists of a reinforced, insulated chamber filled with silica gravel. Factory-installed heating elements are distributed in the bed. Gas ducts are positioned above and below the gravel bed. When VOC-laden air enters the unit through the gas ducts, passes through the silica bed, and approaches the heating elements, its temperature rapidly increases. Due to the high oxygen content of the VOC-contaminated air, complete combustion readily occurs as the ignition point is reached in the middle of the RETOX oxidizer. The combustion by-products are carbon dioxide and water. This technology is applicable to highly volatile contaminants in porous soils.

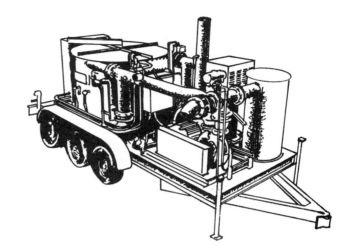

STATUS OF DEVELOPMENT AND USAGE

Commercially available product in general use.

TECHNICAL SPECIFICATIONS

Physical

The RETOX oxidizer is available in sizes of 5.4 m^3/min (200 cfm), 17 m^3/min (600 cfm), 13 m^3/min (1,000 cfm), and 43 m^3/min (1,500 cfm). It is fully pre-wired and assembled. Trailer-mounted units are available on the RETOX 200 and RETOX 600 only. Skid-mounted units are available for all sizes.

Operational

The standard design eliminates more than 95% of the VOCs in the process air stream. Higher levels of reduction can be achieved on an individually evaluated basis. Using a patented design that allows combustion to take place within a high temperature silica gravel bed, virtually no nitrogen oxides (NO_x) are produced. The heat source for the gravel bed contributes no NO_x. The RETOX system requires no catalysts or carbon beds, and requires no additional stoneware to be added or maintained.

The RETOX oxidizer can provide 98% or greater primary heat recovery. The process becomes self-sustaining at 98% primary heat recovery and a hydrocarbon concentration of 1.5% of the lower explosive limit (200 ppm of gasoline). During combustion, the exit portion of the gravel bed is heated by clean exhaust gas. Likewise, the entry portion is cooled by incoming process gas. To maintain a uniform temperature throughout, a valve-switching mechanism changes the direction of the gas flow at regular intervals. Typical incineration temperatures range from 925 to 1,095°C (1,700 to 2,000°F). Process air flows are from 5.4 m^3/min (200 cfm) and up. The RETOX can be used to simultaneously treat both vapours from the soil and groundwater vapours from an air stripper.

As an option, a catalyst pod can be added to the stack of the unit, which guarantees benzene destruction to 99.5%. The catalyst used is a granulated, platinum catalyst on an alumina substrate. The manufacturer states that typical field results are non-detect on benzene through the catalyst (<10 ppb). An automatic influent dilution feature is also available. Modulating automatic inlet dilution valves are computer-controlled and can limit the influent contaminant concentration to a maximum 25% of the lower explosive limit.

Power/Fuel Requirements

230 VAC, 3 phase, 60 Hz electric power.

Transportability

Trailer-mounted units are available on the RETOX 200 and RETOX 600 only. Skid-mounted units are available for all sizes.

PERFORMANCE

Documented performance data were not received. According to the manufacturer's literature, the RETOX oxidizer provides advanced technology with a proven track record for soil remediation using vapour extraction to remove VOCs.

CONTACT INFORMATION

Manufacturer

Adwest Technologies Incorporated
803 West Angus Avenue
Orange, CA 92668
U.S.A.

Telephone (714) 997-8722
Facsimile   (714) 997-8744

OTHER DATA

Options available for the RETOX 200, 600, and 1,000 include TEL-MAX remote telemetry control system, National Fire Protection Association (NFPA) upgrade to Class I, Div. II, Group D, sound-reducing enclosures, stack catalytic pod for high level benzene destruction, and modulating automatic inlet dilution valves.

REFERENCES

(1) Manufacturer's Literature.

ROOTS BLOWERS AND VACUUM PUMPS FOR ENVIRONMENTAL REMEDIATION

No. 60

DESCRIPTION

A complete line of positive displacement blowers and vacuum pumps used to evacuate air in soil contaminated with volatile organic compounds.

OPERATING PRINCIPLE

Wells are drilled into the contaminated soil and cased with slotted vent pipe. The exhauster creates areas of low pressure in the soil around the well casings, which vaporizes the contaminant and the air carries it to the surface where it can be destroyed using an off-gas or vapour treatment system. Applicable to highly volatile contaminants in porous and permeable soils.

STATUS OF DEVELOPMENT AND USAGE

Commercially available products marketed for many environmental applications including soil remediation, landfill degassing, air stripping, radon removal, and bioremediation.

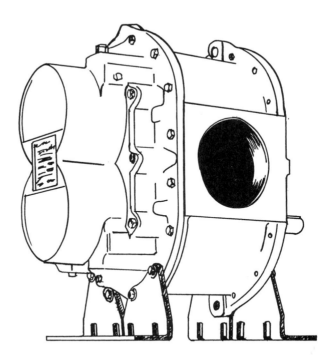

TECHNICAL SPECIFICATIONS

Physical and Operational - A complete line of blowers in a wide range of capacities. Available in models with maximum air flow capacities from 0.054 to 1,380 m^3/min (2 to 48,000 cfm), maximum pressures up to 207 kPa (30 psi), and maximum vacuum ratings of 90 kPa (27 in Hg). An exclusive wraparound discharge plenum and proprietary jets reduce internal pressure pulsations, resulting in lower operating noise levels. Type DVJ vacuum blowers provide vacuum levels to 90 kPa (27 in Hg) and require no sealing water, a significant benefit where water is limited, such as at a remote spill site. Vacuum pumps are available in models with maximum air flow capacities from 0.49 to 600 m^3/min (200 to 22,000 cfm) and maximum vacuum ratings of 95 kPa (28 in Hg). All vacuum pumps are water-sealed and use up to 90% less water than liquid-ring vacuum pumps. Both blowers and vacuum pumps are available as packages, or as blower or vacuum pump only, depending on the model.

Power/Fuel Requirements and Transportability - Power or fuel required depends on the drive selected for the blower or vacuum pump. Smaller blowers are person-portable. Larger blowers have built-in lifting rings and require lifting equipment and a truck for transportation. Package systems are typically skid-mounted with built-in forklift channels.

Manufacturer

Dresser Industries, Incorporated, Roots Division
900 West Mount Street
Connersville, IN 47331, U.S.A.

Telephone (317) 827-9200
Facsimile (317) 825-7669

Roots have specialized in air- and gas-handling equipment since 1854.

Canadian Representative

Beckland Equipment, Ltd.
3250 Beta Avenue
Burnaby, British Columbia V5G 4K4

Telephone (604) 299-8808
Facsimile (604) 299-6162

Prolew-Scott, Ltd.
5859 Chemin St-François
Saint-Laurent, Quebec H4S 1B6

Telephone (514) 332-1010
Facsimile (514) 336-1158

REFERENCES

(1) Manufacturer's Literature.

Section 2.5

Removal - Sorbents

| SPILL SORBENTS - GENERAL LISTING | No. 61 |

DESCRIPTION

Natural or synthetic materials used to absorb and/or adsorb spilled liquids on land and, in some cases, on water.

OPERATING PRINCIPLE

Various forms of sorbents can be used to recover spilled products in areas not accessible by mechanical skimmers, such as on land, in marshes, or in confined watercourses, and to provide a final "polishing" of areas where the bulk of spilled product has already been removed by mechanical skimming. Sorbents are based on *ab*sorption (the penetration of chemicals **into** the material) and *ad*sorption (the adherence of chemicals **onto** the material's surface). To enhance recovery, the sorbent material should be **chemophilic** (chemical-attracting) and **hydrophobic** (water-repelling).

STATUS OF DEVELOPMENT AND USAGE

Most forms of sorbent have been well tested in both oil and chemical response and are commercially available and widely used. Sorbents are applied primarily to small spills in onshore and nearshore environments where accessibility is good and used sorbent can be collected. For recovering larger spills, commercial sorbents are generally inadequate and too expensive.

TECHNICAL SPECIFICATIONS

Physical

Type

Pads	Sheets of sorbent material strong enough to be collected even when saturated with spilled product.
Booms	Sorbent material, enclosed in a cylindrical form, that can be handled and deployed as a boom (see Section 1.3).
Sweeps	Long sheets of sorbent material reinforced with rope and stitching, used to remove thin slicks or rainbow sheens from surface water.
Pillows	Sorbent enclosed in a small sack that can be easily handled and placed in a confined area.
Pom Poms	Bunches of sorbent strips used to recover viscous products, typically on shorelines, and that can be strung together to ease handling and recovery.
Rolls	Continuous sheets of sorbent suitable for lining industrial work areas and chemical storage areas.
Particulate	Loose sorbent material that can be sprinkled over a contaminated area, preferably on land, and recovered with a raking device.

Dimensions

Thickness [<1 cm (<0.4 in) for pads, rolls]
Length [≈3 m (10 ft) for booms]
Width [0.5 to 1 m (1.6 to 3.3 ft) for pads, sweeps, rolls]
Number of units per bale
Dry weight per bale (kg/lb)

Construction

Sorbent type (melt-blown polypropylene, treated cellulose, clay, sawdust, etc.)
Strength and reusability
Special disposal requirements

Operational - The various forms of sorbent can provide spill response capability for a wide range of situations but are ideally suited for removing small spills from confined areas on land. Sorbent can generally be applied in sub-zero temperatures if the products to be recovered are in liquid form and sorbent materials to be used are not prone to freezing, e.g., peat moss, which retains moisture.

Application of sorbent may involve using a broadcast spreader to distribute particulate or a team of workboats to recover a spill on water. Proper safety and breathing protection equipment should be worn when applying loose forms of sorbent that are known to affect bronchial passages or skin. Health and safety information, storage requirements, suitable methods of application, substances that can be recovered by the sorbent, and suitable methods of final disposal should be obtained from the sorbent manufacturer.

Note: When disposing of contaminated sorbents, regulatory authorities should be consulted about proper disposal procedures and to determine whether special approval is required.

Power/Fuel Requirements - None

Transportability - Transportation of most commercial sorbents should not pose any special problems. When transporting contaminated or used sorbents to storage or disposal facilities, applicable packaging and handling requirements must be followed.

PERFORMANCE

The following criteria can be used to evaluate the performance of chemical sorbents.

Chemophilic and Hydrophobic Properties - These properties are the sorbent's ability to attract the spilled chemical and repel surrounding water, thus contributing to recovery efficiency.

Sorption Capacities - Critical factors of sorption capacities are the rate and relative amount of sorption from a given amount of sorbent (measured by weight or volume of chemical over weight or volume of sorbent), and the sorbent's ability to retain collected product during recovery and handling. The best way to ensure that a sorbent meets an organization's spill response needs is to test the product under simulated recovery conditions.

Cost/Reusability - Cheaper transportation, storage, and disposal costs, as well as improved environmental performance, can make the initial expense of reusable sorbents worthwhile when compared to the total cost of acquiring and disposing of non-reusable products.

Buoyancy - Sorbents must remain afloat even when fully saturated with spilled product. Good hydrophobic properties ensure that floating ability remains adequate during spill recovery operations on water.

Disposability - Synthetic sorbents, such as foams, may emit toxic fumes if incinerated, and therefore have special disposal requirements. All contaminated sorbents require special disposal provisions if they contain more than a minimum level of hazardous material.

Storage - Since large quantities of sorbent must be kept on hand for emergencies, their performance should not deteriorate under standard storage conditions. Some polyurethane foams deteriorate in sunlight, and other natural sorbents, such as peat moss, may accumulate moisture if left unsealed.

Ease of Retrieval - Ideally, sorbents should be visible against the colour of the spilled product, should not sink or tear during recovery, and should be easy to collect for disposal. Loose particulate may be easier to apply but harder to recover than other forms of sorbent.

CONTACT INFORMATION

The "World Catalog of Oil Spill Response Products - Fifth Edition", Robert Schulze (ed.), 1995, contains general specifications and contact information for commercial sorbents.

OTHER DATA

Although dated, the U.S. EPA report "Sorbent Materials for Cleanup of Hazardous Materials" (Henick *et al.*, 1982) provides information on the technical specifications and performance of various chemical sorbents.

REFERENCES

(1) Environment Canada, "Selection Criteria and Laboratory Evaluation of Oilspill Sorbents, Update IV", S.L. Ross Environmental Research Limited, for the Emergencies Engineering Division, Environment Canada, Environmental Protection Series Report, EPS 3/SP/3, Ottawa, Ont. (1991).
(2) Henick, E.C., D. Carstea, and G. Goldgraben, "Sorbent Materials for Cleanup of Hazardous Materials", Report EPA-600/2-82-030, United States Environmental Protection Agency (U.S. EPA), Cincinnati, Ohio (March,1982).
(3) Schulze, Robert (ed.), "World Catalog of Oil Spill Response Products", Fifth Edition, World Catalog Joint Venture (JV), Port City Press, Baltimore, MD (1995).

Section 2.6

Removal - Vacuum Systems

2.6 Removal - Vacuum Systems

SUCTION SYSTEMS – GENERAL LISTING   No. 62

DESCRIPTION

Suction systems are designed to recover liquid, solid, or liquid/solid materials in industrial plants, other industrial applications, or at remote spill sites. These systems can be divided into compact or **small vacuum systems** that are wheeled, portable, or mobile and **large suction systems** that are mounted on a skid, trailer, or truck. For specific examples of **small suction systems**, refer to the following entries in this section.

> Minuteman International Vacuum Systems (No. 64)
> Nilfisk Vacuum Systems (No. 65)
> Tornado Vacuum Systems (No. 68)
> Vactagon Vacuum Systems (No. 70)
> Vac-U-Max Vacuum Systems (No. 71)

For specific examples of **large suction systems**, refer to the following entries in this section.

> Max Vac Vacuum Systems (No. 63)
> Powervac Oil Recovery System (No. 66)
> Ro-Vac Vacuum System (No. 67)
> Trans-Vac Oil Recovery Units (No. 69)
> Vactagon Vacuum Systems (No. 70)
> Vac-U-Max Vacuum Systems (No. 71)

OPERATING PRINCIPLE

Small Suction Systems - A vacuum pump, driven by an electric motor, diesel engine, air compressor, or water pump, creates a vacuum in a chamber or small powerhead assembly that is either part of the collection vessel or located away from the suction head. A venturi effect based on air or water flow is sometimes used to establish suction. The induced vacuum allows the operator to recover spilled materials through a hose and suction nozzle. The contaminants are then deposited into a collection canister. As the air moved by the vacuum pump comes into contact with and actually carries the recovered material, it must be filtered so that the contamination is not blown out and redistributed by the vacuum system. Filtration varies from disposal of collected materials into lined or mesh containers, to one-, two-, and three-stage systems including centrifugal separators, linear separators, and cloth dust collectors. High Efficiency Particulate Air (HEPA) filters, designed to retain toxic dusts and other contaminants, are available with industrial-type vacuum cleaners and must be considered when vacuuming hazardous materials.

Large Suction Systems - Vacuum Truck - A low-volume pump or blower directly evacuates a large collection vessel or tank. The free end of a suction hose connected to the tank is then placed in the material to be recovered. Skimming heads can be attached to the end of the hose to increase pickup efficiency in water.

Air Conveyor - A high flow rate of air conveys material into a collection tank through a large diameter hose. The end of the suction hose is held above the material to be recovered to maintain an adequate air flow. The high-capacity blower is protected from liquids and particulates, including dust, by prefilter systems sometimes comprised of steel baffles, as well as primary filtration such as venturi nozzles or other cyclonic separation process, and/or baghouse or multi-stage cloth filters.

STATUS OF DEVELOPMENT AND USAGE

Commercially available products in general use. The use of small suction systems to control spills of hazardous substances other than those in industrial plants is generally not documented. Both vacuum trucks and air conveyors have been used worldwide to recover spilled materials including oil and hazardous substances.

TECHNICAL SPECIFICATIONS

Physical - **Small Suction Systems** - Most devices marketed are electric-powered vacuum cleaners designed to clean up small spills in industrial plants. Other units have been designed specifically for oil spill control or for routine maintenance tasks.

Large Suction Systems - The following are typical dimensions for both vacuum trucks and air conveyors.

	Vacuum Truck	**Air Conveyor**
Suction hose diameter	7.6 to 10 cm (3 to 4 in)	10 to 20 cm (3 to 8 in)
Tank diameter	1 to 2 m (3.3 to 6.6 ft)	1 to 3 m (3.3 to 9.9 ft)
Length (excluding truck)	3 to 10 m (9.9 to 33 ft)	4 to 5 m (13 to 16 ft)
Weight (excluding truck)	2,000 to 8,000 kg (4,400 to 17,600 lb)	5,000 to 8,000 kg (11,000 to 17,600 lb)
Capacity	2 to 23 m^3 (70 to 810 ft^3)	5 to 15 m^3 (180 to 530 ft^3)
Construction materials	Steel and stainless steel with various linings or coatings available	

Operational - **Small Suction Systems** - Operational specifications for vacuum and pressure are provided in a wide variety of units. Typically, however, these suction devices create a vacuum of 14 to 68 kPa (4 to 20 in Hg, or 54 to 268 in WC). A maximum (absolute) vacuum approaching 100 kPa (30 in Hg) would theoretically be possible. Capacity varies widely and is usually expressed in litres per hour (L/h) for liquids, tonnes per hour (t/h) for solids, or litres (L) for receiving capacity. If explosion-proof components and noncorrosive parts are needed for particular applications, their availability must be determined.

Large Suction Systems - The following are typical operating specifications for both vacuum trucks and air conveyors.

	Vacuum Truck	**Air Conveyor**
Vacuum		670 to 730 kPa (98 to 106 psi)
Air flow	3.3 to 83 m^3/min (120 to 590 cfm)	100 to 170 m^3/min (3,530 to 6,000 cfm)
Rotor speed	800 to 1,000 rpm	1,500 to 2,000 rpm
Other	Hydraulically operated tank-end door often featured	

Power/Fuel Requirements and Transportability - Small suction systems require electricity, compressed air, diesel fuel, gasoline, or high-pressure water, depending on the pump drive. Large suction systems are typically powered by a power-take-off (PTO) from the truck, or a dedicated gasoline or diesel power source.

Small vacuum systems are typically wheel-mounted and designed to be easily maneuvered by one person. Large suction devices are mounted on trucks, trailers, or skids for transportability. Railway-mounting kits are available that allow a truck to travel, under its own power, on standard gauge rails to work sites accessible only by railway tracks.

PERFORMANCE AND CONTACT INFORMATION

Evaluation data were not obtained for the application of small suction devices to the cleanup of hazardous materials. For large suction systems, the U.S. EPA tested a vacuum truck and an air conveyor at Leonardo, New Jersey using the Oil and Hazardous Materials Simulated Environmental Test Tank (OHMSETT). For the air conveyor, a maximum recovery rate of 7.8 m^3/h (270 ft^3/h) was achieved in an oil slick with a thickness of 25 mm (1 in). Oil content in the collected liquid reached a maximum of 72%. In one series of tests, it was determined that the vacuum truck could recover a maximum of 3.9 m^3/h (140 ft^3/h) from an oil slick with a thickness of 25 mm (1 in), and in one run, oil content reached 40% but was otherwise much lower. In a second series of tests, the vacuum truck collected more than 10 m^3/h (350 ft^3/h) in a slick of 46 mm (1.8 in) and over 24 m^3/h (850 ft^3/h) when a skimming head was connected to the hose end. Factors that affected performance are summarized in the following recommendations and conclusions.

Spills on Water - Air conveyors are especially recommended for recovering thin layers of highly viscous products on water and are suitable for a variety of forms of debris. Vacuum trucks are better for recovering thick slicks than thin layers of floating materials. Skimming heads should enhance the capability of vacuum trucks to recover floating materials. Blower speed and hose length affect the efficiency with which both systems pick up materials. Blower speed must also be selected consistent with proper truck engine maintenance.

Spills on Land - Large suction systems can effectively remove materials from ditches or other hard-to-reach areas. Once the spilled substance is picked up, it remains confined until its ultimate disposal. If no water body is involved, a highly efficient recovery operation should be possible unless there are other interferences, such as a quickly evaporating substance, a fire hazard, a highly viscous fluid, or physical obstructions that prevent the use of the suction hose.

Refer to specific entries for contact information.

REFERENCES

(1) Manufacturer's Literature.

MAX VAC VACUUM SYSTEMS　　　　　　　　　　　　　　　　　　　　　　　　No. 63

DESCRIPTION

Portable, skid-, trailer- or truck-mounted systems for recovering, vacuum loading, and conveying wet/dry material. Self-contained, require no compressed air or electric power, and provide near-HEPA filtration.

OPERATING PRINCIPLE

A diesel, gasoline, or liquified petroleum gas powerhead powers a direct-drive positive displacement vacuum pump, which draws material through a hose into the vacuum unit where it is separated using a four-stage or two-stage separation system. The recovered material is then deposited into a hopper.

STATUS OF DEVELOPMENT AND USAGE

Commercial products designed to recover and convey hazardous materials in plants and at remote sites.

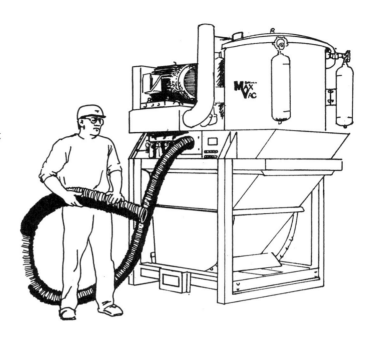

TECHNICAL SPECIFICATIONS

Physical

	P7.5	MV10	MV15	MV20	MV25	XR25	XR30	XR40	XR50	XR75	LD100	LD150	LD200
Height m (ft)	1.2 (4)	1.9 (6.3)	1.9 (6.3)	1.9 (6.3)	1.9 (6.3)	N/A	N/A	N/A	N/A	N/A	N/A	N/A	N/A
Width m (ft)	0.7 (2.3)	1.2 (3.8)	1.2 (3.8)	1.2 (3.8)	1.2 (3.8)	1.5 (5)	1.5 (5)	1.5 (5)	1.6 (5.3)	1.6 (5.3)	2.4 (8)	2.4 (8)	2.4 (8)
Length m (ft)	1.2 (4)	1.3 (4.3)	1.3 (4.3)	1.3 (4.3)	1.3 (4.3)	2 (6.5)	2 (6.5)	2 (6.5)	2 (6.5)	2 (6.5)	2.7 (9)	2.7 (9)	2.7 (9)
Weight kg (lb)	250 (550)	570 (1,250)	590 (1,300)	640 (1,400)	730 (1,600)	1,500 (3,300)	1,500 (3,400)	1,700 (3,800)	1,900 (4,200)	3,000 (4,600)	3,100 (7,000)	3,600 (8,000)	4,090 (9,000)

Operational - Available in portable and manifolded stationary units as well as truck- or trailer-mounted units, ranging from 3.7 to 150 kW (5 to 200 hp) with up to 61 kPa (18 in Hg) positive displacement vacuuming power and conveying capabilities up to 45 t/h (50 tons/hour) over distances of up to 305 m (1,000 ft). Hopper capacities range from 57 L to 1.5 m^3 (2 ft^3 to 2 yd^3). All systems can provide near-HEPA filtration at 99.97% of 0.33 µm or are available with optional final absolute HEPA filtration. The Max Vac incorporates a reverse air pulse filter-cleaning system that provides continuous vacuuming. The Max Vac systems also operate at low noise levels in the range of 85 dB.

Power/Fuel Requirements and Transportability - Available with electric, diesel, liquified petroleum gas, gasoline, or compressed air drive options. 240/480 VAC, 3 phase, 60 Hz electric power is required for the electric-driven models. Skid-mounted units are transported by forklift; other units are either trailer- or truck-mounted for transporting by road.

Manufacturer

DeMarco MaxVac Corporation
P.O. Box 46129
Chicago, IL 60646-0129　　　　Telephone (312) 685-5957
U.S.A.　　　　　　　　　　　　Facsimile　 (312) 283-0307

Canadian Licensee/Manufacturer

Ward Ironworks Limited
123 Victoria Street
P.O. Box 511
Welland, Ontario　　　　Telephone (905) 732-7591
L3B 5R3　　　　　　　　Facsimile　 (905) 732-3310

REFERENCES

(1) Manufacturer's Literature.

MINUTEMAN INTERNATIONAL VACUUM SYSTEMS

No. 64

DESCRIPTION

Small industrial vacuum cleaners designed to pick up wet or dry hazardous wastes, including nuclear, mercury, chemical, asbestos, and other materials.

OPERATING PRINCIPLE

A vacuum is created in the collection tank, either by an electric motor integral to the unit or by a remote source of compressed air. When used for dry pickup, up to five filters, including HEPA filter, impact, cloth bag, and protector bag, entrap contaminants and treat all air flowing through the system.

STATUS OF DEVELOPMENT AND USAGE

Commercial products specifically designed to recover hazardous materials in plants.

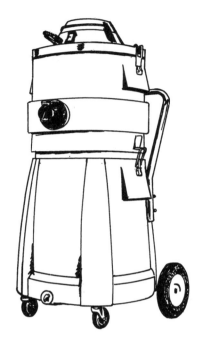

TECHNICAL SPECIFICATIONS

Physical

Series 800 Critical Filter Vacuum Systems - This series offers many sizes and configurations of critical filter vacuums.

	X-100		X-1700		X-703			X-1000					MX-1000	
	-15	-30	-4	-6	-15	-30	-55	-4	-6	-15	-30	-55	Wet/Dry	Dry
Height, m	0.9	1.2	.48	0.6	0.8	1.1	1.3	N/A	0.6	0.9	1.2	1.3	1.1	1.1
(ft)	(2.9)	(4)	(1.6)	(2)	(2.8)	(3.8)	(4.3)		(2)	(2.9)	(3.9)	(4.4)	(3.8)	(3.8)
Width, m	0.5	0.6	0.4	0.4	0.5	0.6	0.6	0.4	0.4	0.5	0.6	0.6	0.5	0.5
(ft)	(1.8)	(2)	(1.2)	(1.2)	(1.8)	(2)	(2)	(1.2)	(1.2)	(1.8)	(2)	(2)	(1.8)	(1.8)
Weight, kg	40	58	10	11	23	44	51	10	11	31	48	55	39	38
(lb)	(88)	(130)	(23)	(24)	(51)	(98)	(112)	(22)	(24)	(69)	(107)	(120)	(86)	(83)
Filter Area	3.3	3.3	1.3	1.8	3.6	3.0	3.0	1.3	1.8	3.0	3.0	3.0	3.0	3.0
m^2 (ft^2)	(36)	(36)	(14)	(19)	(39)	(32)	(32)	(14)	(19)	(32)	(32)	(32)	(32)	(32)

Series X 829 Wet/Dry HEPA Filter Vacuum Systems - Two models of portable vacuums designed to recover hazardous materials. Both models have polyethylene tanks, use a non-woven, embossed, polyester filter bag that provides 50% more filtration area than standard bags, and are designed for wet and dry applications. They are available with tank sizes of 15 L (4 gal) and 23 L (6 gal) and weigh about 12 kg (26 lb).

Series X 839 Asbestos Vacuum Systems - Eighteen models of asbestos vacuums, available with polypropylene, stainless steel, or painted tanks. Three separate vacuum motor assemblies are matched to six tank sizes. Each 57 L (15 gal) and 76 L (20 gal) Hako Asbestos Vacuum is equipped with a disposable collector bag, inside plastic bag liner, disposable filter protector, cloth filter bag, impact pre-filter, and HEPA filter, which is 99.99% effective at 0.12 µm, for a total of 5 filter media for dry recovery and 4 filter media for wet recovery.

X 380-1 Back Pak Vac - A backpack-mounted vacuum for removing hazardous materials. It weighs 4.5 kg (10 lb), has 15 m (50 ft) of electrical cord, and uses a HEPA filter for highly efficient removal of particulate. This unit provides access to wherever the operator can walk.

Operational

Series 800 Critical Filter Vacuum Systems - Equipped with HEPA filters with a copolly-laminated medium for increased durability. HEPA filter is 99.99% effective at 0.12 µm. Includes a modular 4-element system for adapting critical filter vacuum systems to recover liquid mercury and vacuums rated for asbestos pickup and removal.

2.6 Removal - Vacuum Systems

	X-100		X-1700		X-703			X-1000					MX-1000	
	-15	-30	-4	-6	-15	-30	-55	-4	-6	-15	-30	-55	Wet/Dry	Dry
Wet Capacity L (gal)	45 (12)	95 (25)	N/A N/A	N/A N/A	45 (12)	100 (27)	180 (47)	N/A	N/A	45 (12)	100 (27)	180 (47)	45 (12)	N/A
Dry Capacity L (ft³)	53 (1.9)	100 (3.6)	6 (0.21)	44 (1.6)	23 (0.82)	100 (3.6)	204 (7.2)	6 (0.21)	18 (0.62)	53 (1.9)	100 (3.6)	204 (7.2)	59 (2.1)	59 (2.1)
Vacuum kPa (in H₂O)	32 (130)	32 (130)	45 (180)	45 (180)	45 (180)	45 (180)	45 (180)	45 (180)	22 (88)	32 (130)	32 (130)	32 (130)	32 (130)	32 (130)
Air Flow m³/min (cfm)	2.8 (100)	2.8 (100)	4.7 (170)	4.7 (170)	4.7 (170)	4.7 (170)	4.7 (170)	2.7 (95)	2.7 (95)	2.8 (100)	2.8 (100)	2.8 (100)	2.8 (100)	2.8 (100)
Power Watts	1,480	1,480	*	*	*	*	*	930	930	1,480	1,480	1,480	1,480	1,480
Air Pressure kPa (psi)	N/A	N/A	620 (90)	620 (90)	620 (90)	620 (90)	620 (90)	N/A	N/A	N/A	N/A	N/A	N/A	N/A

* Vacuums powered by compressed air 1.2 m³/min (42 cfm).

Series X 829 Wet/Dry HEPA Filter Vacuum Systems - The filter is 99.9% efficient at 3.0 µm and the HEPA filter is 99.99% effective at 0.12 µm.

Wet capacity	9.5 and 17 L (2.5 and 4.5 gal)
Dry capacity	5.9 and 13 L (0.21 and 0.46 ft³)
Vacuum	21 kPa (85 in H₂O)
Air flow	2.7 m³/min (95 cfm)
Power	930 W

Series X 839 Asbestos Vacuum Systems - The HEPA filter is 99.99% effective at 0.12 µm.

Wet capacity	45, 49, 72, 100, and 180 L (12, 13, 19, 27, and 47 gal)
Dry capacity	23, 37, 100, and 200 L (0.82, 1.3, 3.6, and 7.2 ft³)
Vacuum	21, 26, and 33 kPa (85, 105, and 130 in H₂O)
Air flow	2.7, 3.3, and 2.8 m³/min (95, 115, and 100 cfm)
Power	745 and 1,715 W

X 380-1 Back Pak Vac - The HEPA filter is 99.97% effective at 0.3 µm.

Dry capacity	4.2 L (0.15 ft³)
Vacuum	22 kPa (88 in H₂O)
Air flow	2.8 m³/min (100 cfm)
Power	930 W

Power/Fuel Requirements and Transportability

Series 800 Critical Filter Vacuum Systems - 115V or 220V AC/DC or 620 kPa (90 psi) compressed air at 1.2 m³/min (42 cfm).
Series X 829 Wet/Dry HEPA Filter Vacuum Systems - 115V or 220V AC/DC
Series X 839 Asbestos Vacuum Systems - 115V or 220V AC/DC
X 380-1 Back Pak Vac - 115V or 220V AC/DC

All vacuum systems are person-portable. All models have wheels except the X-1700, the two smaller X-1000 units, and the Back Pak Vac, which is mounted on a backpack frame.

Manufacturer

Minuteman International Inc.
111 South Route 53
Addison, IL 60101, U.S.A.

Telephone (708) 627-6900
Facsimile (708) 627-1130

Canadian Representative

Minuteman Canada, Inc.
84 E. Brunswick Blvd.
Dollard des Ormeaux, Quebec H9B 2C5

Telephone (514) 683-3838
Facsimile (514) 683-0809

REFERENCES

Manufacturer's Literature.

NILFISK VACUUM SYSTEMS

No. 65

DESCRIPTION

Small industrial vacuum cleaners designed to pick up wet or dry hazardous wastes, including nuclear, mercury, chemical, asbestos, and other materials.

OPERATING PRINCIPLE

Suction, created in the vacuum unit by an electric motor or a remote source of compressed air, draws the contaminant through a hose into the vacuum unit where it is caught by the appropriate filter media and contained within the vacuum tank.

STATUS OF DEVELOPMENT AND USAGE

Commercial products specifically designed to recover hazardous materials in plants.

TECHNICAL SPECIFICATIONS

Physical

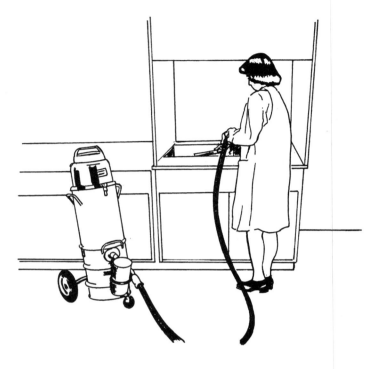

VT 60 and VT 60A Wet/Dry HEPA Vacuums - The VT 60 vacuum is a wet/dry HEPA vacuum with a 57 L (15 gal) polyethylene tank. The VT 60A is similar to the VT 60 except that it is powered by a remote compressed air source. Both models are 0.9 m (3.1 ft) high, 0.5 m (1.8 ft) wide, have a filter area of 1.4 m^2 (15 ft^2), and weigh about 30 kg (65 lb).

GS 80 Series - This series includes the GS 80 with a disposable bag capacity of 8.5 L (2.3 gal) dry bulk; the GS 81 with a disposable bag capacity of 15 L (4 gal) dry bulk; the GS 82 with a capacity of 45 L (12 gal) dry bulk; and the GS 83 with a disposable bag capacity of 68 L (18 gal) dry bulk.

GS 90 Asbestos Vac - A lightweight, heavy-duty, dry-only HEPA vacuum cleaner, which is 40 cm (16 in) high, 30 cm (12 in) wide, with a filter area of 0.9 m^2 (9.7 ft^2), and weighs 6 kg (14 lb).

GS 90 Mercury Collection Vacuum - Recovers liquid mercury spills and granular mercury compounds, while eliminating mercury vapours in the process. No physical specifications available.

SCV Spray-Cleaner Vac - A HEPA-filtered, spray-cleaning vacuum with corrosion-resistant polyethylene supply and collection tanks, designed to apply Tri-sodium phosphate or other cleaning solutions with simultaneous or delayed solution recovery. The SCV is suited for lead abatement and other wet-cleaning applications that require particle control. This unit is 0.9 m (3.1 ft) high, 0.5 m (1.8 ft) wide, has a filter area of 1.4 m^2 (15 ft^2), and weighs 55 kg (120 lb).

Operational

All Nilfisk vacuums use an "absolute filtering system", which begins with first-stage separation, during which the centrifugal airflow pattern of the cleaner separates the heavier dust from the collected fines. The collected material is then filtered by the main filter, and then by a microfilter which provides pre-filtering protection for the motor at an efficiency of 99.5% at 2 microns. HEPA or ULPA filters further increase the retention efficiencies to absolute standards of up to 99.9995% at 0.3 or 0.12 microns.

VT 60 and VT 60A Wet/Dry HEPA Vacuums - The VT 60 offers three collection/disposal options: a 19 L (5 gal) disposable pail, a polyliner, or direct collection into the 57 L (15 gal) vacuum tank.

2.6 Removal - Vacuum Systems

	VT60	**VT60A**
Capacity [L (gal)]	38 (10)	57 (15)
Vacuum [kPa (in H$_2$O)]	29 (117)	43 (172)
Air flow [m^3/min (cfm)]	2.8 (100)	3.0 (140)
Power	1,000 W	15,000 W (20 hp) compressor required

GS 90 Asbestos Vac and GS 80 Series

Dry capacity	12 L (0.4 ft^3)
Vacuum	19 kPa (75 in H$_2$O)
Air flow	2.5 m^3/min (87 cfm)
Power	700 W

GS 90 Mercury Collection Vacuum - In addition to the HEPA or ULPA filters, this vacuum is equipped with an activated carbon adsorbent filter cartridge that adsorbs mercury vapour. The 15 kg (33 lb) charge of activated carbon has a shelf life of up to five years. A transparent, smooth-lined Tygon hose reduces buildup of droplet residue.

Capacity	1.4 L (0.38 gal)
Vacuum	19 kPa (75 in H$_2$O)
Air flow	2.5 m^3/min (87 cfm)

SCV Spray-Cleaner Vac - The SCV offers three collection/disposal options: an 19 L (5 gal) disposable pail, a polyliner, or direct collection into the 57 L (15 gal) vacuum tank.

Capacity	57 L (15 gal)
Waterlift	29 kPa (117 in H$_2$O)
Air flow	2.8 m^3/min (100 cfm)
Power	1,000 W

Power/Fuel Requirements

VT 60 and VT 60A Wet/Dry HEPA Vacuums - 115 VAC, 8.3 amps, or 15,000 W (20 hp) air compressor
GS 80 Series - 115 VAC
GS 90 Asbestos Vac - 115/220 VAC, 5.9/3.0 amps
GS 90 Mercury Collection Vacuum - Specifications not available
SCV Spray-Cleaner Vac - 120 VAC, 8.3 amps

Transportability

All vacuum systems have wheels and are person-portable.

CONTACT INFORMATION

Manufacturer

Nilfisk Limited
396 Watline Avenue
Mississauga, Ontario
L4Z 1XZ

Telephone (905) 712-3260
Facsimile (905) 712-3255
Toll-free   (800) 668-0400 (Canada)

REFERENCES

(1) Manufacturer's Literature

POWERVAC OIL RECOVERY SYSTEM

No. 66

DESCRIPTION

A trolley-mounted, engine-driven suction unit which includes an integral pressure washer. The vacuum head is designed for use with a standard oil drum or with a Vikoma Hopper unit which is offered as an option.

OPERATING PRINCIPLE

A vacuum is created in the collection tank by a vane-type vacuum pump powered by a diesel motor. The vacuum draws the contaminant through a hose into the vacuum unit and then deposits it directly into a standard drum or the optional hopper. A water pump that is tolerant of fresh or salt water is coupled to the diesel to power a pressure washer jet lance.

STATUS OF DEVELOPMENT AND USAGE

Commercial product specifically designed to recover very viscous materials.

TECHNICAL SPECIFICATIONS

Physical - The unit is 1.5 m (5 ft) long, 1.1 m (3.7 ft) wide, 1.4 m (4.5 ft) high, and weighs 330 kg (730 lb). It has a 5.8 hp, 2-cylinder, water-cooled, electric start, 4-stroke diesel engine. The frame is made of marine-grade aluminum channel box section and sheet and both the vacuum head and fuel tank are made of marine-grade aluminum sheet. The covers are made of glass-reinforced plastic. All metal parts are cleaned with metal conditioner and painted with two coats of epoxy primer and orange polyurethane.

Operational - The vane-type vacuum pump produces extremely high air velocities in the recovery lance hoses to carry entrained sludge lumps and solids to recovery and generate high vacuums. Vacuum levels are adjustable, and recovered sludges can either be air carried/sucked to, or dumped in, standard open-top oil drums or, at much higher vacuum levels, to a recovery hopper with a discharge shoot. The PowerVac generates free air velocities to 100 m/s (328 ft/s) in the suction lance, but will, for example, in the vacuum-operating mode, continue to recover viscous muds with the tube/lance completely filled with 12 m (39 ft) of sand/mud. Typical sludge recovery is at a rate of 30 t/h. The water jet operates at a pressure of 6,900 kPa (1,000 psi).

Power/Fuel Requirements and Transportability - The unit operates on diesel fuel. The system is trolley-mounted with wheels for movement on hard-packed surfaces. The unit can be transported in a pickup truck.

CONTACT INFORMATION

Manufacturer

Vikoma International Limited
88 Place Road, Cowes
Isle of Wight, P031 7AE
England

Telephone 983 296021
Facsimile 983 299035

Canadian Representative

Can-Ross Environmental Services Limited
2270 South Service Road West
Oakville, Ontario
L6L 5M9

Telephone (905) 847-7190
Facsimile   (905) 847-7175

REFERENCES

(1) Manufacturer's Literature.

2.6 Removal - Vacuum Systems

RO-VAC VACUUM SYSTEM   No. 67

DESCRIPTION

A trolley-mounted, vacuum system with an integral pressure washer for recovering oils, sludges, and waterborne solids from areas not accessible to vacuum trucks. The vacuum head is designed for use with a standard oil drum or with the optional hopper unit.

OPERATING PRINCIPLE

A vacuum is created in the collection tank by a rotary sliding vane vacuum pump powered by a diesel motor. The vacuum draws the contaminant through a hose into the vacuum unit and deposits it directly into a standard drum or into the Ro-Vac hopper. A water pump that is tolerant of fresh or salt water is coupled to the diesel to power a pressure washer jet lance.

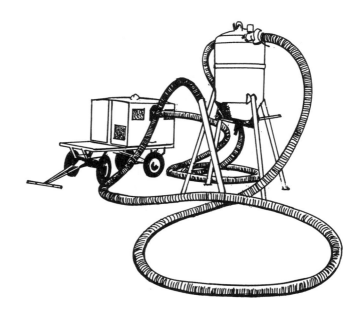

STATUS OF DEVELOPMENT AND USAGE

Commercial product specifically designed and used to recover very viscous materials and solids from water or ground surfaces.

TECHNICAL SPECIFICATIONS

Physical

Power Unit
Length	0.9 m (3 ft)
Width	0.8 m (2.8 ft)
Height	0.7 m (2.3 ft)
Weight	360 kg (790 lb)

Engine   12 kW (16 hp), 2 cylinder, air-cooled, diesel or electric motor - standard or explosion-proof to CSA, ISO, etc.

Vacuum Hopper
Material	Aluminum
Height	1.9 to 2.6 m (6 to 8.4 ft) adjustable
Diameter	1.2 to 1.6 m (4 to 5 ft) adjustable
Weight	38 kg (84 lb)

Trolley Unit
Flat platform	Four wide-tread, low-pressure tires
Length	1.5 m (4.9 ft)
Width	0.7 m (2.6 ft)
Height	0.5 m (1.7 ft)
Weight	60 kg (130 lb)

Operational

The vacuum pump generates a maximum air flow of 7.2 m³/min (250 cfm) through a 7.6 cm (3 in) diameter hose and a vacuum of up to 80 kPa (24 in Hg). Pumps of the same type and specification are used in industrial vacuum trucks. When the suction nozzle is immersed, the Ro-Vac fills an oil drum in less than 10 seconds. Coupled with the powerful vacuum lift, the Ro-Vac transfers materials over hose lengths of up to 90 m (295 ft). The pressure washer pump, which tolerates salt or fresh water, is driven by a toothed belt and provides pressures up to 10,300 kPa (1,500 psi) through a 10-m (33-ft) hose to a hand-held spray lance. The optional recovery hopper allows the highest vacuum levels to be used to recover very viscous materials. With its extending legs and base-dumping facility, the hopper can transfer recovered materials to skips, dump trucks, or other storage containers in an almost continuous operating cycle. The vacuum head and hopper unit feature a vacuum gauge and adjustable vent and upper limit vacuum valves. An automatic shutoff is provided for drum/hopper fill and fully automatic open-close dumping of recovered product.

Power/Fuel Requirements

Diesel fuel or electric mains supply - three-phase.

Transportability

The Ro-Vac is designed with a low centre of gravity and is easy to transport and move on site. The power unit has a space volume of 0.6 m^3 (21 ft^3) and weighs 360 kg (790 lb). It is mounted on a trolley which has four large diameter pneumatic tires and is fitted with Ackerman steering for easy maneuverability and stability when turning. A central lifting eye and shackle are provided. All components, including the power unit, vacuum head, hopper, and hoses, can be transported by pickup truck.

PERFORMANCE

As the Ro-Vac diesel engine and vacuum pump operate significantly below their maximum ratings and the unit is equipped with built-in fans and vents, air flow is good and there is no excess buildup of heat. The Ro-Vac was used during a 40-day beach-cleaning trial in Saudi Arabia conducted for IMO/MEPA and the Danish government after the Gulf War. The objective was to remove an average 9 m (30 ft) wide strip of heavily weathered oil from beaches. The Ro-Vac was used to recover the weathered crude from the beach zones, pressure wash the beaches, and clean other equipment used in the beach cleanup. The trial results are published in a report available from Ro-Clean International.

In conjunction with a U.K.-U.S. chemical company, a range of special suction head tools has been developed for recovering caustic, crystallized wastes from tanks.

CONTACT INFORMATION

Manufacturer

Ro-Clean International A/S
Hestehaven
DK-5260 Odenses
Denmark

Telephone (+45) 65 91 02 01
Facsimile   (+45) 65 90 88 77

Canadian Representative

Navenco Marine Incorporated
350 Ford Boulevard, Suite # 120
Chateauguay, Quebec
J6J 4Z2

Telephone (514) 698-2810
Facsimile   (514) 698-1008

OTHER DATA

A versatile floating suction head is also available to enhance the efficiency of the Ro-Vac. The **Ro Weir** can be deployed in many ways to recover substances floating on water. Made of aluminum, the head unit has a manually adjustable weir and two sealed buoyancy compartments.

REFERENCES

(1) Foulds, M., "New Methods to Remove Oil Penetrated into the Sand in a Tidal Zone", Ro-Clean International, Odenses, Denmark (1991).
(2) Manufacturer's Literature.

TORNADO VACUUM SYSTEMS No. 68

DESCRIPTION

Small industrial vacuum cleaners for picking up wet or dry hazardous wastes, including chemicals, asbestos, and nuclear waste, in industrial plants and laboratories.

OPERATING PRINCIPLE

A vacuum created in the collection tank, either by an electric motor integral to the unit or by a remote source of compressed air, draws the contaminants through a hose into the vacuum unit where they are caught by the filter media and contained within the vacuum tank.

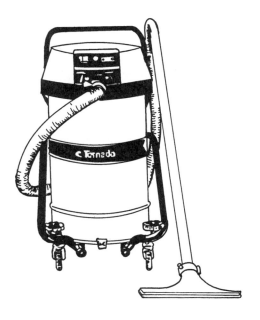

STATUS OF DEVELOPMENT AND USAGE

Commercial products specifically designed to recover hazardous materials in industrial plants.

TECHNICAL SPECIFICATIONS

Physical

The following are brief descriptions of the five
Tornado vacuum systems that are suitable for recovering hazardous materials, including a table of specifications for each model.

Headmaster™ ToxVac7000 - A hazardous material vacuum with a 68-L (18-gal) stainless steel tank. The all-welded "tip and pour" carrier has 10-cm (4-in) swivel ball bearing casters, 25-cm (10-in) semi-pneumatic rear tires, and two 10-cm (4-in) rubber bumper wheels. A utility basket is mounted between the back supports.

Air ToxVac - An air-powered hazardous material vacuum with a 68-L (18-gal) stainless steel vacuum tank, mounted on an all-steel, four-wheel dolly, with tilt handle and easy-roll casters. Optional Manometer filter gauge and 3.8 cm (1.5 in) tool sets.

Jumbo AirToxVac (ATV5) - An air-powered hazardous material vacuum that uses a 208-L (55-gal) steel drum as a vacuum tank. The system is mounted on an all-steel, four-wheel dolly, with tilt handle and easy-roll casters.

Mini ToxVac - A sub-compact hazardous material vacuum with a Lift-Grip handle and a 28-L (7.4-gal) stainless steel tank mounted on a bumper-guarded carrier with four 6.4-cm (2.5-in) ball bearing swivel casters and an optional Manometer filter gauge.

Portable ToxVac - A compact, portable, backpack-mounted hazardous material vacuum with a 7.6-L (2-gal) ABS plastic vacuum tank and weighing only 8 kg (18 lb). Includes shoulder harness, 1.5 m (5 ft) of vinyl hose, 7.6 cm (3 in) dusting tool, 38 cm (15 in) crevice tool, and 12 disposable paper filters.

		ToxVacs			
	Headmaster 7000	Air	Jumbo Air (ATV5)	Mini	Portable
Height	0.9 m (3 ft)	1 m (3.5 ft)	1.5 m (4.8 ft)	0.6 m (2 ft)	0.4 m (1.3 ft)
Width	0.5 m (1.7 ft)	0.5 m (1.7 ft)	0.6 m (2 ft)	0.4 m (1.4 ft)	0.2 m (0.8 ft)
Diameter	0.6 m (2 ft)	not available	0.6 m (2 ft)	0.4 m (1.4 ft)	0.3 m (1.1 ft)
Weight	28 kg (60 lb)	24 kg (53 lb)	59 kg (130 lb)	16 kg (35 lb)	7.9 kg (18 lb)
Filter area	1.8 m^2 (19 ft^2)	2.1 m^2 (22 ft^2)	1.5 m^2 (16 ft^2)	1.3 m^2 (14 ft^2)	0.8 m^2 (8.6 ft^2)
Motor	1.6 kW (2.1 hp)	air-driven	air-driven	0.8 kW (1.1 hp)	0.8 kW (1.1 hp)
Cord	15 m (50 ft)	N/A	N/A	15 m (50 ft)	15 m (50 ft)

Operational

Headmaster™ ToxVac7000 - Features injection-molded plastic power modules designed to accept the HEPA, SCAFITEX, and paper filtration systems. The powerhead also includes a carbon brush motor filter, for use in Class 100 "Clean Rooms". The HEPA filter is tested to assure a minimum removal efficiency of 99.97% at 0.3 µm. The SCAFITEX filter is designed to protect the HEPA filter and entrap standard dust with efficiency of 99.4 %. The primary paper filter, which fits around both the SCAFITEX and HEPA filters, is designed as a disposable fail-safe filter with 0.4 m^2 (650 in^2) of filtration. A standard manometer light gauge indicates when the HEPA filter system needs to be cleaned or replaced. The hose intake on the Headmaster vacuum is located in the powerhead unit and not in the tank, which provides the following advantages: 1) all filters stay attached to the power module; 2) a plastic liner can be placed in the tank for self-bagging of debris; 3) the intake deflector sets up a cyclonic action (in-air separation) that forces the debris down into the tank; 4) the powerheads are readily adaptable to 208-L (55-gal) drums; and 5) the intake accepts 3.8 and 5 cm (1.5 and 2 in) attachments.

Air ToxVac and Jumbo Air ToxVac (ATV5) - A Venturi powerhead runs on compressed air and has no moving parts, for trouble-free operation. This design permits recovery of hazardous materials where a spark-free vacuum system is required. The unit must be grounded when used in a potentially volatile environment. Both Air ToxVacs feature Absolute filter systems that incorporate a mounting frame, fail-safe paper filter, polypropylene filter, and a HEPA filter. The HEPA filter is capable of 99.97% removal at 0.3 µm.

Mini ToxVac - A small, sub-compact vacuum designed primarily for recovering dry material. It has a four-stage filter system with a HEPA filter capable of 99.97% removal at 0.3 µm.

Portable ToxVac - A compact, backpack-mounted, hazardous material vacuum, designed to recover dry material only. It features a three-stage filter system with a HEPA filter capable of 99.97% removal at 0.3 µm.

		Toxvacs			
	Headmaster 7000	Air	Jumbo Air (ATV5)	Mini	Portable
Wet capacity L (gal)	19 (5)	19 (5)	152 (40)	15 (4)	N/A
Dry capacity L (bu)	46 (1.3)	35 (1)	190 (5.4)	35 (1)	8 (0.23)
Vacuum kPa (in H$_2$O)	30 (120)	60 (240)	60 (240)	20 (80)	20 (80)
Air flow m^3/min (cfm)	3 (107)	2.3 (83)	2.3 (83)	2.1 (75)	2 (69)
Air pressure kPa (psi)	N/A	620 (90)	620 (90)	N/A	N/A

Power/Fuel Requirements

Headmaster™ ToxVac7000 - 115 VAC, 60 Hz, 12 amps, 230 VAC models available.

Air ToxVac and Jumbo Air ToxVac (ATV5) - 620 kPa (90 psi) compressed air at 1.3 m^3/min (47 cfm). An 11 kW (15 hp) compressor with 1.3 cm (0.5 in) airline and quick connectors is recommended.

Mini ToxVac and Portable ToxVac - 115 VAC, 60 Hz, 6.9 amps, 230 VAC models available.

Transportability

All vacuum systems are person-portable. All models have wheels, except the Portable ToxVac which is mounted on a back-pack frame.

CONTACT INFORMATION

Manufacturer

Breuer/Tornado Corporation
7401 W. Lawrence Avenue
Chicago, IL 60656-3489
U.S.A.

Telephone (708) 867-5100
Toll-free (800) 822-8867
Facsimile (708) 867-6968

REFERENCES

(1) Manufacturer's Literature.

TRANS-VAC OIL RECOVERY UNITS

No. 69

DESCRIPTION

High-capacity, self-contained high-suction/pressure pump systems incorporating a vacuum pump, liquid-air-solids separator, and positive displacement rotary pump.

OPERATING PRINCIPLE

These units combine the air-handling capacity of a vacuum pump and the high transfer capability of a positive displacement pump. The **Trans-Vac 500D** uses the vacuum created by a high capacity vacuum pump to move oil, water, debris, and air into a receiving vessel on the unit. The **Trans-Vac 300DH and 220D** convey oil, water, and debris from skimming heads through floating hoses into a receiving vessel. Inside the vessel, air and liquid are continuously evacuated through different outlets, while debris is screened and retained by a grate inside the tank. On all models, the vacuum pump and transfer pump are each driven by a separate diesel engine for independent speed control.

STATUS OF DEVELOPMENT AND USAGE

Commercial product specifically designed to recover and convey spilled oil at remote sites or floating chemicals from water or land depending on whether skimmers or suction heads are used on hoses.

TECHNICAL SPECIFICATIONS

Physical

Models	500D	300DH	220D
Length	3.6 m (12 ft)	2.3 m (7.5 ft)	2.3 m (7.5 ft)
Width	2.1 m (7 ft)	1.5 m (4.8 ft)	1.5 m (4.8 ft)
Height	2.1 m (7 ft)	2 m (6.5 ft)	2 m (6.5 ft)
Weight	3,100 kg (7,000 lb)	1,100 kg (2,500 lb)	1,100 kg (2,500 lb)
Engines	Two 40 hp	Hydraulic Power Pack	9 and 22 hp
Capacity	1,500 L (400 gal)	790 L (210 gal)	790 L (210 gal)

Operational

Trans-Vac units are self-contained, requiring only a floating or land-based holding tank or tank trucks to operate. All components except hoses and skimmer heads are mounted on a skid. The pumps are driven by independent diesel engines. Once the pumps are set at the operating speed, suction rate is controlled with a single lever control valve near the control panel. As the pumps are driven by separate engines, the operator can set any combination of suction and discharge rates. Trans-Vac uses the vacuum pump exhaust and a connection on the unit to make blow-back of lines a simple operation. By manipulating a few valves, positive air pressure is used to force obstructions from the suction line. The Trans-Vac is equipped with an overfill alarm and vacuum pump shutdown which is activated if the alarm continues unchecked. Relief valves are fitted on all lines where positive pressure is possible, including the receiver vessel. Exhaust gas from the blower is monitored continually and an alarm indicates when temperatures become excessive. The Trans-Vac is equipped with three spotlights for night operations. Custom trailers are also available.

The following are recovery and pumping specifications.

Models	500D	300DH	220D
Suction lift	5 m (16 ft)	5 m (16 ft)	5 m (16 ft)
Discharge head	35 m (115 ft)	21 m (70 ft)	21 m (70 ft)
Discharge rate	32 L/s (500 gpm)	19 L/s (300 gpm)	14 L/s (220 gpm)
Air capacity	24 m^3/min @ 51 kPa Hg	8.5 m^3/min @ 51 kPa Hg	8.5 m^3/min @ 51 kPa Hg

Power/Fuel Requirements

Diesel fuel.

Transportability

All components of the Trans-Vac except hoses and skimmer heads are mounted on a heavy structural steel skid which assures rigidity and durability under adverse field conditions. Trans-Vac units can be moved with a forklift and transported by helicopter.

PERFORMANCE

Trans-Vac Oil Recovery Units have been used successfully in many oil spill cleanups including the San Jacinto River Spill in 1994, the *Exxon Valdez* spill in Alaska in 1989, and the *Texaco Panama* spill in 1982.

CONTACT INFORMATION

Manufacturer

Slickbar Products Corporation
18 Beach Street
Seymour, CT 06483
U.S.A.

Telephone (203) 888-7700
Toll-free (800) 921-2221
Voice Mail System (800) 322-BOOM (2666)
Facsimile (203) 888-7720

OTHER DATA

Slickbar Products Corporation has been manufacturing spill response systems and products for 35 years. These include a range of **Oil Booms**, **Special Purpose Booms**, other **Oil Recovery Products** such as pumps and skimmer heads, and **Boats, Trailers, and Accessories**, such as Folding Oil Spill Workboats. The company also offers both classroom and on-site training and engineering advice for special applications or to solve difficult spill response problems.

REFERENCES

(1) Manufacturer's Literature.

VACTAGON VACUUM SYSTEMS

No. 70

DESCRIPTION

Small industrial vacuum cleaners designed to pick up wet or dry hazardous wastes, including nuclear waste, chemicals, and asbestos, in industrial plants and laboratories, and a mid-sized vacuum system that provides high suction, wet/dry versatility with total portability and HEPA filtration.

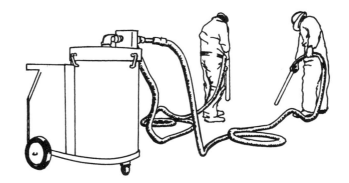

OPERATING PRINCIPLE

A vacuum is created in the collection tank by either an electric motor integral to the unit or a remote source of compressed air. The contaminant is drawn through a hose, caught by the appropriate filter media, and contained within the vacuum tank.

STATUS OF DEVELOPMENT AND USAGE

The small vacuums are commercial products designed to recover hazardous materials in plants, while the mid-sized vacuum is designed for environmental contractors and fills the gap between small "in-plant" vacuums and large truck-mounted systems.

TECHNICAL SPECIFICATIONS

Physical

Vactagon vacuums use a two-stage filter system. The primary filter is an exclusive, Porthane filter element consisting of hundreds of porous plastic tubes. The HEPA filters are located ahead of the motor(s) to prevent contamination and prolong service life and are rated at 99.99% efficient at 0.3 µm. A suspended plate seal (SPS) liquid shutoff system cuts off the suction and protects the final filter from liquids if the vacuum overflows.

Drum Top Vacuum Series - Eight models of drum-top vacuum units are available, which vacuum materials directly into 208-L (55-gal) drums, or into a plastic bag placed within a drum. Available as either electric or air-powered, the units operate simultaneously as wet and dry vacuums. The vacuum heads are constructed of either aluminum or steel. A drum trolley is available as an option.

TecVac Vacuum - A small, low profile vacuum for vacuuming wet or dry materials into plastic bags for one-step cleanup and disposal. The 76-L (20-gal) polyethylene tank has a collection capacity of 35 L (9.2 gal). It is 0.6 m (1.8 ft) in diameter, 0.7 m (2.4 ft) high, and has a net weight of 25 kg (55 lb) and a shipping weight of 29 kg (63 lb).

Gold Series Vacuums - Larger than the TecVac, with a capacity of 0.07 m^3 (2.3 ft^3) and constructed of either carbon steel or stainless steel. Three models are available with a choice of carbon brush, turbine, or positive displacement blower. Reverse pulse filtration is optional. All models are 0.8 m (2.8 ft) long, 0.6 m (1.9 ft) wide, and 1 m (3.2 ft) high. The AU200 weighs 73 kg (160 lb) and the AU400 and AU500 weigh 82 kg (180 lb).

Vaculoader Vacuum Loading System - A mid-sized system that provides high suction, wet/dry versatility with total portability and HEPA filtration. It fills the gap between small "in-plant" vacuums and large, truck-mounted, positive displacement or centrifugal blower-driven vacuums. Four models with power ratings from 4 to 38 kW (5 to 50 hp).

Operational

Due to plastic filtration media which never rot and are washed instead of thrown away, Vactagon vacuums can handle a wide range of chemicals, solvents, caustics, and even some acids. All vacuums operate in a simultaneous wet/dry mode. Each unit is supplied with a full complement of wet/dry cleaning tools. Quick-disconnect couplers provide positive locking connections. Designed with a low centre of gravity so that units do not fall over even when pulled aggressively by the hose. Special hoses are available for chemicals, as well as static and temperature-resistant hoses.

Drum Top Vacuum Series - Vacuum materials directly into 208-L (55-gal) drums or a plastic bag within a drum.

	DT55	DT55-H	DT255	DT255-H	DT55-AP	DT55-APH	DT255-AP	DT255-APH
Air flow m^3/min (cfm)	3.2 (112)	3.2 (112)	6.4 (225)	6.4 (225)	2.4 (83)	2.4 (83)	4.7 (166)	4.7 (166)
Vacuum kPa (in H$_2$O)	26 (106)	26 (106)	26 (106)	26 (106)	52 (220)	52 (220)	52 (220)	52 (220)
Filtration	Standard	HEPA	Standard	HEPA	Standard	HEPA	Standard	HEPA

TecVac Vacuum - The TecVac has a capacity of 35 L (9.2 gal) and generates an air flow of 2.8 m^3/min (100 cfm) and a vacuum of 24 kPa (95 in H$_2$O). The polyethylene tank is dent- and corrosion-resistant.

Gold Series Vacuums - Gold Series vacuums have a capacity of 65 L (2.3 ft^3) and can support two operators and hoses simultaneously. A gauge indicates when the HEPA filter needs changing. Three models are available with air flows ranging from 3.2 to 8.5 m^3/min (110 to 300 cfm) and vacuums ranging from 24 to 35 kPa (95 to 140 in H$_2$O).

Vaculoader Vacuum Loading System - Four models are available, with air flow ranging from 5.7 to 28 m^3/min (200 to 1,000 cfm), vacuum ranging from 30 to 75 kPa (120 to 300 in H$_2$O), recovery distance from 23 to 152 m (75 to 500 ft), and using from 2 to 10 hoses. Combines high suction and high airflow with wet/dry versatility, portability, and HEPA filtration. It pulls materials weighing up to 1,600 kg/m^3 (100 lb/ft^3) from 69 m (225 ft) and yet rolls easily through a standard doorway. This is made possible by the system's high-efficiency Roots positive displacement blower which brings full horsepower into use only when hose-loading occurs, when it is needed most. The patented bag-loading system allows the operator to collect wet or dry bulk materials into standard, off-the-shelf plastic bags. The collection hopper tilts forward so that full bags can easily be removed. Volumes of fine powders, solids, liquids, or sludges can be collected into poly bags that are pneumatically retained inside steel drums. This method, called speed loading, permits an operator to load up to a bag per minute while continuously vacuuming. Larger collection hoppers are available.

For applications that require extremely long hoses, an optional Twin Package provides an inexpensive alternative to truck- or trailer-mounted, or stationary vacuum equipment. Designed for environmental contractors, this system connects two units together for twice the power of a single machine and pulls material from up to 137 m (450 ft) horizontally. A single system can also be operated with multiple hoses. Up to six operators using 38 mm (1.5 in) diameter hoses can be used. The HEPA filter is continuously monitored and a gauge indicates when the filter needs changing. Optional equipment includes explosion-proof electric motors and controls and a silencer package.

Power/Fuel Requirements and Transportability

Drum Tops - 110 VAC, 11 or 20 amp; or 690 kPa (100 psi) compressed air at 1.7 to 3.4 m^3/min (60 or 120 cfm).
TecVac Vacuum - 120 VAC, 15 amp (220 VAC optional).
Gold Series Vacuums - 120 VAC, 11, 15, or 20 amp.
Vaculoader Vacuum Loading System - 020 VAC or 220/440 VAC, gas or diesel.

All vacuum systems are mounted on a trolley and most can be handled by one person. All systems will pass through a standard doorway. The larger Gold Series and Vaculoader systems require more than one person to lift.

PERFORMANCE AND CONTACT INFORMATION

The Vaculoader has been widely used in the United States for removing contaminated soil from crawl spaces and tunnels and for removing asbestos. For more demanding applications, Vactagon builds custom vacuum equipment, central in-plant units, and diesel-powered units to meet customers' specifications. A free video on Vactagon's systems is available from the manufacturer.

Manufacturer

Vactagon Pneumatic Systems, Inc.
P.O. Box 11583
Rockhill, SC 29730　　　　　　　　　　Telephone (803) 328-5252
U.S.A.　　　　　　　　　　　　　　　　Facsimile　(803) 328-5255

REFERENCES

(1)　　Manufacturer's Literature.

VAC-U-MAX VACUUM SYSTEMS

No. 71

DESCRIPTION

Industrial vacuum cleaners designed to pick up hazardous wastes in wet or dry form, including nuclear waste, mercury, chemical, and asbestos, in industrial plants and laboratories, and a mid-sized vacuum system that provides high suction, wet/dry versatility with total portability and HEPA filtration.

OPERATING PRINCIPLE

A vacuum created in the collection tank by either an integral electric motor powered vacuum pump or a remote source of compressed air draws the contaminant through a hose into the vacuum unit where it is caught by the appropriate filter media and contained within the vacuum tank.

STATUS OF DEVELOPMENT AND USAGE

Commercially available vacuums. Drum vacuum and liquid recovery system are designed for spill recovery.

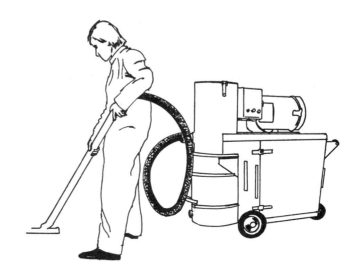

TECHNICAL SPECIFICATIONS

Physical

Model 55 - Air-operated, drum-top vacuums that vacuum directly into 208-L (55-gal) drums. Covers are equipped with a Venturi-type vacuum developer and connection for vacuum hose. Dust filters, automatic full-drum shutoffs, and units with drum and four-wheel dolly for portable use are all available. This unit is 0.7 m (2.2 ft) in diameter, 1.1 m (3.8 ft) high, and weighs 57 to 77 kg (125 to 170 lb).

Model 450 - A HEPA-rated electric vacuum cleaner, powered by a 3.7 kW (5 hp) motor. The unit is liquid- and solid-compatible and has a 76-L (20-gal) castered debris container.

Model 747 - An air- or gasoline-powered unit for recovering liquid or sludge, consisting of a vacuum device with a 1,040-L (275-gal) tank, mounted on a steerable, four-wheel chassis with towbar and handbrake. This unit is 3.4 m (11 ft) long with towbar extended, 1.8 m (5.8 ft) wide, 1.8 m (5.8 ft) high, and weighs 790 kg (1,750 lb) when empty.

Model 1010 - A HEPA-rated, 7.5 kW (10 hp) electric unit with positive displacement rotary vane vacuum pump and a 95-L (25-gal) debris container. Also equipped with a high air flow rotary suction blower for fast material pickup.

Model 1500 - A HEPA-rated, 7.5 kW (10 hp), electric unit available in several versions for portable or stationary use. Designed as a high capacity cleaner for general cleanup and for continuous use as a vacuum source in central systems.

Model 3300 - A HEPA-rated, 22 kW (30 hp) electric unit designed for use with an interceptor hopper for large-scale removal of sand, cement, tailings refuse, sludges, or liquids. A rotary piston vacuum pump provides strong suction and high air flow to remove up to 9 t (10 tons) of debris per hour. Available on a rugged, steerable four-wheel cart.

Operational

Model 55 - Drum-top vacuums shipped with the inlet orifice set to develop 27 kPa (8 in Hg) vacuum and adjustable to 54 kPa (16 in Hg). A compressed air supply of 410 kPa (60 psi) is required in volumes of 1 to 4 m^3/min (35 to 140 cfm), depending on the number of venturies and produces induced air flows of 2.8 to 11 m^3/min (100 to 400 cfm).

Model 450 - This unit has a maximum vacuum of 22 kPa (87 in H$_2$O), a maximum air flow of 10 m^3/min (355 cfm), and a sound level of 72 dB.

Model 747 - The unit powered with compressed air is shipped with the inlet orifice set to develop 27 kPa (8 in Hg) vacuum and is adjustable to 54 kPa (16 in Hg). It requires an air supply of 410 kPa (60 psi) at a flow rate of 1 to 2 m^3/min (35 to 70 cfm), and produces a maximum induced air flow of 2.8 m^3/min (100 cfm). The gasoline-powered unit develops 41 kPa (12 in Hg) vacuum and a maximum airflow of 5.6 m^3/min (200 cfm). An optional static bonding reel ensures that all components are properly bonded to prevent buildup of static electricity.

Model 1010 - A maximum vacuum of 30 kPa (120 in H_2O) and air flow of 12 m^3/min (410 cfm).

Model 1500 - A maximum vacuum of 35 to 42 kPa (140 to 170 in H_2O) and air flow of 19 m^3/min (670 cfm).

Model 3300 - This unit has a maximum vacuum of 54 kPa (215 in H_2O) and air flow of 22 m^3/min (750 cfm)

Power/Fuel Requirements and Transportability

For **Model 55**, a compressed air supply of 410 kPa (60 psi) is required in volumes of 1 to 4 m^3/min (35 to 140 cfm). For **Model 747**, a compressed air supply of 410 kPa (60 psi) at a flow rate of 1 to 2 m^3/min (35 to 70 cfm), **or** gasoline, is required, depending on the model. All other models require a standard 120 VAC electric power supply.

The drum-top vacuums are available with a four-wheel dolly for portability on hard surfaces such as floors, tarmac, or concrete and are light enough to be person-portable. Model 747 is mounted on a steerable, four-wheel chassis with towbar and handbrake. The trailer is not equipped for highway travel. All other models are available on steerable carts, suitable for hard surfaces such as floors, tarmac, or concrete, and have slim contours for maneuvering in tight areas.

PERFORMANCE AND CONTACT INFORMATION

Vac-U-Max vacuum systems, which have been developed and marketed for more than 30 years, are used by the United States Air Force, chemical companies, and airlines.

Manufacturer

Vac-U-Max
37 Rutgers Street
Belleville, NJ 07109-3196
U.S.A.

Telephone (201) 759-4600
Facsimile (201) 759-6449

Canadian Representative

Machine Control B.A.A. Canada Incorporated
701 Meloche Avenue
Dorval, Quebec
H9P 2S4

Telephone (514) 631-4587
Facsimile (514) 631-4588

Shadrack Engineering Limited
501 Passmore Avenue, Unit 13
Scarborough, Ontario
M1V 5G4

Telephone (416) 293-3100
Facsimile (416) 293-2370

Tes-Tech Engineering Systems Limited
Suite 250
2221 West 4th Avenue
Vancouver, British Columbia
V6J 1N3

Telephone (604) 737-2217
Facsimile (604) 737-8874

OTHER DATA

Vac-U-Max also produces an **oil skimmer attachment** for its drum-top vacuum systems. The skimmers operate on high pressure air which eliminates the potential hazards of electric motors. They are portable, generate a vacuum of up to 61 kPa (18 in Hg) at 410 kPa (60 psi), and are capable of handling up to #4 weight oil, along with most entrained material able to pass through the 51 mm (2 in) inlet. Unit includes jet suction unit, Auto Vac automatic liquid cutoff, hose adapter, quick disconnect assembly, static conductive high-pressure air hose, and oil skimmer wand and hose.

REFERENCES

(1) Manufacturer's Literature.

Section 2.7

Removal - Dredging

DREDGING EQUIPMENT - GENERAL LISTING   No. 72

DESCRIPTION

Dredges are devices for removing bottom sediments from bodies of water. They are typically used for underwater excavation to deepen water courses for navigation, for underwater mining, or for collecting benthos. Dredging is also used to remove sediments contaminated by discharges of hazardous and harmful substances. See the following entries in this section.

> Crisafulli Dredges (No. 73)
> H&H Dredges (No. 74)
> Mud Cat Model 370 PDP Dredge (No. 75)
> Pit Hog Auger Dredges (No. 76)
> Trac Pump Dredging System (No. 77)
> VMI Mini Dredges and Piranha Auger Dredge Attachment (No. 78)

OPERATING PRINCIPLE

Dredges are categorized according to three operating principles: mechanical, hydraulic, or pneumatic.

Mechanical dredges apply direct mechanical force to dislodge and excavate material. Bottom sediment is usually scooped up with a bucket and carried to the surface where it is placed in a barge, scow, truck, or impoundment for treatment or disposal.

Hydraulic dredges remove and transport sediment in liquid slurry form. They are usually barge-mounted systems that remove and transport the liquid slurry using centrifugal pumps. Pumps are either barge-mounted or submersible. The suction end of the dredge is mounted on a moveable ladder that can be adjusted to a specific dredging depth. A cutterhead is often fitted to the suction end of a dredge to assist in dislodging bottom materials. The slurries, usually containing 10 to 20% solids by weight, are often transported several hundred metres to the disposal site through pontoon-supported pipelines or floating hose. Other handling methods include sidecasting, loading into barges or scows, and direct loading on board hoppers.

Pneumatic dredges are a special category of hydraulic dredge that use compressed air and/or hydrostatic pressure instead of centrifugal force to draw sediments to the collection head and through the transport piping. Pneumatic dredges are usually barge-mounted and dredged material is discharged to hopper barges or scows.

STATUS OF DEVELOPMENT AND USAGE

Commercially available products originally developed for maintaining navigable waterways. Small dredge systems are now marketed specifically for maintaining industrial ponds/lagoons. Dredging equipment and practices have focused primarily on maximizing the rate of sediment removal, although recently the environmental effects of dredging have also been a consideration. Equipment and operating methods are now being modified to accommodate the removal of contaminated sediments.

TECHNICAL SPECIFICATIONS

Physical

Size, weight, and configuration of dredging equipment vary considerably with each dredge, even within the three categories. Specific data should therefore be obtained for any particular dredge. Physical specifications have a direct bearing on selecting appropriate equipment for spill situations since temporary storage, transportation, and treatment and disposal of dredged material must be considered. See also the specifications for No. 73 to 78 in this section.

Operational

Mechanical Dredges

These include clamshell, dragline, dipper, and bucket ladder dredges, as well as conventional earth excavation equipment, such as backhoes and front-end loaders. Mechanical dredges can be either vessel-mounted for offshore use or track-mounted and land-based. Mechanical dredges can remove sediments at nearly in-situ densities, i.e., the water content of sediments is not increased through the dredging process. When the sediment has the maximum possible solids content, the scale of facilities required for transport, treatment, and disposal is reduced. Mechanical dredges, however, are capable of only modest production rates [<500 m³/h (<650 yd³/h)] and require separate disposal vessels and equipment. These dredges cause a great deal of sediment re-suspension and are ineffective for free or unabsorbed liquid contaminants.

Clamshell or grab dredges are crane-operated dredges mounted on flat-bottomed barges or pontoons, on other seagoing vessels, or on track-mounted, land-based vehicles. Material is removed with large, hinged buckets ranging in capacity from 0.75 to 10 m³ (1 to 13 yd³). Open buckets are lowered on a control cable to the bottom surface and sediment is scooped up as the two halves of the bucket come together. Clamshell dredges operate at 20 to 30 cycles per hour, depending on working depth and the characteristics of the sediment. Because standard clamshell buckets are open at the top, considerable sediment is lost through spillage and leakage when the bucket breaks the water surface.

Clamshell dredges can be used to excavate at in-situ densities in all types of material except highly consolidated sediments and solid rock, and can excavate to depths of 30 m (100 ft) or more. These dredges are well suited for use in confined areas, as position and depth can be carefully controlled. As with other types of mechanical dredges, clamshells are generally used for relatively small-scale operations (up to a few hundred thousand cubic metres).

The major disadvantages of clamshell and other mechanical dredges are the relatively low rate of production and the relatively high degree of sediment re-suspension compared to other dredges. Sediment re-suspension occurs as the bucket impacts and pulls free from the bottom and also as hoisting drag forces wash away part of the load. Clamshells generally excavate a heaped load and more material may be lost through rapid drainage of entrapped water and slumping of the material above the rim of the bucket as the bucket clears the water surface. Concentrations of suspended solids in the vicinity of clamshell dredging may be as great as 500 mg/L, compared with background levels, which do not generally exceed 50 mg/L. A watertight bucket has been designed in Japan that reduces the problem of fines and loose material escaping. The rubber gaskets used to create the seal may not stand up to continuous use in a full-scale dredging operation, however, and may present compatibility problems with spilled chemicals in certain spill situations. Watertight clamshell dredge buckets have not been used to date in North America. The re-suspension of bottom materials may also warrant the use of containment measures, such as curtains or pneumatic barriers.

Draglines use the same basic equipment as clamshell dredges and share many of the applications and limitations. Material is excavated as the buckets are pulled by a drag cable toward the crane. Draglines are operated by the same type of crane as clamshell dredges and can also be mounted on seagoing vessels or operated from land. A 2 m³ (2.6 yd³) bucket removes approximately 230 m³ (300 yd³) of loosely consolidated material per hour. Draglines can be used in the same circumstances as clamshell dredges, however, with less control of depth and position. In addition, sediment re-suspension is a problem as it is with clamshell and all other mechanical dredges. A watertight bucket could not be used for a dragline dredge due to the dredge's operating characteristics.

Dipper dredges can exert great mechanical force and are well suited to excavating soft rock and dense, highly consolidated sedimentary deposits such as clay and glacial tills. Material is excavated with a bucket attached to a long boom, which is forcibly thrust into the material to be removed. Dipper dredges are commonly mounted on flat-bottomed barges or other vessels. Vertical columns called spuds are anchored into the bottom sediments to hold the dredge in position. Dipper Dredges have a maximum working depth of 15 m (50 ft), and can usually achieve a production rate of 30 to 60 cycles per hour. Dipper buckets typically hold 6 to 9 m³ (8 to 12 yd³). The extreme mechanical action of a dipper dredge results in worse sediment re-suspension than that caused by clamshell dredges. The use of dipper dredges for removing spilled hazardous materials is, therefore, probably limited.

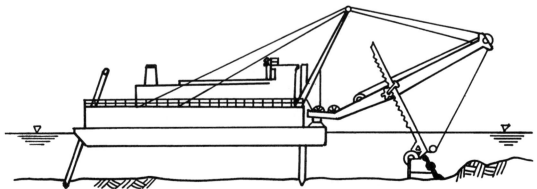

2.7 Removal - Dredging

Bucket ladder dredges have an inclined ladder that supports a continuous chain of buckets that rotate around two tumbler pivots located at each end of the ladder. As the buckets revolve around the lower tumbler, material is scooped, then transported up the ladder and dumped into a storage area as the buckets round the top pivot. Most bucket ladder dredges are mounted on pontoons and are not self-propelled. These dredges can store dredged material on board and can also load barges or other vessels for transport. Production rates are generally higher for bucket ladder dredges than for other mechanical dredges. Bucket volumes range from .075 to 1 m^3 (0.1 and 1.3 yd^3) and several hundred cubic metres of sediment per hour can be excavated under calm water conditions. Normal working depth for a bucket ladder dredge is approximately 18 m (60 ft) with some able to operate at depths to 30 m (100 ft).

The bucket ladder dredge is limited by the high degree of re-suspension of sediments and bucket leakage common to all mechanical dredges and the fact that they require support equipment, including tow boats and barges, and a complicated configuration of mooring lines to hold them in place.

Conventional earth excavation equipment can be used on a limited scale to remove sediments in shallow waterways. Backhoes and front-end loaders are sometimes used to remove bottom sediments. Both pieces of equipment scoop up material to be excavated and transfer the material to a vehicle or holding area. Backhoes are normally used for trench and other subsurface excavation and can reach 12 m (40 ft) or more below the level of the machine. Front-end loaders are normally used to excavate loose or soft materials in a narrow vertical range of operation a few feet above or below grade. Loaders are useful in removing sediments from dewatered portions of water bodies where track-mounting of equipment allows mobility in soft materials. Specially fabricated wide tracks (called "low-earth-pressure" tracks) provide added support and traction under such conditions.

Conventional earth excavation equipment is very limited for removing contaminated sediments from water bodies as its lateral and vertical reach is restricted by the length of the boom to which the excavation bucket is attached. This requires that the equipment be close to the material being dredged, which is often not practical. Also, submerged sediments and soils around water bodies have a low load-bearing capacity, making it difficult or impossible to maneuver equipment.

Hydraulic Dredges

Hydraulic dredges include portable dredges, handheld dredges, plain suction dredges, cutterhead dredges, dustpan dredges, and hopper dredges. They are usually barge-mounted and use centrifugal pumps to remove and transport sediment in a liquid slurry state. Production rates are generally higher and re-suspension rates lower than with mechanical dredges. As hydraulic dredges can also remove liquid contaminants, they are susceptible to damage by debris and clogging with weeds and thus more extensive pre-dredging work is required than with mechanical dredges. A major disadvantage of hydraulic dredges is the large amount of water recovered with the sediment which calls for a large receiving and de-watering area, as well as potentially extensive waste treatment systems to treat the recovered water and any contaminated drainage.

Portable hydraulic dredges are dredge vessels that can be moved easily over existing roadways without major dismantling. Dredging capabilities range from 3 to 15 m (10 to 50 ft). Vessel draft is generally less than 1.5 m (5 ft) with many less than 0.6 m (2 ft). Production rates average between 39 to 385 m^3/h (50 to 500 yd^3/h) depending on model, size, and site conditions. Methods of launching portable dredges vary and include amphibious/self-launching, launching by crane, launching from transport trailers, and launching at boat ramps from transport trailers. Most portable dredges are positioned using a cable and winch arrangement anchored on land.

Portable dredges are particularly suitable for many contaminated sediment remediation projects because of their transportability and shallow operating draft. Their primary limitations are their relatively low production rates and limited depth capabilities. They are also limited to conditions where waves do not exceed 30 cm (1 ft) in height.

Handheld hydraulic dredges remove a mixture of sediment and water in a slurry in much the same manner as larger hydraulic dredges. Handheld dredges can be operated either underwater or above water. Underwater handheld dredges are operated by divers who manipulate handheld suction devices that are connected to a pump and storage tank on board either a barge, a boat, or a land-based truck. Above-water handheld dredges are normally used by operators wading into shallow waterways or from small water crafts.

Handheld dredges cannot be operated in strong currents. They are suited for small-volume spills in calm water where precision dredging is important. They feature high mobility, manual positioning, which allows for meticulous cleanup work and effectiveness in vacuuming identifiable masses of pure contaminant, particularly liquids and free-flowing solids. Some diver-operated dredges can operate to depths of 305 m (1,000 ft) with a removal rate of 190 m^3/h (250 yd^3/h). Above-water handheld dredges are generally limited to use in waters no deeper than 1.2 m (4 ft), with bottom sediments firm enough to allow wading by the operator. The solids concentration of recovered slurry from a handheld dredge is typically 5 to 10%.

The major disadvantage of diver-operated handheld dredges is the risk of contaminant exposure for the divers and/or surface support personnel. Experiences in these environments have often resulted in injuries, primarily chemical burns. Problems with diving equipment also occur in chemically-contaminated water environments, primarily due to petroleum products. Equipment deterioration and failure caused by contaminant exposure has been responsible for fatalities and for incidents of diver exposure to contaminants. The U. S. government has extensively evaluated diver safety at spills of hazardous materials.

Plain suction dredges are the simplest form of hydraulic dredges, relying solely on the suction created by a centrifugal pump to dislodge and transport sediments. Suction dredges are only suitable for removing free-flowing materials such as sand or unconsolidated sediments. The dredge has no digging devices and therefore is not useful for hard or cohesive bottom materials.

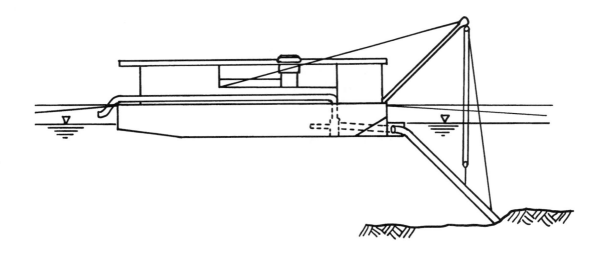

The dredge head is attached to the end of a ladder and is positioned vertically and horizontally by adjusting cables attached to the ladder. Material of 10 to 15% solids is drawn in through the dredge head, transported up the suction line, and discharged into a scow or through a pipeline for treatment. Recovery rates vary with the size and horsepower of the pump and the type of material being dredged, but typically range from 765 to 7,645 m^3/h (1,000 to 10,000 yd^3/h). Plain suction dredges are moved along a straight line fixed by a cable and winch arrangement anchored on land or on the bottom of the water body. Suction dredge vessel draft is typically from 1.5 to 1.8 m (5 to 6 ft). Hard and cohesive materials, such as clays or firm native bottom soils, are not readily removed by plain suction dredges as mechanical dislodging devices are not used. Generally, these dredges cannot be used in waters with waves of 0.9 m (3 ft) or higher.

Cutterhead dredges have a similar configuration and principle of operation as the plain suction dredge; however, a mechanical device is added to dislodge material. The device, called the cutterhead, is located at the intake of the suction pipe and rotates to break up hard and cohesive materials to create a slurry that can be recovered by the suction. When the cutterhead is shut off, the dredge operates in the same manner as a plain suction dredge. Slurries of 10 to 20% solids are typically achieved when the cutterhead is operating. Production rates vary according to pump size and can be as large as 1,910 m^3/h (2,500 yd^3/h). Vessels range from 15 to 69 m (50 to 225 ft) in length with a typical draft of between 0.9 and 1.5 m (3 and 5 ft).

Cutterhead dredges are propelled in a different pattern than other hydraulic dredges by alternately anchoring on one of two spuds. The anchored spud is used as a pivot and the vessel is drawn along an anchored cable, swinging the cutterhead in a short horizontal arc about the spud. When the cutterhead repeatedly swings in arcs while alternating anchoring spuds, the resulting partially overlapping cuts form a wide swath through the area being dredged. Cutterhead dredges can reach materials 15 m (50 ft) below the water surface. While they are best suited for use in calmer waters, their large size and self-propulsion make them more suitable for use in rough waters than the smaller dredges or plain suction dredges.

2.7 Removal - Dredging

Dustpan dredges are similar in configuration and operation to plain suction dredges. The sediment collection head, called a "dustpan", is a widely flared head containing high-pressure water jets that loosen and agitate sediments which are then captured in the dustpan as the dredge is winched forward into the bottom materials. The high-pressure jets allow cohesive sediments to be dredged but also produce considerable re-suspension of the bottom material into the water column. Slurries of 10 to 20% solids are typically recovered. Production rates range from 380 to 11,500 m^3/h (500 to 15,000 yd^3/h). Vessel draft varies from 1.5 to 4.3 m (5 to 14 ft). Dustpan dredges can reach materials up to 15 m (50 ft) below the water surface.

Hopper dredges, unlike other suction dredges, are mounted on self-propelled, seagoing vessels instead of barges. The suction pipes are hinged on each side of the ship with the intake extending downward toward the stern of the vessel. The head is dragged along the bottom as the vessel moves forward at speeds of up to 13 km/h (7.0 knots). The material moves up the suction pipe and is stored in the hold or hopper of the ship. Excess water containing a high level of suspended solids is normally allowed to flow over weirs in the hoppers and back into the waterbody. When excavating contaminated sediments, however, this overflow of the hoppers would not be acceptable, and would not be practiced. The material is normally unloaded by dumping it back into the waterbody at a different location which is also not acceptable for contaminated sediments. Contaminated dredged material would have to be pumped out of the hoppers to a safe disposal site. Some hopper dredges are equipped with built-in pump-out capability while others require auxiliary equipment.

Hopper dredges are most efficient for excavating loose, uncohesive materials. It is the only dredge that can operate in rough open waters, relatively high currents, in and around marine shipping traffic, and in adverse weather conditions. Slurries from hopper dredges have a solids content of 10 to 20%. The vessel drafts range from 3.6 to 9.5 m (12 to 31 ft) with production rates ranging from 380 to 1,530 m^3/h (497 to 2,000 yd^3/h). The minimum dredging depth varies from 3 to 8.5 m (10 to 28 ft) with a maximum dredging depth of 20 m (66 ft).

Pneumatic Dredges

These are a special category of hydraulic dredges that use compressed air and/or hydrostatic pressure instead of centrifugal force to draw sediments to the collection head and through the transport piping. They are often barge-mounted and dredged material is discharged to hopper barges or scows. Pneumatic dredges generally produce slurries with a higher concentration of solids than those produced by hydraulic dredges and there is less re-suspension of bottom materials. They can dredge both solids and liquids.

Airlift dredges use compressed air to dislodge and transport sediments. Compressed air is introduced into the bottom of an open vertical pipe that is usually supported and controlled by a barge-mounted crane. The air expands and rises as it is released, creating an upward current that carries both water and sediment up the pipe. Air can also be introduced through a special transport head that is vibrated or rotated to dislodge more cohesive sediments. The dredge is controlled laterally by swinging the boom of the crane in a manner similar to mechanical dredging. Vertical control is achieved by raising and lowering the open end of the vertical transport pipe, and by varying the pressure of the air released at the end of the pipe.

Pneumatic dredges were developed in Italy ("Pneuma" is a trade name). They consist of a pneumatic pump suspended on cables from a barge-mounted crane. The pump consists of three cylinders that are alternately filled with sediment by hydrostatic pressure. When full, the cylinders are serially emptied by closing the inlet and applying compressed air to force the contents into a common header line. Pneuma systems can dredge bottom materials at near in-situ densities. The major advantage is that the sediment does not have to be in liquid slurry form, which minimizes material-handling problems and turbidity. Pneuma dredges can discharge directly to disposal or treatment areas through a discharge pipeline.

Theoretically, there is no depth limitation to the pneumatic pump and because it is crane-mounted, it offers accurate dredging, which is useful in and around port structures. Pneuma dredges cannot be effectively used in shallow water because water depth and corresponding hydrostatic pressure are required to induce the flow into the dredgehead. While the entire system is not small, its components can be transported by truck or by air.

Oozer dredges were developed in Japan ("Oozer" is a trade name). These pumps are similar in concept to the Pneuma dredge. Oozer dredges use negative pressure (vacuum) to fill the chambers in addition to the difference between water pressure at the depth of operation and atmospheric pressure, which drives the Pneuma dredge. This allows dredging to be done in shallower waterways.

The Oozer pump is usually mounted at the end of a ladder. The pump body consists of two cylinders to which a vacuum is applied to increase the differential pressure and induce the flow between the sediment and cylinders. Positive air pressure is used to discharge, similar to the Pneuma dredge. Sediment thickness detectors, underwater television cameras, and a turbidimeter are attached near the suction mouth. Suspended oil can be collected by an attached hood and cutters can be attached for dislodging hard soils. Oozer dredges can pump slurries at near in-situ densities, at rates of 380 to 610 m^3/h (500 to 800 yd^3/h), while minimizing re-suspension of sediments. Dredged material is usually discharged to a hopper barge or scow.

The table, "Summary of Dredging Techniques for Contaminated Material" is adapted from the "EPA's Handbook Responding to Sinking Hazardous Substances", (U.S. EPA, 1988). This table summarizes information presented in this section and can be used to select dredges for specific applications.

Power/Fuel Requirements

Power or fuel requirements vary with the type and model of dredge used. Power and fuel sources typically include gasoline or diesel, with any electrical requirements being generated on-board by the prime mover.

Transportability

Most dredges cannot be transported by road or air, with the exception of the portable hydraulic dredges, handheld dredges, and some mechanical dredge heads, all of which can be transported by road, and some of which can be transported by air.

PERFORMANCE

Dredging is often the logical and perhaps the only feasible means of recovering spilled materials from the bottom of a waterway. Dredging alone, however, can seldom totally recover spilled chemicals and should be considered in conjunction with other measures, such as burial or in-situ chemical treatment. These techniques are particularly applicable around the periphery of a spill where contaminant levels are too low to justify continued dredging. Dredges can also be used to build barriers, such as trenches or dykes, to prevent dispersion. Dredging should always be considered as a follow-up technique after burial or use of sorbents.

The environmental effects of dredging and disposal operations have been studied at the United States Army Engineers Waterways Experiment Station. Documented effects include increases in turbidity, suspended solids concentration, and biochemical oxygen demand, and decreases in concentration of dissolved oxygen and distribution of benthic organisms. Less direct effects are alterations in bottom topography due to changes in benthic communities and current flow patterns. The degree of adverse effect varies with the dredge used and the type of sediment being removed. Dredges that cause the greatest turbidity cause the most contaminant re-suspension and pose the greatest threat to aquatic organisms.

Maximizing the rate of sediment removal has been the major focus in the development of dredging equipment and practices. More recently, the environmental effects of dredging have also been considered. Existing equipment and operating methods are being modified to accommodate the removal of contaminated sediments. Both in North America and abroad, new specialized dredging equipment is being developed to reduce environmental effects. Factors that are being considered include the need for precise delineation of the area to be dredged; precise lateral and vertical control of the dredgehead; special precautions tailored to specific chemicals; reduction or elimination of the re-suspension of contaminated sediments; prediction of any damage to aquatic and benthic organisms; and temporary storage, transport, and treatment before disposal of the dredged material.

Experience relating to these problems is minimal as only a few dredging operations of contaminated sediments are actually documented.

2.7 Removal - Dredging

Summary of Dredging Techniques for Contaminated Material

Technique	Applications	Limitations	Secondary Impacts	Availability/ Transportability	Vessel Length/Draft (m)	Production (m³/h)	Maximum Depth of Use (m)	Relative Cost
Mechanical Dredges								
Clamshell	• small volumes of sediments; • confined areas and near structures; • removal of bottom debris; • unconsolidated sediments; • interior waterways; harbours.	Low production rates; cannot excavate highly consolidated sediments or solid rock.	Considerable re-suspension of sediments.	Dredgehead can be moved over existing roads as-is and mounted on conventional crane; widely available.	Depends on support vessel.	23 to 460	30	Low
Dragline	• small volumes of sediments; • confined areas; • non-consolidated sediments; • harbours and interior waterways.	Low production rates; cannot excavate highly consolidated sediments or solid rock.	Considerable re-suspension of sediments.	Dredgehead can be moved over existing roads as-is and mounted on conventional crane; widely available.	Depends on support vessel.	46 to 535	30	Low
Conventional Excavation Equipment	• small volumes of sediments in shallow or dewatered areas.	Restricted capacities and reach; limited to very shallow water.	Considerable re-suspension of sediments.	Can be moved over existing roads; widely available.	Highly variable.	Up to 535	9	Low
Dipper	• small volumes of sediments; • up to consolidated sediments; • confined areas; • harbours and interior waterways.	Low production rates.	Considerable re-suspension of sediments.	Not easily transported over roads; limited availability.	Depends on support vessel.	23 to 460	15	Low
Bucket Ladder	• small volumes of sediments; • up to highly consolidated sediments; • interior waterways.	Low production rates.	Considerable re-suspension of sediments.	Not easily transported over roads; very limited availability.	30/1.5	230	18	Medium
Hydraulic Dredges								
Portable Hydraulic	• moderate volumes of sediments; • lake and inland rivers; • very shallow depths.	Limited to waves of less than 30 cm; depending on model, low production rates and limited depth.	Moderate re-suspension of sediments.	Readily moved over existing roads, may require some disassembling; widely available.	6 to 15/ 0.6 to 1.5	38 to 1400	15	Low

Summary of Contaminated Material Dredging Techniques (continued)

Technique	Applications	Limitations	Secondary Impacts	Availability/ Transportability	Vessel Length/Draft (m)	Production (m³/h)	Maximum Depth of Use (m)	Relative Cost
Hand-held Hydraulic	• small volumes of solids or liquids in calm waters; • for precision dredging.	Operated from above-water units only in shallow water; underwater units require diver operation.	Moderate re-suspension of sediments.	Easily moved over existing roads; can be assembled using commonly available equipment.	N/A	8 to 190	305	Low
Cutterhead	• large volumes of solids and liquids; • up to very hard and cohesive sediments; • in calm waters.	Dredged material is 80 to 90% water; cannot operate in rough, open waters; susceptible to debris damage and weed clogging.	Moderate re-suspension of sediments.	Transport in navigable waters only; wide availability.	15 to 76/1 to 4	19 to 7,650	15	Medium
Dustpan	• large volumes of free-flowing sediments and liquids; • in calm interior waterways.	Dredged material is 80 to 90% water; cannot operate in rough, open waters; susceptible to debris damage.	Moderate re-suspension of sediments.	Transport in navigable waters only.	30/1.5 to 4	19 to 7,650	18	Medium
Hopper	• solids and liquids in deep harbours; • rough, open water and high currents; • in adverse weather; • in and around shipping.	Not for highly consolidated sediments.	Moderate re-suspension of sediments; normal operation involves large overflow of dredged material.	Can only be moved in deep waters.	60+/3.5 to 10	380 to 1,500	20	High
Pneumatic Dredges								
Airlift	• deep dredging of loose sediment and liquids; • in interior waterways.	Not for consolidated sediments; dredged material is 75% water.	Re-suspension of sediment is low.	Dredgehead can be moved over existing roads; not widely available in North America.	30/1 to 1.8	45 to 300	None	Medium
Pneumatic	• nonconsolidated solids and liquids in interior waterways.	Not for consolidated sediments; not for shallow waters; may obstruct water traffic.	Re-suspension of sediment is low.	Dredgehead can be moved over existing roads; not widely available in North America.	30/1.5 to 1.8	45 to 300	46	High
Oozer	• soft sediments and liquids from river beds or harbour bottoms; • in relatively shallow depths.	Modest production rates; may obstruction traffic.	Re-suspension of sediment is low.	Dredgehead can be moved over existing roads; not widely available in North America.	35/2	380 to 610	None	High

2.7 Removal - Dredging

CONTACT INFORMATION

Manufacturers and suppliers of dredging equipment or marine/dredging contractors are listed in industrial registers and local telephone directories. Refer also to No. 73 to 78 in this section.

OTHER DATA

Because of the high capital cost of dredge systems, cleanup contractors who own and operate such equipment are usually called upon in the event of a chemical spill.

The following are recommendations for dredge use during spill cleanup (Bonham, 1989).

1) In flowing watercourses, redeposition of the contaminant from upstream will be minimized by initiating cleanup at the furthest upstream point and proceeding downstream.

2) In non-flowing water bodies, the spread of material will be minimized by initiating cleanup at the area with the heaviest contamination.

3) To prevent over-dredging and under-dredging, analytical capabilities appropriate to the spilled material should be maintained at the site to determine the presence of contaminant in the dredged material.

4) Before cleanup is started, the safe residual concentration of the chemical, its persistence in the environment, and the location of the spill and uses of the affected waterbody should be determined.

5) Large-scale dredging should not be done in small streams unless it is the only feasible method, because of flow, terrain, or other circumstances. This prevents permanent scarring of the stream bank or the need for expensive restorative landscaping.

REFERENCES

(1) Bonham, N., "Response Techniques for the Cleanup of Sinking Hazardous Materials", Environmental Protection Series Report, EPS 4/SP/1, Environment Canada, Ottawa, Ont. (December, 1989).
(2) Cullinane, M.J., D.E. Averett, R.A.Shafer, J.W. Male, C.L.Truitt, and M.R Bradbury, "Contaminated Dredged Material - Control, Treatment and Disposal Practices", Noyes Data Corporation, Park Ridge, NJ (1990).
(3) Manufacturer's Literature.
(4) U.S. EPA (United States Environmental Protection Agency), "EPA's Handbook Responding to Sinking Hazardous Substances", Pudvan Publishing Company, Northbrook, IL (January, 1988).

CRISAFULLI DREDGES   No. 73

DESCRIPTION

The **Rotomite High-Solids Dredge** and the **Sludge Dredge** are portable hydraulic dredges equipped with a rototiller cutterhead and the power to pump silts, sands, and heavy sludges. The Rotomite is self-propelled and the Sludge Dredge is propelled by an onboard winch attached to a traversing cable. The **Flump Pond Dredge** is a remote, electric- or hydraulic-controlled lagoon/pond dredger with a cutterhead, also propelled by an onboard winch attached to a traversing cable.

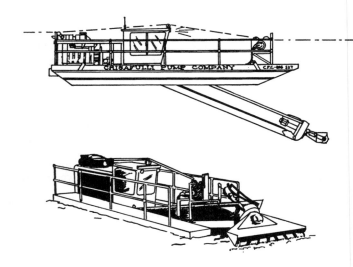

OPERATING PRINCIPLE

A pump and cutterhead are mounted on a catamaran hull. The pump and cutterhead are swung down between the hulls into the water to the desired depth. The cutterhead agitates the sediments and the pump recovers and transfers them through a discharge hose to the surface, where they are discharged to shore through a floating discharge hose.

STATUS OF DEVELOPMENT AND USAGE

Commercially available products used for removing sludge from industrial and municipal waste ponds or lagoons.

TECHNICAL SPECIFICATIONS

Physical

	Rotomite Dredge	Sludge Dredge	Flump Pond Dredge
Length	8.2 m (27 ft)	12 m (38 ft)	4.6 m (15 ft)
Width	2.4 m (8 ft)	2.4 m (8 ft)	2.3 m (7.5 ft)
Height	1.5 m (5 ft)	2.4 m (8 ft)	1.4 m (4.5 ft)
Weight	5,400 kg (12,000 lb) on 180P	7,700 to 8,600 kg (17,000 to 19,000 lb)	1,500 kg (3,400 lb)
Draft	58 cm (23 in) on 180P	0.6 m (2 ft)	
Engine	120 kW (160 hp)	Cummins Diesel Engine	
Fuel Capacity	300 L (80 gal)	10-hour fuel tank with sight gauge	
Flotation	Foam-filled steel pontoons	Foam-filled, airtight, 10-gauge steel pontoons	UHMW polyethylene pontoons, foam-filled with integral mounting lugs

The rototiller on the **Rotomite Dredge** is 2.4 m (8 ft) wide, bi-rotational style with 508 N·m (4,500 in/lb) torque. The **Sludge Dredge** has a 2.4 m (8 ft) wide, auger/tiller style cutterhead with 1,017 N·m (9,000 in/lb) torque, and the **Flump Pond Dredge** has 2.3 m (7.5 ft) wide auger/tiller style, single-motor drive with 283 N·m (2,500 in/lb) torque. The following are additional specifications for the **Flump Pond Dredge**.

2.7 Removal - Dredging

Pump	25 cm (10 in) long, severe-duty, centrifugal pump with an enclosed drive shaft and dual mechanical seals
Pump coupling	Flexible three-piece coupling
Depth winch	0.6 kW (0.75 hp) reversing-gear motor and brake with 7.6 m (25 ft), 5 mm (3/16 in) stainless steel cable
Traversing winch	1 kW (1.5 hp), 15 cm (6 in) triple-sheave reversing winch with greasable idler bearings, constant speed 2.1 m/min (7 ft/min)
Frame	Mild carbon steel with four channel beams, two crossbeams, double A-frames, end caps, pump pivot mount, and four lifting lugs

Operational

Rotomite - Dredges to a depth of 4.5 m (15 ft) and 2.4 m (8 ft) wide. It is propelled with an outboard hydraulic drive and the steering is hydraulically controlled. It is equipped with a steering wheel, engine throttle, key start, depth control, pump speed control, on/off/reverse switch for both the rototiller and winch, as well as a speed control switch for the winch. A tachometer, oil pressure gauge, hydraulic pressure gauge, ammeter, and flow meter are optional.

Sludge Dredge - This unit must be towed or moved by winch and traversing cables attached to shore. It has a maximum dredging depth of 4.5 m (15 ft) and a dredging width of 2.4 m (8 ft).

Flump Pond Dredge - This unit must be towed or moved by cables attached to shore. Its dredging depth is 2.4 m (8 ft) standard, 9.1 m (30 ft) maximum, and it has a pump discharge of 15 cm (6 in) irrigation quick-coupling (other discharges are available).

Power/Fuel Requirements

Both the Rotomite and the Sludge Dredge have an electrical system that requires an automotive style, 12-V, DC battery and engines that run on diesel. The Flump Pond Dredge requires 460-V, 3-phase, 60-cycle electric power.

Transportability

These Crusafulli dredges are equipped with four lifting lugs for loading and/or transportation purposes and can be moved on a trailer for highway transport.

CONTACT INFORMATION

Manufacturer	*Canadian Representative*
SRS Crisafulli Inc.	ENVIROCAN
Crisafulli Drive	26 McCauley Drive
P.O. Box 1051	Bolton, Ontario
Glendive, MT 59300	L7E 5R8
U.S.A.	
Telephone (406) 365-3393	Telephone (905) 880-2418
Facsimile (406) 365-8088	Facsimile (905) 880-2327

OTHER DATA

Delivery time is 4 to 8 weeks. Custom designs are available. SRS Crisafulli Inc. also produces hydraulic-driven, submersible, variable speed centrifugal dredge pumps.

REFERENCES

(1) Manufacturer's Literature.

H&H DREDGES

No. 74

DESCRIPTION

Portable, hydraulic dredges, equipped with cutterhead and the power to pump silts, sands, and heavy sludges. Propelled by a windlass or winch travel system attached to traversing cables.

OPERATING PRINCIPLE

A pump and cutterhead are mounted on a catamaran hull. The pump suction and cutterhead are swung down between the hulls into the water to the desired depth. The cutterhead agitates the sediments and the pump recovers and transfers them through a suction hose to the surface where they are discharged to shore through a floating discharge hose.

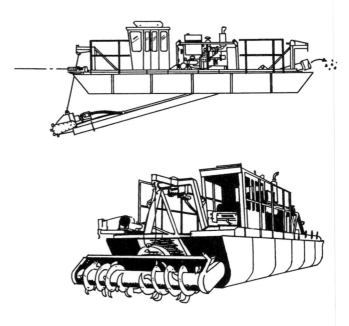

STATUS OF DEVELOPMENT AND USAGE

Commercially available products for removing sludge from industrial and municipal waste ponds or lagoons. Application to chemical spills not documented.

TECHNICAL SPECIFICATIONS

Physical

Little Monster Dredge - Can be constructed with stainless steel dredge pump, cutterhead, pivot tube, and decking with polyethylene pontoon floats for operation in caustic materials. An operator's cab is available only on dredges with polyethylene pontoon floats.

Length	7.9 m (26 ft)
Width	2.4 m (8 ft)
Height	3.0 m (10 ft)
Weight	4,300 kg (9,500 lb)
Draft	43 cm (17 in)
Engine	Cummins Model 4BT3.9, 100 hp, diesel engine
Fuel capacity	340 L (90 gal)
Cutterhead	2.4 m (8 ft) H&H Auger/Tine cutterhead or H&H Jetting Ring
Flotation	Two 6.7 × 2.4 m (22 × 8 ft) pontoons, with four sealed compartments, constructed of 10-gauge steel. Total displacement 8,100 kg (18,000 lb); polyethylene floats also available

Monster of the Mudway (M-150 and M-177) - Features an enclosed cab with a 360° view, and heating and air conditioning.

Length	10 m (34 ft)
Width	2.4 m (8 ft)
Height	2.7 m (9 ft)
Weight	5,900 kg (13,000 lb)
Draft	46 cm (18 in)
Engine	150 hp or 177 hp Cummins diesel engine
Fuel capacity	340 L (90 gal)
Cutterhead	2.4 m (8 ft) H&H Auger/Tine cutterhead or H&H Jetting Ring
Flotation	Two 9.5 m (31 ft) pontoons, with six sealed compartments, constructed of 10-gauge steel. Total displacement 11,700 kg (25,800 lb)

2.7 Removal - Dredging

Operational

Little Monster Dredge

Propulsion	Must be towed or moved by winch and traversing cables attached to shore
Dredging depth	5.5 m (18 ft) maximum
Dredging width	2.4 m (8 ft)
Dredging speed	0 to 30 m/min (0 to 100 ft/min)
Production rate	23 to 409 m^3/h (30 to 532 yd^3/h) heavy sludge; 114 to 500 m^3/h (148 to 650 yd^3/h) water; 38 to 60 m^3/h (50 to 80 yd^3/h) sand

A variable discharge of the 15 cm (6 in) hydraulic submersible steel pump delivers sludge from 23 to 408 m^3/h (30 to 530 yd^3/h through a 15-cm (6-in) pipe at distances up to 305 m (1,000 ft). The H&H cutterhead can cut through sand, silt, gravel, fly ash, and municipal, industrial, and animal wastes.

Monster of the Mudway (M-150 and M-177)

Propulsion	Must be towed or moved by winch and traversing cables attached to shore
Dredging depth	7.6 m (25 ft) maximum
Dredging width	2.4 m (8 ft)
Dredging speed	0 to 30 m/min (0 to 100 ft/min)
Production rate	227 to 568 m^3/h (295 to 738 yd^3/h) heavy sludge; 410 to 1,100 m^3/h (530 to 1,430 yd^3/h) water; 57 to 230 m^3/h (75 to 300 yd^3/h) sand

Power/Fuel Requirements

Electrical system requires an automotive style, 12-V, DC battery. Engine runs on diesel.

Transportability

Both the Little Monster and the Monster of the Mudway can be transported on a trailer without permits or escorts and can be loaded or unloaded in tight spaces.

CONTACT INFORMATION

Manufacturer

H&H Pump Company
P.O. Box 431
Holden, LA 70744
U.S.A.

Telephone (504) 567-2408
Facsimile (504) 567-3091

Canadian Representative

R.M.S. Enviro Solv Incorporated
Suite 102, 456 Alliance Avenue
Toronto, Ontario
M6N 2J2

Telephone (416) 766-7471
Facsimile (416) 766-7399

R.M.S. Enviro Solv Incorporated
2200 - 46th Avenue
Lachine, Quebec
H8T 2P3

Telephone (514) 631-3533
Facsimile (514) 631-8224

REFERENCES

(1) Manufacturer's Literature.

MUD CAT MODEL 370 PDP DREDGE

No. 75

DESCRIPTION

A portable, hydraulic, lagoon/pond dredger with a cutterhead. The dredger is propelled by an onboard winch attached to a traversing cable.

OPERATING PRINCIPLE

Pump and cutterhead are mounted on a catamaran hull. The pump suction and cutterhead are swung down into the water to the desired depth. The cutterhead agitates the sediments and the pump recovers and transfers them through a suction hose to the surface, where they are discharged to shore through a floating hose.

STATUS OF DEVELOPMENT AND USAGE

Commercially available product for removing sludge from industrial and municipal waste ponds or lagoons. Application to chemical spills not documented.

TECHNICAL SPECIFICATIONS

Physical

Length	17 m (57 ft)
Width	7.9 m (26 ft)
Height	4.3 m (14 ft)
Weight	43,500 kg (96,000 lb)
Draft	26 cm (22 in)
Engine	280 kW (375 hp) Caterpillar 3406
Dredge head	Dynamic force (vibrating) excavator with two spring-mounted, material-separation screens
Flotation	Two pontoons constructed of 0.47 cm (3/16 in) steel with internal bulkheads and stiffeners
Pump	Two positive displacement pumps

Operational

The dredging depth is 7.6 m (25 ft), dredging speed is 16 m/min (53 ft/min), and production rate is 60 m^3/h (78 yd^3/h). This dredge must be towed or moved by cables attached to shore. It pumps high solids concentrations of up to 80% volume, potentially eliminating the need to dewater the recovered material. The dredge pumps slurries with viscosities of up to 1,080 cSt (5,000 SSU). Typically, no dilution is required when excavating and pumping in-situ material. There is little or no re-suspension of excavated material in the water column. A deck crane is available for maintenance and trash removal from the excavator screens. An environmentally controlled cab filters most toxic gases, heats or cools the filtered air, and maintains the inside noise level below 85 dB.

Power/Fuel Requirements and Transportability

The Mud Cat 370 PDP runs on diesel fuel and can be moved on a trailer for highway transport.

Manufacturer

Mud Cat, A Division of Ellicott Machine Corporation International
1611 Bush Street
Baltimore, MD 21230
U.S.A.

Telephone (410) 837-7900
Facsimile (410) 752-3294

REFERENCES

(1) Manufacturer's Literature.

2.7 Removal - Dredging

PIT HOG AUGER DREDGES — No. 76

DESCRIPTION

Manned or remote-controlled diesel- or electric-powered, lagoon/pond dredges with cutterheads. The dredges are propelled by an onboard winch attached to a traversing cable.

OPERATING PRINCIPLE

A pump and cutterhead are mounted on a catamaran hull. The pump suction and cutterhead are swung down between the hulls into the water to the desired depth. The cutterhead agitates the sediments and the pump recovers and transfers them through a suction hose to the surface, where they are discharged to shore through a floating discharge hose.

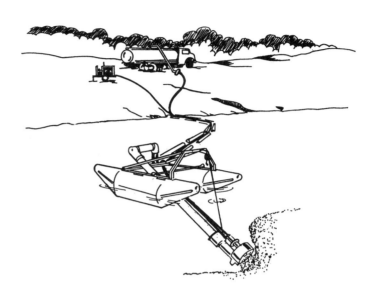

STATUS OF DEVELOPMENT AND USAGE

Commercially available products used for removing sludge from industrial and municipal waste ponds or lagoons. Application to chemical spills has not been documented.

TECHNICAL SPECIFICATIONS

Physical

Pit Hog Series 7MAU Dredges - Compact and portable hydraulic dredges with 10-gauge steel, foam-filled pontoons that maintain minimum operational draft requirements. Exact dimensions vary as these dredges are built to meet customers' specific needs.

Draft	46 cm (18 in)
Cutterhead	Variable speed, horizontal auger head with chopper and vortex pump
Engine	John Deere industrial diesel, or totally enclosed, fan-cooled electric motors
Pump	Variable speed, cast-iron, centrifugal pump

Remote Control Lagoon Pumpers - The remote controlled pumper systems are available in several sizes and styles depending on the intended use. They can be built to be controlled from shore and/or by an operator on the pumping unit. The hydraulic system is available in several horsepower sizes and driven by diesel engine or electric motor.

Basic Lagoon Pumpers - A simple lagoon pumping system with no traversing cable or propulsion capability. The system features a manually adjustable depth winch, hydraulically operated Pit Hog submersible hydraulic chopper pump, and slurry agitation gate. The pumps are powered by electric, diesel, or hydraulic power-take-off shore power.

Operational

Pit Hog Series 7MAU Dredges - The horizontal auger head and submerged chopper or vortex pump allow high pumping rates of viscous, heavy solids, slurry residuals, pulp, silts, and sands. For hazardous materials, the units can be remotely controlled from shore. The variable-speed auger cuts and removes material from the solids bed and conveys it to the submerged pump head where it is induced into the pump. The totally enclosed shroud can be adjusted to the proper approach angle with tilting mechanisms to minimize solids bypass and re-suspension. Pit Hog dredges can discharge directly to shore or to a booster pump for extended transfer distances.

Propulsion	Must be towed or moved by cables attached to shore
Dredging depth	4.6 m (15 ft) maximum
Dredging width	2.4 m (8 ft)

Remote Control Lagoon Pumpers - Both the Pit Hog manned and remote-controlled lagoon pumping systems are designed for applications where it is necessary to travel into the material to be removed and it is not physically or economically feasible to use a full-sized portable hydraulic dredge. The systems offer submersible hydraulic chopper pumps with variable-speed control, 2.4 m (8 ft) auger cutterhead, pump depth control, slurry agitation gate, travel winch, and shore-mounted or handheld remote controls.

Basic Lagoon Pumpers - As the float system is not self-propelled or equipped for traversing cable propulsion, these units are only suitable for pumping small ponds or when the material to be pumped is homogeneous and free-flowing.

Power/Fuel Requirements

All Pit Hog dredges and lagoon pumping systems are available in either diesel- or electric-powered configurations.

Transportability

All Pit Hog dredges can be moved on a trailer for highway transport.

CONTACT INFORMATION

Manufacturer

Liquid Waste Technology Incorporated
Box 250, 422 Main Street
Somserset, WI 54024
U.S.A.

Telephone (715) 247-5464
Facsimile (715) 247-3934

OTHER DATA

Liquid Waste Technology Incorporated offers custom construction to meet specific requirements, with special coatings or materials for explosion-proof, acidic, caustic, or toxic operations.

REFERENCES

(1) Manufacturer's Literature.

TRAC PUMP DREDGING SYSTEM

No. 77

DESCRIPTION

A remote-controlled, mobile dredging system that excavates and pumps slurries and sludges from ponds, tanks, digesters, ice-covered lagoons and canals, pipelines, sewers, and other locations that are difficult to reach.

OPERATING PRINCIPLE

Trac Pump travels through the sludge by crawling along the bottom on individually controlled dual tracks. Dual, rotating cutterheads excavate, slurry, and feed material into the sludge pump. Powered and controlled by hydraulic hose connected to a diesel-powered hydraulic unit. The operator controls the unit from the remote power unit.

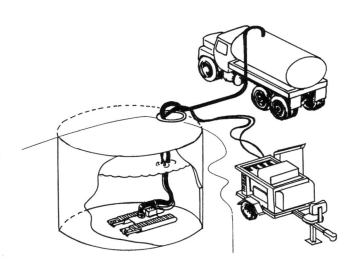

STATUS OF DEVELOPMENT AND USAGE

Commercially available product used by industry to remove accumulated sludges from ponds, tanks, digesters, lagoons, canals, pipelines, sewers and other locations that are difficult to reach.

TECHNICAL SPECIFICATIONS

Physical

A complete system includes Trac Pump, diesel/hydraulic power unit with controls and operator station mounted on a highway trailer, 61 m (200 ft) of hydraulic power hose assembly, and discharge tubing with line floats.

Operational

The Trac Pump travels through the sludge by crawling along the bottom on individually controlled dual tracks, which enable it to "turn on a dime". By using a rotating cutterhead to excavate, slurry, and feed material into the 10 cm (4 in) sludge pump, the system pumps heavy sludges at rates averaging 25 to 32 L/s (400 to 500 gpm), through tubing to an external discharge location. The operator, sitting at the remote power unit, controls all unit functions, including steering, forward/reverse, and pump and cutterhead speed. The unit fits through a standard 61 cm (24 in) manhole. The mobility and remote control capabilities of the Trac Pump enable personnel to work away from the hazardous area.

Power/Fuel Requirements and Transportability

The remote power supply unit is diesel-powered. The entire system can be transported either by road on a trailer or airlifted.

CONTACT INFORMATION

Manufacturer

H&H Pump Company
P.O. Box 431
Holden, LA 70744
U.S.A.

Telephone (504) 567-9200
Facsimile (504) 567-3091

REFERENCES

(1) Manufacturer's Literature.

VMI MINI DREDGES AND PIRANHA AUGER DREDGE ATTACHMENT

No. 78

DESCRIPTION

Mini Dredges are portable, hydraulic dredges, equipped with cutterheads and the power to pump silts, sands, and heavy sludges. The dredges are propelled by a windlass or winch travel system attached to traversing cables. The **Piranha Auger Dredge Attachment** is a hydraulic, cutterhead dredge attachment that fits onto an excavator/backhoe, converting it to a portable hydraulic dredge.

OPERATING PRINCIPLE

Mini Dredges consist of a pump and cutterhead mounted on a catamaran hull. The pump suction and cutterhead are swung down between the hulls into the water to the desired depth. The cutterhead agitates the sediments and the pump recovers and transfers them through a suction hose to the surface where they are discharged to shore through a floating discharge hose. The **Piranha Auger Attachment** consists of a pump and cutterhead which are mounted on a large backhoe or excavator, replacing the standard bucket and converting the machine into a portable dredge.

STATUS OF DEVELOPMENT AND USAGE

Commercially available products for removing sludge from industrial and municipal waste ponds or lagoons.

TECHNICAL SPECIFICATIONS

Physical

Six models of **VMI Mini Dredges** are available. All models are 2.4 m (8 ft) high and 2.7 m (9 ft) wide, except the MDE-415HS which is 2.4 m (8 ft) wide. They range in length from 7.6 m (25 ft) to 13 m (41 ft). The various models weigh from 5,700 kg (12,500 lb) to 10,200 kg (22,500 lb). All models have a 56 cm (22 in) draft except the MDE-415HS which has a 51 cm (20 in) draft. The Cummins Model engines range from 116 to 243 hp and Caterpillar, Detroit, and Deutz diesel engines are also available. Cutterheads on all models range from .37 x 2.6 m (1.2 x 8.5 ft) to 0.4 x 2.6 m (1.3 x 8.5 ft) and have a variable speed head with 52 removable teeth, with full width, flow-through suction. Most are capable of 790 N·m (7,000 in/lb) torque, while both the MDE-615HN and -620HN models are capable of 1,073 N·m (9,500 in/lb) torque. Pumps on all dredges are Gould or Hazelton centrifugal, recessed impeller pumps, constructed of either 28% chrome or 316 stainless steel. All models have foam-filled pontoons constructed of either 10-gauge or 12-gauge steel with integral bulkheads and stiffeners.

Piranha Auger Attachment - Two models are available. The smaller model, MDE-4HN, requires an 18 t (40,000 lb) or greater excavator, while the larger model, MDE-6HN, requires a 36 t (80,000 lb) or greater excavator. Both models have two 61-cm (24-in) diameter augers, 1.2 m (4 ft) long on the smaller model and 1.8 m (6 ft) long on the larger model, each driven by an independent 7.5 kW (10 hp) motor. Attachments have a Hazelton Hi-chrome centrifugal pump, with 40 cm (16 in) impeller and 15 cm (6 in) suction and discharge. The following are additional specifications.

	MDE-4HN	MDE-6HN
Width	3.4 m (11 ft)	4.3 m (14 ft)
Weight	2,580 kg (5,700 lb)	3,860 kg (8,500 lb)
Engine	Cummins 6BT5.9, 152 hp, diesel	Cummins 6CTA8.3. 234 hp, diesel
Fuel capacity	470 L (125 gal)	1,100 L (290 gal)

2.7 Removal - Dredging

Operational

All **VMI Mini Dredges** must be towed or moved by winch and traversing cables attached to shore.

Model	Dredging depth	Dredging width	Average dredging speed	Pump capacity
MDE-415HN	4.6 m (15 ft)	2.6 m (8.5 ft)	1.5 to 4.6 m/min (5 to 15 ft/min)	3,800 L/min (1,000 gpm) @ 40 m (130 ft) head
MDE-415HS	4.6 m (15 ft)	2.4 m (8 ft)	1.5 to 4.6 m/min (5 to 15 ft/min)	3,000 L/min (800 gpm) @ 27 m (90 ft) head
MD-E610HS	3 m (10 ft)	2.6 m (8.5 ft)	2.4 to 3.6 m/min (8 to 12 ft/min)	5,680 L/min (1,500 gpm) @ 37 m (120 ft) head
MD-E615HS	4.6 m (15 ft)	2.6 m (8.5 ft)	2.4 to 3.6 m/min (8 to 12 ft/min)	5,680 L/min (1,500 gpm) @ 37 m (120 ft) head
MDE-615HN	4.5 m (15 ft)	2.7 m (9 ft)	1.5 to 4.6 m/min (5 to 15 ft/min)	7,570 L/min (2,000 gpm) @ 43 m (142 ft) head
MDE-620HN	6.1 m (20 ft)	2.7 m (9 ft)	1.5 to 4.6 m/min (5 to 15 ft/min)	7,570 L/min (2.000 gpm) @ 43 m (142 ft) head

Piranha Auger Attachment - Conventional excavation equipment is limited for removing contaminated sediments from water bodies. As both lateral and vertical reach are restricted by the length of the boom, the equipment must be close to the material being dredged. Also, because of the low load-bearing capacity of submerged sediments and soils surrounding water bodies, it is often difficult or impossible to maneuver conventional equipment. The Piranha system can be used for dredging small areas where other portable dredges may not be able to operate. Its small size makes it easy to transport and store, and the excavation equipment required to operate the system is readily available. It can be operated from shore, in shallow waters with sediments that will support the excavator, from docks and wharfs, or from a barge in shallow water. For both models of the attachment, the dredging depth varies with the size of the excavator.

Power/Fuel Requirements and Transportability

Diesel fuel is required for mini dredges, dredge unit, and excavator. All VMI dredges and the Piranha dredge attachment can be moved by trailer for highway transport. Transport trailers are available from VMI.

CONTACT INFORMATION

Manufacturer

VMI Incorporated
1125 North Maitlen Drive
Cushing, OK 74023
U.S.A.

Telephone (918) 225-7000
Facsimile (918) 225-0333
Toll-free (800) 762-2257

OTHER DATA

Another attachment available from VMI is the **Sludgifier**, which is a solidifier that attaches to the excavator's boom instead of the bucket. It has a manifold induction system that feeds a reagent evenly over the mixer drum. The reagent solution is mixed into the material by a 0.8 × 2.4 m (2.5 × 8 ft) cutter drum driven by two planetary gear drives and two axial piston motors at each end. All controls to the diesel engine and hydraulics are located in the excavator's cab.

Options available for all VMI dredges include air conditioner, heater, radio, flow meter, discharge pipe, flexible hose, floats, cable kit, transport trailers, booster pump, and a jet propulsion system for moving the dredge into the working area. Additional attachments available for the Piranha Auger attachment are a hydraulic attachment for controlling a variable width cut and an attachment to solidify hazardous waste.

REFERENCES

(1) Manufacturer's Literature.

Section 3.1

Temporary Storage - Flexible Containers

FLEXIBLE CONTAINERS - GENERAL LISTING No. 79

DESCRIPTION

Flexible or collapsible containers designed for storing and/or transporting chemicals, hazardous materials, and petroleum products. Types of design include bladders, often called "pillow tanks", for storage on land; open-top collapsible reservoirs; water-towable or air-liftable tanks, some of which float on water, while others are for use only on land; and folding tanks consisting of a fabric liner and metal support structure. Both transportable and stationary tanks require very little space when collapsed or folded for storage or transport. See also the following entries in this section.

Canflex Sea Slug and Open-Top Reservoir (No. 80)
Helios Flexible Bulk Containers (No. 81)
Port-A-Tank (No. 82)
Terra Tank (No. 83)

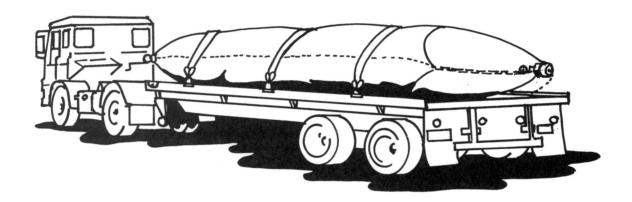

OPERATING PRINCIPLE

"Pillow Tanks" - Large volume, collapsible containers for storing recovered liquids on land. Self-supporting containers with a low profile, they lie flat when empty and assume a "pillow" shape when filled with liquid. They typically have a single inlet/outlet that can be fitted with a hose connector of choice. They are foldable and require minimal space when stored or transported.

Open-Top Collapsible Reservoirs - Self-supporting, open-top containers for storing recovered liquids on land. These tanks have a large opening at the top and a floating collar so the tank expands as it fills. The tanks are drained through an outlet drain fitted to the bottom of the tank.

Towable or Airliftable Tanks - Containers designed specifically for storing recovered liquids at sea. Inlet and outlet fittings are typically placed at each end of the bladder. The bladder floats just below the water surface when full and can be pumped out while in the water or lifted out by crane-lifting straps, which are built into the structure of the bladder to equalize the loading and decrease stress on the fabric. Most towable bladders can be towed at speeds of up to 5 knots.

Folding Frame Flexible Tanks - A portable, folding storage reservoir that can be set up quickly without tools to store both liquids and solids. Typically, a metal frame is assembled or unfolded to support a flexible fabric liner. A bottom drain allows differentially floating liquids to be separated.

STATUS OF DEVELOPMENT AND USAGE

Commercially available products in general use. Most are marketed specifically for spill response, with some pillow tanks marketed for storage and supply of potable water, firefighting water, or fuel in remote areas.

TECHNICAL SPECIFICATIONS

Physical

Construction materials vary and include PVC-coated polyester or nylon, Neoprene, Hypalon, urethane-coated nylon, PVC-coated polyamide, and a choice of many custom materials and coatings. While physical specifications also vary between companies and models, storage containers are generally available with the following specifications.

	Capacity	Dimensions	Shipping Weight
Pillow Tanks	380 to 380,000 L (100 to 100,000 gal)	1.7 × 1.2 m to 26 × 10.7 m (5.5 × 4 ft to 85 × 35 ft)	18 to 1,070 kg (40 to 2,350 lb)
Open-Top Collapsible Reservoirs	1,900 to 45,000 L (500 to 12,000 gal)	1.9 × 1 m to 5.3 × 1.5 m (6 × 3 ft to 17 × 5 ft)	15 to 58 kg (33 to 130 lb)
Towable or Airliftable Tanks	1,900 to 79,500 L (500 to 21,000 gal)	4.3 × 0.5 m to 28 × 2.7 m (14 × 1.5 ft to 92 × 9 ft)	68 to 500 kg (150 to 1,100 lb)
Folding Frame Flexible Tanks	450 to 21,200 L (120 to 5,600 gal)	0.9 × 0.9 × 0.6 m to 6.1 × 6.1 × 0.9 m (3 × 3 × 2 ft to 20 × 20 × 2 ft)	14 to 130 (30 to 280 lb)

Operational

Operational specifications also vary among containers. Some systems can be used on both land and water, while some can be used only on land or only on water. Some require a hard, level surface for use. Fill and discharge procedures vary, as well as the size and type of any inlet and outlet fittings. Characteristics of the fabric that must also be considered include chemical compatibility, range of operating temperature, puncture-resistance, capability of being repaired in the field, and sensitivity to ultraviolet light. Other operational considerations include the need for a protective ground cover, tow rate or maximum lift capacities, ease of cleaning after use, and folded or collapsed size for storage and transport.

Power/Fuel Requirements

All storage containers listed in this section are self-supporting and require no outside source of power or fuel.

Transportability

Flexible storage containers are easy to transport because they fold or collapse to a compact size.

PERFORMANCE

Most flexible storage containers have been used in spill situations, although documentation of performance is limited. Most towable or liftable containers have been tested and have safety factors of up to 5:1 built into the lift/tow specifications and handle well within the recommended towing speeds and wave heights.

CONTACT INFORMATION

Several representative manufacturers and their products and Canadian distributors are listed alphabetically here. Selected products from each category of flexible storage containers are discussed in more detail in specific entries in this section.

Manufacturer	Product(s)	Comments
ABASCO P.O. Box 38573, 363 West Canino Houston, TX 77037 U.S.A. Telephone (713) 931-4400 Toll-free (800) 242-7745 Facsimile (713) 931-4406	Kwik-Tank Kwik Pool Towable Oil Bag	Folding frame flexible tanks Open-top collapsible reservoirs Towable tanks

3.1 Temporary Storage - Flexible Containers

Aero Tec Laboratories Incorporated Spear Road Industrial Park Ramsey, NJ 07446 U.S.A. Telephone (201) 825-1400 Toll-free (800) 526-5330 Facsimile (201) 825-1962	Flex-Tanks Petro-Flex Tanks Chem-Flex Tanks Aqua-Flex Tanks	Pillow tanks Standard and custom designs, including designs for collecting chlorofluorocarbons (CFCs).
Aqua-Guard Technologies Incorporated Suite 200 - 1130 West Pender Street Vancouver, British Columbia V6E 4A4 Telephone (604) 681-3773 Facsimile (604) 681-6825	Flexible Oil Storage Tanks	Open-top collapsible reservoirs Pillow tanks
Can-Ross Environmental Services 2270 South Service Road, West Oakville, Ontario L6L 5M9 Telephone (905) 847-7190 Facsimile (905) 847-7175	Portatank and Floating Tanks Sanator Tanks Pillow Tanks	Towable tanks Open-top collapsible reservoirs Pillow tanks A Canadian distributor for the Eldred Corporation (listed below).
Canflex Manufacturing Incorporated #2, 1742 Marine Drive West Vancouver, British Columbia V7V 1J3 Telephone (604) 925-2003 Toll-free (800) 544-8356 Facsimile (604) 925-3740	Open-Top Reservoirs Pillow Tanks Truck-mountable Pillow Tanks "Sea Slug" Series	Open-top collapsible reservoirs Towable or airliftable tanks Pillow tanks for mounting on flatbed trucks Towable tanks See No. 80 in this section.
The Eldred Corporation P.O. Box 110 Milan, IL 61264 U.S.A. Telephone (309) 787-3500 Facsimile (309) 787-3635	Liquid Containment Storage Tanks	Folding frame flexible tanks Open-top collapsible reservoirs Towable tanks Pillow tanks
Helios Container Systems Incorporated 251 Covington Drive Bloomingdale, IL 60108 U.S.A. Telephone (708) 529-7590 Toll-free (800) 336-3422 Facsimile (708) 529-8695	Flexible Bulk Containers	Woven polypropylene bags with lifting straps, used for any dry free-flowing material. See No. 81 in this section.
Kepner Plastics Fabricators, Incorporated 3131 Lomita Boulevard Torrance, CA 90505 U.S.A. Telephone (310) 325-3162 Facsimile (310) 326-8560	Sea Container	Towable tanks

Montgomery Environmental 290 Mitchell Avenue Oshawa, Ontario L1H 2V9 Telephone (905) 432-2531 Toll-free (800) 668-7523 Facsimile (905) 404-0791		Pillow tanks
Navenco Marine Incorporated 350 Ford Boulevard, Suite # 120 Chateauguay, Quebec J6J 4Z2 Telephone (514) 698-2810 Facsimile (514) 698-1008	Ro-Tank	Towable tanks A Canadian distributor for Ro-Clean International.
Oil Pollution Environmental Control Ltd. 1, NAB Lane, Birstall Bately, West Yorkshire WF17 9NG UK Telephone 44 1 924 442701 Facsimile 44 1 924 471925	Flexi Tank Storage and Transport Tank	Towable tanks
SEI Industries Limited 7400 Wilson Avenue Delta, British Columbia V4G 1E5 Telephone (604) 946-3131 Facsimile (604) 940- 9566	Terra Tank Buoywall	Pillow tanks Open-top collapsible reservoirs See No. 83 in this section.
Sopers Engineered Fabric Products P.O. Box 277 144 Chatham Street Hamilton, Ontario L8N 3E8 Telephone (905) 528-7936 Toll-free (800) 263-8334 Facsimile (905) 528-8128	Port-A-Tank	Foldable storage reservoir See No. 82 in this section.

OTHER DATA

Consult local telephone directories or industrial trade registers for other manufacturers or local distributors.

REFERENCES

(1)   Manufacturer's Literature.

CANFLEX SEA SLUG AND OPEN-TOP RESERVOIR

No. 80

DESCRIPTION

Canflex manufactures welded products made of coated high strength polyester, nylon, and kevlar fabrics. The **Sea Slug** is an ocean-towable bladder for oil spill recovery and bulk liquid storage/transportation by water. It can also be used to store liquid on land. The **Open-Top Reservoir** is a collapsible tank with a floating collar used for storing oil, potable water, sewage, and other liquids.

OPERATING PRINCIPLE

The **Sea Slug** is a rugged, ocean-towable bladder, used primarily on water although it can also be used for storing liquids on land. The **Open-Top Reservoir** is completely self-supporting and has a large opening at the top and a floating collar that allows the tank to expand as it is filled. The tank is emptied through an outlet drain fitted to the bottom.

STATUS OF DEVELOPMENT AND USAGE

Commercially available products developed for heavy-duty use in emergency response applications. The design is based on experience in designing, testing, and supplying oil booms to government and industry from the Arctic to equatorial regions throughout the world, and supplying underwater recovery bags.

TECHNICAL SPECIFICATIONS

Physical

The Canflex **Sea Slug** comes in a range of sizes. Replaceable inner liners are available to assist with cleaning or to avoid long-term chemical exposure. The following are specifications for the various models of Sea Slug.

Model No.	Capacity	Dimensions - Filled (length/diameter)	Storage/Shipping Dimensions	Shipping Weight
FCB-12	3,900 L (1,040 gal)	6 m/1 m (20 ft/3.3 ft)	1.5 × 1.2 × 1.2 m (5 × 4 × 4 ft)	68 kg (150 lb)
FCB-25	8,300 L (2,200 gal)	9 m/1.3 m (30 ft/4.3 ft)	1.5 × 1.5 × 1.2 m (5 × 5 × 4 ft)	140 kg (315 lb)
FCB-60	20,800 L (5,500 gal)	12 m/1.8 m (38 ft/6 ft)	1.8 × 1.5 × 1.2 m (6 × 5 × 4 ft)	245 kg (540 lb)
FCB-125	41,600 L (11,000 gal)	14 m/2.3 m (47 ft/7.5 ft)	3 × 2.1 × 1.2 m (9.8 × 6.9 × 4 ft)	820 kg (1,800 lb)
FCB-250	83,300 L (22,000 gal)	20 m/2.8 m (66 ft/9.2 ft)	3.4 × 2.7 × 1.6 m (11 × 8.8 × 5.2 ft)	1,270 kg (2,800 lb)
FCB-650	208,000 L (55,000 gal)	44 m/2.8 m (144 ft/9.2 ft)	3.4 × 3.1 × 1.8 m (11 × 10 × 6 ft)	1,800 kg (4,000 lb)

The **Open-Top Reservoir** is a sturdy but lightweight response unit requiring no frame setup. It is available in a range of sizes from 1,900 to 378,500 L (500 to 100,000 gal). Custom-made tank sizes can be ordered. Options include 2.5 to 20 cm (1 to 8 in) threaded flanges, 2.5 to 7.6 cm (1 to 3 in) ball valves, collar covers to reduce evaporation of the contained liquid, tie-down pegs and straps, ground pads, and transport bags.

Operational

The Canflex **Sea Slug** is constructed using state-of-the-art radio frequency welding machines. High quality, foldable closed cell foam flotation panels welded along each side of the bladder provide auxiliary flotation. Because of the streamlined design and inherent flotation, the Sea Slug can be used in ocean environments while filled with fluids, including high-density, weathered oils. It can also be used for emergency storage of fluid on land. The Sea Slug is constructed with high tenacity oil grade coated fabric materials with tensile strengths up to 78,600 kg/m (4,400 lb/in) width. Units come with all equipment and accessories required for operation. Access ports for lowering submersible pumps into the Sea Slug are standard with the larger units.

The **Open-Top Reservoir** is self-supporting and self-inflating (the collar rises with the fluid level), easily deployed, and can be filled by one person.

Power/Fuel Requirements

None.

Transportability

All Canflex storage products are collapsible and can be folded for easy storage and transportation.

PERFORMANCE

Canflex collapsible storage products are used by the Canadian and U.S. Coast Guard, Alaska Clean Seas, the Clean Caribbean Cooperative, Larco Environment, Brown and Root Services Corporation, and others.

CONTACT INFORMATION

Manufacturer and Distributor

Canflex Manufacturing Inc.
#2, 1742 Marine Drive
P.O. Box 91727
West Vancouver, British Columbia
V7V 1J3

Telephone (604) 925-2003
Toll-free   (800) 544-8356
Facsimile (604) 925-3740

OTHER DATA

Other collapsible storage products available from Canflex are **Canflex Pillow Tanks**, which are used for a wide variety of liquids including potable water, oil, jet fuel, diesel fuel, wastewater, and sewage and come in capacities from 1,900 to 378,500 L (500 to 100,000 gal); **Canflex Truck-mountable Pillow Tanks**, which have welded tiedown grommets and reinforcing webbing; and the **Canflex Model FL Balloon Style**, which is an oil spill response bag designed for use with a disc-type skimmer in harbour spills.

REFERENCES

(1) Manufacturer's Literature.

HELIOS FLEXIBLE BULK CONTAINERS — No. 81

DESCRIPTION

Woven polypropylene bags with lifting straps for collecting and storing dry free-flowing material or as single-use bags for non-flowable or moist products.

OPERATING PRINCIPLE

The bags are filled through a "full open, duffel top", or a top-loading "charge spout" and then closed with tie strings. They are emptied through bottom-dispensing shoots on some models or by simply cutting the bottom out of single-use bags. The bags are moved with a forklift using polyester lifting straps.

STATUS OF DEVELOPMENT AND USAGE

Commercially available product marketed as intermediate bulk shipping containers.

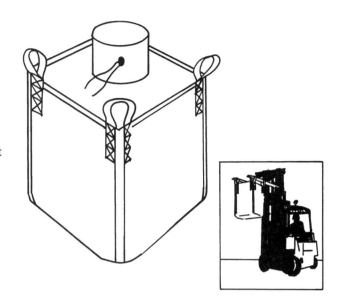

TECHNICAL SPECIFICATIONS

Physical - These bags are designed to accommodate dry or semi-moist flowable materials in capacities up to 2.3 m^3 (80 ft^3) and 2,000 kg (4,400 lb). Most bags are 0.9 m (2.9 ft) wide by 0.9 m (2.9 ft) long and vary in height from 0.5 m (1.5 ft) to 1.7 m (5.7 ft). The bags are made of woven polypropylene with polyester lifting straps that extend down and under the complete length of the bag and are then connected by a separate strap to create a "bucket lift" which ensures maximum lifting security. Several different loop designs are available to meet special needs.

Operational - Every container has a top-loading spout and bottom-dispensing capability. To save costs for a "single-use", closed-bottom bags are available which are cut at the bottom for dispensing material and are quickly and easily loaded and unloaded with a forklift or other conventional material-handling system. The containers collapse and fold to 10% of their full size for easy storage or shipment. The containers are sturdy enough for repeated use, with the actual number of uses depending on the care used in handling and the nature of the material being stored.

Power/Fuel Requirements and Transportability - There are no power or fuel requirements. The bags, which fold to a compact size for ease of storage and transport, are person-portable.

PERFORMANCE

Although no documentation of use in chemical spills was received, the design appears to lend itself to use in spill cleanup operations. Every container must pass rigorous strength and quality testing, including shock and drop testing of filled containers and fabric-tear testing.

CONTACT INFORMATION

Manufacturer and Distributor

Helios Container Systems Incorporated
251 Covington Drive
Bloomingdale, IL 60108
U.S.A.

Telephone (708) 529-7590
Toll-free (800) 336-3422
Facsimile (708) 529-8695

REFERENCES

(1) Manufacturer's Literature.

PORT-A-TANK No. 82

DESCRIPTION

A portable, foldable storage reservoir that can be quickly set up without tools.

OPERATING PRINCIPLE

A flexible, open-top tank for quick, temporary storage of recovered materials. It features simple drainage systems that can be easily adapted to specific requirements and can be opened by one person.

STATUS OF DEVELOPMENT AND USAGE

Commercially available product used by fire departments oil spill cooperatives, and emergency response teams across Canada.

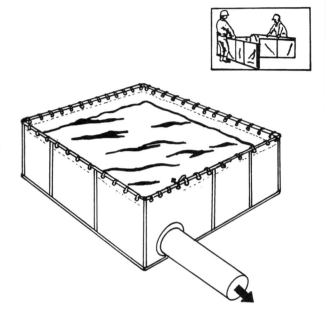

TECHNICAL SPECIFICATIONS

Physical

The following are the specifications for the five standard sizes. Custom sizes are also available.

Capacity	3,785 L (1,000 gal)	5,680 L (1,500 gal)	1,900 L (500 gal)	950 L (250 gal)	380 L (100 gal)
Size (open)	2.5 × 2.5 × 0.8 m	3 × 2.5 × 0.8 m	0.6 × 2 × 2.1 m	0.6 × 1.5 × 1.6 m	0.6 × 0.9 × 1 m
	(8.2 × 8.2 × 2.5 ft)	(10 × 8.2 × 2.5 ft)	(2 × 6.8 × 7 ft)	(2 × 4.9 × 5.1 ft)	(2 × 3 × 3.3 ft)
Size (closed)	2.5 × 0.8 × 0.2 m	3 × 0.8 × 0.2 m	0.6 × 2.1 × 0.2 m)	0.6 × 1.8 × 3 m	0.6 × 1 × 0.2 m
	(8.2 × 2.5 × 0.8 ft)	(10 × 2.6 × 0.8 ft)	(2 × 7 × 0.8 ft)	(2 × 5.9 × 10 ft)	(2 × 3.3 × 0.8 ft)
Weight	50 kg (110 lb)	65 kg (140 lb)	30 kg (70 lb)	28 kg (60 lb)	20 kg (50 lb)

Operational - A 522 g/m² (22 oz/yd²) yellow vinyl-coated polyester liner is standard. A special XR-5, 710 g/m² (30 oz/yd²) black vinyl-coated polyester liner is available specifically for the chemical and oil industry. Liners are constructed with welded seams, are flame- and tear-resistant and inert to mildew, can be stored when wet, and will not absorb oil or grease. Punctures can be repaired on-site with a patch kit or heat gun.

Power/Fuel Requirements and Transportability - There are no power or fuel requirements. Person-portable.

PERFORMANCE

Frequently used for oil spill cleanup operations in Canada. The manufacturer has thoroughly tested the liner materials for chemical resistance and permeability, strength, water absorption, wicking, high temperature blocking (sticking), dead load, abrasion resistance, ease of field repair, and shelf life. Test results are available from the manufacturer.

Sopers Engineered Fabric Products
P.O. Box 277
144 Chatham Street
Hamilton, Ontario
L8N 3E8

Telephone (905) 528-7936
Toll-free   (800) 263-8334
Facsimile (905) 528-8128

REFERENCES

(1) Manufacturer's Literature.

TERRA TANK No. 83

DESCRIPTION

A portable, "pillow tank" that can be quickly set up without tools for storing recovered liquids on land.

OPERATING PRINCIPLE

Low profile, self-supporting containers lie flat when empty and assume a "pillow" shape when filled with liquid. A single inlet/outlet can be fitted with a hose connector of choice.

STATUS OF DEVELOPMENT AND USAGE

Commercially available product used in mineral and oil exploration, helicopter and fixed wing aviation, forestry, logging, and by the military in many countries.

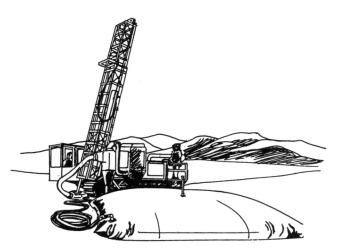

TECHNICAL SPECIFICATIONS

Physical - Available in small to very large sizes to contain fuel, water, or virtually any chemical or solution. The smallest holds 380 L (100 gal) and weighs about 45 kg (100 lb) including crate. When empty, it is 1.6 × 1.5 m (5 × 4.9 ft) in size and its shipping crate is 0.9 × 10 × 0.4 m (3 × 33 × 1.4 ft). The largest Terra Tank holds 189,300 L (50,000 gal) and weighs about 680 kg (1,500 lb) including crate. When empty, it is 9.2 × 18 m (30 × 60 ft) in size and its approximate shipping dimensions are 1.2 × 2.1 × 1 m (4 × 7 × 3.3 ft). Larger Terra Tanks are also available if required.

The **Petro Guard Terra Tank** is made of urethane-coated nylon fabric and stores JP1 jet fuel, JP4 jet fuel, diesel fuels, regular gasoline, isopropyl alcohol (with optional corrosion-proof fittings), phosphoric acid (10%), and sodium hydroxide (60%). The **Aqua Guard Terra Tank** is made of vinyl-coated polyester fabric (approved by the U.S. Food and Drug Administration) and is suited for storing potable water, some acids, and certain other liquids. The **Chem Guard Terra Tank** is constructed from urethane-coated nylon fabric and is resistant to a wide variety of chemicals and solutions. All Terra Tanks have two 51-mm (2-in) female NPT fittings, one for the fill/drain fitting and the other at the centre of the tank for vent/overfill protection. Many additional valves, flanges, and fittings can also be supplied, as well as camouflage covers, berm liners for secondary containment, ground sheets, flow meters, and pumping systems.

Operational - Terra Tanks are extremely lightweight and fully collapsible, offering a liquid containment capacity many times larger than their transportable size. They require almost no site preparation, can be easily and quickly installed and used immediately, and can be quickly repaired in the field in almost any weather. As the fabric of the tank hugs the surface of the contained liquid, only a few square centimetres of the liquid are exposed to air and no dangerous buildup of vapours occurs. The tank continuously adjusts itself to whatever volume of liquid is stored, and does not have to breathe as it has no vapour space. Terra Tanks have been proven in the most remote and harsh environments, including the Amazon Basin, Ecuador. The manufacturer's "torture test" includes a tracked front end loader crawling over the top of a filled tank. Industry endorsements of the Terra Tank are available from the manufacturer.

Power/Fuel Requirements and Transportability - No power or fuel is required and they are person-portable.

Manufacturer

SEI Industries Limited
7400 Wilson Avenue
Delta, British Columbia, V4G 1E5

Telephone (604) 946-3131
Facsimile (604) 940-9566

REFERENCES

(1) Manufacturer's Literature.

Section 3.2

Temporary Storage - Rigid Wall Containers

RIGID WALL STORAGE CONTAINERS - GENERAL LISTING — No. 84

DESCRIPTION

Containers and tanks designed to store and transport chemicals, hazardous materials, and petroleum products. Types of design include overpacks or salvage drums, modular tanks, roll-offs, bulk intermediate shipping containers, and standard steel, plastic, and fibreglass tanks. See also the following entries in this section.

> Custom Metalcraft Transportable Storage Systems (No. 85)
> Enviropack Polyethylene Overpacks/Salvage Drums (No. 86)
> Greif Containers - Steel Overpacks/Salvage Drums (No. 87)
> ModuTank Modular Storage Tanks (No. 88)
> Skolnik Steel Overpacks/Salvage Drums (No. 89)

OPERATING PRINCIPLE

Overpacks/Salvage Drums - Open-top steel or polyethylene drums used as overpack shipping containers for damaged or leaking drums, or for secondary containment and collective packaging of small quantities of containerized hazardous materials. Many such drums are nestable while not in use, reducing space requirements during storage and transport.

Modular Tanks - A variety of lined tanks for storing solids, liquids, or slurries consisting of portable modular components ranging from circular or rectangular bolt-together steel or concrete units to free-standing, smaller packaged systems that can be quickly assembled using panels and support frames. Modular components are delivered to the site in a compact package and are easily and quickly bolted together with ordinary hand tools. The structures often incorporate flexible membrane liners and can be easily taken down for compact storage, relocation, or resizing with a new liner.

Roll-Offs - Containers designed to be "rolled on and rolled off" of specially equipped train or truck trailers. The system is standardized so that containers can be delivered, dropped off, filled, and later picked up by any roll-off compatible train or truck. Available in many shapes suitable for storing and transporting solids, sludges, and sometimes liquids.

Bulk Intermediate Shipping Containers - Bulk liquid or solid shipping containers designed to be transported, when full, by standard lifting and transport equipment. They are available in many shapes and all are equipped with lifting fittings and are approved for transporting hazardous goods.

Standard Steel, Plastic, or Fibreglass Tanks - Standard product tanks constructed of steel, plastic, or fibreglass, available with a variety of coatings to allow the storage of any liquid.

STATUS OF DEVELOPMENT AND USAGE

Commercially available products in general use.

TECHNICAL SPECIFICATIONS

Physical

Physical specifications vary from company to company and model to model. Construction materials also include varying grades of steel, polyethylene, fibreglass, and fibreglass-reinforced plastic (FRP), with a choice of many custom materials, liners, and coatings. Storage containers are usually available in the following range of specifications.

	Capacity	Dimensions	Shipping Weight
Overpacks/Salvage Drums	30 to 416 L (8 to 110 gal)	0.4 × 0.4 m to 1.9 × 2.6 m (1.2 × 1.2 ft to 6.3 × 8.6 ft)	5 to 46 kg (12 to 100 lb)
Modular Tanks	≥ 9,120 L (≥ 2,500 gal)	≥ 3.2 × 1.4 m (≥ 11 × 4.5 ft)	≥ 72 kg (≥ 160 lb)
Roll-Off Containers	15,300 to 76,450 L (20 to 100 yd^3)	7.3 × 2.6 × 1.4 m to 14.6 × 2.6 × 2.6 m (24 × 8.5 × 4.5 ft to 48 × 8.5 × 8.5 ft)	≥ 2,270 kg (≥ 5,000 lb)
Bulk Intermediate Shipping Containers	890 to 2,000 L (235 to 525 gal)	1 × 0.9 m to 1.3 × 1.5 m (3.5 × 3.1 ft to 4.3 × 5.1 ft)	36 to 230 kg (80 to 500 lb)
Standard Tanks	114 to 62,000 L (30 to 16,400 gal)	0.6 × 0.5 m to 4.3 × 5 m (1.9 × 1.8 ft to 14 × 17 ft)	8 to 1,900 kg (18 to 4,200 lb)* *for polyethylene

Operational

Operational specifications also vary among containers. Major considerations when choosing an appropriate tank include whether the material to be stored in the container is compatible with it and will satisfactorily contain the material, transportation to the site, whether the container will need to be transported when full, site conditions, including ground stability, slope, and access, and preparation required to use the container.

Power/Fuel Requirements

All storage containers listed in this section are self-supporting and require no outside source of power.

Transportability

The transportability of rigid wall storage systems varies greatly.

PERFORMANCE

Documentation of performance in chemical spill situations is limited. Tanks of appropriate durability, chemical compatibility, and shape should prove useful for isolating and transporting a wide variety of spilled hazardous materials.

CONTACT INFORMATION

The following is an alphabetical list of some representative manufacturers, their products, and Canadian distributors. Selected rigid storage containers are discussed in more detail in the individual entries in this section.

Manufacturer	Product(s)	Comments
21st Century Container Limited 150 Selig Drive Atlanta, GA 30336 U.S.A. Telephone (404) 699-1308 Toll-free (800) 772-3745 Facsimile (404) 699-7912	Atlas Intermediate Bulk Container	Bulk intermediate shipping container Open top; Xurex plastic; 1,360 L (360 gal) capacity
Aco-Asmann of Canada Limited 794 McKay Road Pickering, Ontario L1W 2Y4 Telephone (905) 683-8222 Facsimile (905) 683-2969	Pallet Tanks Polyethylene Tanks	Bulk intermediate shipping container A complete range of polyethylene tanks and bulk intermediate shipping containers
Air Plastics Incorporated 1224 Castle Drive Mason, OH 45040 U.S.A. Telephone (513) 398-8081 Facsimile (513) 398-8082	Corrosion-resistant Fibreglass and Plastic Tanks	A complete range of plastic and fibreglass tanks and bulk intermediate shipping containers
Canbar Incorporated Box 280, One Canbar Street Waterloo, Ontario N2J 4A7 Telephone (519) 886-2880 Facsimile (519) 886-5546	Polyethylene Tanks	A complete range of polyethylene tanks and bulk intermediate shipping containers

3.2 Temporary Storage - Rigid Wall Containers

Can-Ross Environmental Services 2270 South Service Road, West Oakville, Ontario L6L 5M9 Telephone (905) 847-7190 Facsimile (905) 847-7175	Big Mouth Fast Tank	Steel overpack/salvage drum Modular tank
Cassier Engineering Sales Ltd. 11 Progress Avenue Unit 3, Suite 206 Scarborough, Ontario M1P 4S7 Telephone (416) 298-1628 Facsimile (416) 298- 9584	TOTE Systems and TranStore® Systems	Bulk intermediate shipping containers Canadian distributor for Custom Metalcraft Inc. and TOTE Systems (both listed below)
Chem-Tainer Industries Inc. 361 Neptune Avenue West Babylon, NY 11704 U.S.A. Telephone (516) 661-8300 Toll-free (800) 275-2436 Facsimile (516) 661-8209	Polyethylene Tanks	A complete range of polyethylene tanks, overpacks/salvage drums, and bulk intermediate shipping containers
Custom Metalcraft Inc. 3232 East Division P.O. Box 10587 Springfield, MO 65808 Telephone (417) 862-0707 Facsimile (417) 864-7575	TranStore® Systems	Portable, stackable bulk-handling units and transportable storage units See No. 85 in this section
ENPAC Corporation 121 Industrial Parkway P.O. Box 1100 Chardon, OH 44024-5100 U.S.A. Telephone (216) 286-9222 Toll-free (800) 936-7229 Facsimile (216) 286-9297	Poly Labpack Poly Overpack	Polyethylene overpacks/salvage drums
Enviropack Company 465 Hamilton Road Bossier City, LA 71111 U.S.A. Telephone (318) 742-1100 Facsimile (318) 742-7424	Generation III Polyethylene Overpacks Chem-Tank	Polyethylene overpacks/salvage drums Bulk intermediate shipping container See No. 86 in this section
Galbreath Incorporated Rosser Drive Post Office Box 220 Winimac, IN 46996 U.S.A. Telephone (219) 946-6631 Facsimile (219) 946-4579	Roll-Off Sludge Containers	Roll-offs

Greif Containers Incorporated 370 Millen Road Stoney Creek, Ontario L8E 2H5 Telephone (905) 664-4433 Facsimile (905) 664-4491	PLAST-I-LINED and Tapered NEST-ALL Steel Drums SAF-T-DRUM steel salvage drums LOK-RIM Fibre Drums	A complete line of all types of steel drums from 35 to 320 L (9 to 85 gal), as well as tapered/nestable steel drums, and plastic and fibre drums See No. 87 in this section
J & L Industrial Tanks 32351 Huntingdon Road, R.R. # 5 Abbotsford, British Columbia V2T 5Y8 Telephone (604) 854-6776 Facsimile (604) 854-1992	Polyethylene Tanks Permaglas Storage Systems	A complete range of polyethylene tanks and bulk intermediate shipping containers Modular tanks made of steel panels with a silica glass layer chemically bonded to inside and outside steel surface for corrosion resistance
May Fabricating Company Incorporated P.O. Box 1029 Beeville, TX 78104 U.S.A. Telephone (512) 358-7022 Toll-free (800) 242-0122 Facsimile (512) 358-0418	Roll-Offs	Roll-off and intermodal containers for both hazardous and non-hazardous materials
McDaniel Tank Manufacturing 714 North Saginaw Street Holly, MI 48442 U.S.A. Telephone (810) 634-8214 Toll-free (800) 627-6508 Facsimile (810) 634-2240	Safety-Tote Tanks	Bulk intermediate shipping container Forklift-portable, stackable, steel
Milepost Manufacturing Inc. R.R. # 2 St. Albert, Alberta T8N 1M8 Telephone (403) 459-1030 Facsimile (403) 458-6377	Portable Concrete Systems	Easily assembled modular tanks with pre-formed structural walls constructed of concrete and a synthetic liner
ModuTank Inc. 41-04 35th Avenue Long Island City, NY 11101 U.S.A. Telephone (718) 392-1112 Facsimile (718) 786-1008	Lined Modular Storage Tanks	Ranging from bolt-together steel units either circular or rectangular in shape to smaller, free-standing packaged systems See No. 88 in this section
Poly Processing Co. P.O. Box 80, 8055 South Ash Street French Camp, CA 95231, U.S.A. Telephone (209) 982-4904 Facsimile (209) 982-0455	Polyethylene Tanks	A complete range of polyethylene tanks and bulk intermediate shipping containers

3.2 Temporary Storage - Rigid Wall Containers

Skolnik Industries Incorporated 4900 South Kilbourn Avenue Chicago, IL 60632-4593 U.S.A. Telephone (312) 735-0700 Facsimile (312) 735-7257	Overpack/Salvage Drums	A complete line of steel overpack/salvage drums
Stelfab Niagara Limited 8594 Earl Thomas Avenue Niagara Falls, Ontario L2E 6X8 Telephone (905) 356-8683 Facsimile (905) 356-8728	Steltight Intermediate Bulk Containers	Steel and stainless steel intermediate bulk shipping containers for liquid and dry materials
Techstar Plastics Incorporated 15400 Old Simcoe Road Port Perry, Ontario L9L 1L8 Telephone (905) 985-8479 Toll-free (800) 263-7943 Facsimile (905) 985-0265	Bulk Frame Containers	Steel-framed polyethylene bulk intermediate shipping containers for both liquid and dry materials
TOTE Systems (A Division of Thomas Conveyor Company) 2042-A South Brentwood Springfield, MI 65804 U.S.A. Telephone (417) 889-8683 Facsimile (417) 889-7911	TOTE Transportable Storage Systems	Tote bins for shipping bulk and dispensing product at the destination
Turner Equipment Company Incorporated P.O. Box 1260, Hwy. 117 South Goldsboro, NC 27530 U.S.A. Telephone (919) 734-8345 Toll-free (800) 672-4770 Facsimile (919) 736-4550	Custom Fabricated Steel Tanks	Standard steel tanks

OTHER DATA

Consult local telephone directories or industrial trade registers for additional manufacturers and local distributors.

REFERENCES

(1) Manufacturer's Literature.

CUSTOM METALCRAFT
TRANSPORTABLE STORAGE SYSTEMS

No. 85

DESCRIPTION

Portable, stackable, bulk-handling containers suitable for storing or shipping hazardous materials.

OPERATING PRINCIPLE

Square-shaped, stackable, all-metal bulk containers, featuring built-in pallets, top-filling ports, and bottom discharge outlets with butterfly and ball valves.

STATUS OF DEVELOPMENT AND USAGE

Commercial products designed for storing and transporting hazardous liquids or flowable solids in bulk.

TECHNICAL SPECIFICATIONS

Physical - TranStore® Systems - Eleven standard sizes are available in either steel, 304/316 stainless steel, or aluminum, ranging from 415 to 2,160 L (110 to 570 gal) in capacity and from 0.5 to 1.8 m (1.7 to 5.9 ft) in height.

Operational - The tanks require 44% less storage space than equivalent volumes of drum storage systems. All corners have a 6.3 cm (2.5 in) radius for strength and easy cleaning. Tanks come with heavy-duty combination lifting lugs and leg positioners for stacking. A 0.5 m (1.8 ft) drum cover on tanks provides easy access for cleaning and inspection. A side door bin for storing solids, a variety of closures and valves, and a one-piece sump for complete drainage are also available. Tanks are tested and tagged U.S. DOT Specification 56 for dry material, U.S. DOT Specification 57 for liquids, and are UN-rated.

Transportability - All containers come with lifting lugs and a unitized pallet so units can be lifted by forklifts or other hoisting devices. The containers can be shipped by standard flat bed, tractor trailer, or rail. Twenty to twenty-two containers can be loaded in a standard transport truck and 26 to 28 containers can be loaded in a standard box car.

PERFORMANCE

According to the manufacturer, this system of containerization provides safer storage and transportation, greater ease of handling, and more efficient use of space than conventional drum storage. Cost and frequency of use, however, might affect the decision to use such a system. Similar systems have been used by Canadian offshore oil and gas exploration companies in the Arctic to supply lubricating oil to support/supply ships and to collect, store, and tranship waste oils. Specializing in custom, stainless steel processing equipment, transportable storage units, and complex bulk-handling systems since 1977, Custom Metalcraft Inc. has developed a reputation for excellence in designing and manufacturing high quality equipment for the food, beverage, chemical, and pharmaceutical industries.

Manufacturers

Custom Metalcraft
2332 East Division
P.O. Box 10587
Springfield, MI 65808, U.S.A.

Telephone (417) 862-0707
Facsimile (417) 864-7575

Canadian Distributor

Cassier Engineering Sales Limited
11 Progress Avenue
Scarborough, Ontario
M1P 4S7

Telephone (416) 298-1628
Facsimile (416) 298-9584

REFERENCES

(1) Holden, Nikki, personal communication, Custom Metalcraft Inc., Springfield, MI (February 2, 1996).
(2) Manufacturer's Literature.

ENVIROPACK POLYETHYLENE OVERPACKS/SALVAGE DRUMS No. 86

DESCRIPTION

Open-top polyethylene drums for overpacking hazardous waste drums, designed to be used as overpack salvage drums for 208-L (55-gal) and smaller containers, and as lab packs.

OPERATING PRINCIPLE

Designed to fully enclose various sizes of drums and pails to contain leaks and to facilitate safe transport.

STATUS OF DEVELOPMENT AND USAGE

Commercial product developed to control leakage of chemicals from drums, or for co-packaging small quantities of containerized chemicals (lab packs).

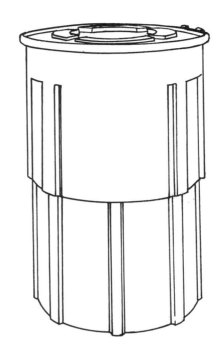

TECHNICAL SPECIFICATIONS

Physical

The salvage drums have a capacity of 322 L (85 gal) and accept all drums 208-L (55-gal) and smaller. They weigh 25 kg (55 lb), outside height is 1 m (3.5 ft), inside height is 0.9 m (3.2 ft), and outside diameter is 0.7 m (2.5 ft). They are made of 100% UV-stabilized, seamless, rotationally molded polyethylene and are sealed with a Neoprene gasket with steel locking ring and bolt.

Operational

Enviropack lids are double-walled for added leak protection. The lid and drum base are designed to interlock so drums can be stacked. Empty drums nest at 50% of the total height for space savings. Enviropack overpacks meet the following specifications: the United States Department of Transport (DOT) E-9775, the United States Code of Federal Regulations (CFR) 49, 173.3, and the United Nations Specification 1H2/Y454/S.

Power/Fuel Requirements and Transportability

No power or fuel are required. Overpacks are designed for transporting chemicals. A drum hoist, hand truck, fork lift and pallet, or other drum-handling equipment may be required when handling full drums. Empty overpacks can be nested when empty for a 50% height reduction, thereby reducing shipping bulk.

CONTACT INFORMATION

Manufacturer

Enviropack Co.
465 Hamilton Road
Bossier City, Louisiana 71111
U.S.A.

Telephone (318) 742-1100
Facsimile (318) 742-7424

Enviropack overpack drums can be shipped out of Bossier City, Louisiana; North Las Vegas, Nevada; or Knoxville, Tennessee.

REFERENCES

(1) Manufacturer's Literature.

GREIF CONTAINERS - STEEL OVERPACKS/SALVAGE DRUMS No. 87

DESCRIPTION

Heavy-duty, HI-bake coated, fully removable head steel drum for use as overpack shipping containers for damaged or leaking drums of hazardous waste.

OPERATING PRINCIPLE

To safely assist in the containment of hazardous materials during transportation.

STATUS OF DEVELOPMENT AND USAGE

Commercially available product, manufactured in four sizes capable of enclosing industry standard type containers ranging from 20 to 208 L (5 to 55 gal).

TECHNICAL SPECIFICATIONS

Physical

Greif overpack/salvage drums are made of 16-gauge, 18-gauge, or 20-gauge stainless steel throughout, with other gauges and combinations available for customizing salvage drum requirements. All drums have a corrosion-resistant lining and are painted yellow with "SALVAGE DRUM/FÛT DE RÉCUPÉRATION" printed on the outside. The following are other specifications for the overpack/salvage drums.

	Q85-BR-16	U55-BR-16	S30-BR-18	B15-BR-20
Capacity	320 L (85 gal)	208 L (55 gal)	110 L (30 gal)	60 L (15 gal)
Inside diameter	0.7 m (2.1 ft)	0.6 m (1.9 ft)	0.5 m (1.5 ft)	0.4 m (1.2 ft)
Inside height	1 m (3.2 ft)	0.9 m (2.8 ft)	0.7 m (2.3 ft)	0.6 m (1.9 ft)
Tare weight	36 kg (80 lb)	28 kg (62 lb)	16 kg (35 lb)	7 kg (16 lb)
Cover Gasket	tubular rubber	tubular Neoprene	tubular rubber	flowed latex

Operational

All salvage drums (except B15-BR-20) come with a 12-gauge Drop Forged closing ring, complete with 1.5 cm (5/8 in) bolt and locking nut. The Q85-BR-16 has a 1.9 cm (3/4 in) NPT fitting installed in the cover to relieve any internal pressure before opening.

Transportability

A drum hoist, hand truck, forklift and pallet, or other drum-handling equipment may be required when handling full drums. To reduce transportation and storage space, salvage drums are also available in quad packs which consist of four sizes of drums nested into one another.

PERFORMANCE

Greif Salvage Drums meet the requirements of the *Transportation of Dangerous Goods Regulations,* Section 7.29 for the use of salvage drums for domestic shipments. All Canadian manufacturing locations are registered to **ISO 9002-94** quality standards by the International Organization for Standards.

3.2 Temporary Storage - Rigid Wall Containers

CONTACT INFORMATION

Manufacturer

Greif Containers Inc.
370 Millen Road
Stoney Creek, Ontario
L8E 2H5

Telephone (905) 664-4433
Oakville   (905) 825-0296
Facsimile (905) 664-4491

OTHER DATA

Greif Containers Inc./Contenants Greif Inc is a wholly owned subsidiary of Greif Bros. Corporation. The company operates 95 locations in the United States and Canada. Greif also manufactures other heavy- and light-gauge steel drums, overpacks, kegs, and composites in capacities from 30 to 320 L (8 to 85 gal), straight-sided or tapered (nestable), as well as a complete range of plastic and fibre drums. These containers meet the following standards:

- **steel drums and composites** meet the Transport Canada TC - 17H, 17E, 37A, 37B, 6D, 37M, and United Nations 1A1, 1A2, 6HA1 (these can be used as overpacks/salvage drums for international shipping);
- **fibre drums and composites** meet the Transport Canada TC - 21C, 21P, and United Nations 1G, 6HG1;
- **plastic drums** meet the Transport Canada TC - 34 and United Nations 1H1 and 1H2.

REFERENCES

(1)   Manufacturer's Literature.
(2)   Tweedle, Ross, personal communication, Greif Containers Inc., Stoney Creek, Ont. (February, 1996).

MODUTANK MODULAR STORAGE TANKS — No. 88

DESCRIPTION

A variety of lined tanks of portable modular components, ranging from bolt-together steel units either circular or rectangular in shape to smaller, freestanding packaged systems that can be quickly assembled using panels and support frames.

OPERATING PRINCIPLE

Modular components are delivered to the site in a compact package and are easily and rapidly bolted together with simple hand tools. The structures are made of galvanized or epoxy-coated steel with flexible membrane liners and can be easily taken down for compact storage, relocation, or resizing with a new liner.

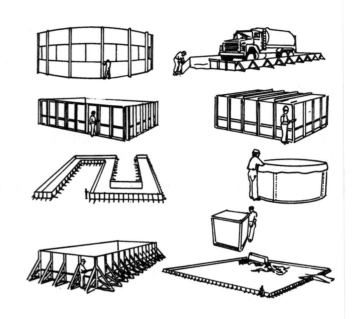

STATUS OF DEVELOPMENT AND USAGE

Commercial products specifically conceived as highly portable, modular systems. They are widely used for temporarily storing or treating a range of liquids including many chemicals.

TECHNICAL SPECIFICATIONS

Physical

Modutank/EconoTank - Designed for heavy-duty permanent or standby containment, ModuTanks and EconoTanks are rectilinear shaped tanks with individual panels, support posts, liners, covers, and optional piping arrangements. An exterior system of angled supports maintains the sidewalls in a vertical position. Modutanks and EconoTanks are available in any desired rectilinear size from 1.2 × 1.2 m (4 × 4 ft) upward, based on 1.1 m (3.7 ft) long modules and 1.4 m (4.7 ft) maximum wall height. A 1.9 m (6.2 ft) high model called the HiStor, is available in sizes from 946,350 L (250,000 gal) and greater. The following are some typical dimensions for various models.

Capacity	Dimensions [1.4 m (4.7 ft) deep]
30,300 L (8,000 gal)	4.7 × 4.7 m (16 × 16 ft)
113,600 L (30,000 gal)	9.3 × 9.3 m (31 × 31 ft)
189,300 L (50,000 gal)	12 × 12 m (39 × 39 ft)
378,500 L (100,000 gal)	17 × 17 m (57 × 57 ft)
1,135,600 L (300,000 gal)	29 × 29 m (94 × 94 ft)
1,892,700 L (500,000 gal)	38 × 38 m (124 × 124 ft)
3,785,400 L (1,000,000 gal)	53 × 53 m (173 × 173 ft)

Wall panels	16-gauge galvanized steel
Support frames	5 × 5 × 0.3 cm (2 × 2 × 1/8 in) and 5 × 5 × 0.5 cm (2 × 2 × 3/16 in) steel angle, hot dip galvanized
Rails	7.6 × 5 cm (3 × 2 in) steel angle, hot dip galvanized after fabrication
Fasteners	
ModuTank	Stainless steel
Tension Cables	
ModuTank	Stainless steel
EconoTank	Galvanized steel

3.2 Temporary Storage - Rigid Wall Containers

Liners
- ModuTank — A choice of fitted liners and floating covers in various gauges and materials is available. Materials include XR-5, Hypalon, PVC, HDPE, and reinforced polypropylene membrane.
- EconoTank — A 0.5 mm (20 mil) HDPE liner is standard.

ModuStor - Similar to ModuTanks and EconoTanks in design but circular in shape. No external supports required.

Size	18,900 to 3,785,400 L (5,000 to 1,000,000 gal) capacity
	4.6 to 37 m (15 to 120 ft) diameter
	1.2 to 4.6 m (4 to 15 ft) wall heights
Wall panels	18 to 10 gauge, mill galvanized steel
Girths	Structural steel angle, hot dip galvanized
Hardware	Grade 5 plated steel or better
Liners	A choice of fitted liners and floating and supported covers is available in various gauges and materials. Available materials include XR-5, Hypalon, PVC in thicknesses from 0.8 to 1.1 mm (30 to 45 mil), and reinforced polypropylene membrane in thicknesses from 0.0 to 1.1 mm (20 to 45 mil).

EconoStor/ComPakt/ChemStor - Small capacity rectilinear tanks for permanent or standby storage, indoors or outdoors. The ChemStor tanks are engineered for storing liquid chemicals.

Size	950 L (250 gal) and greater capacity
Wall panels	
EconoStor	16-gauge galvanized steel
Compakt	16-gauge galvanized steel
Chemstor	Epoxy-coated steel
Support frames	
EconoStor	7.6 × 5 × 0.6 cm (3 × 2 × 1/4 in) steel angle
Compakt	7.6 × 5 × 0.6 cm (3 × 2 × 1/4 in) steel angle
Chemstor	Epoxy-coated 7.6 × 5 × 0.6 cm (3 × 2 × 1/4 in) steel angle
Hardware	High-tensile steel
Liners	PVC liners are Standard, with XR-5, Hypalon, HDPE, reinforced polypropylene membrane, and other materials available
Covers	Modular galvanized steel with or without hatches

QuickStor - Circular tanks designed for temporary or emergency containment and featuring very fast setup.

Size	8,300 to 132,500 L (2,200 to 35,000 gal) capacity
	3.0 to 12 m (10 to 40 ft) diameter
Wall panels	1.2 m (4 ft) high flat wall panels are made from 20-gauge galvanized sheet steel, 2.0 m (6.7 ft) long
Hardware	Corrosion-resistant
Liners	0.5 mm (20 mil) PVC

TerraStor - Designed for temporarily storing and treating hazardous earthen materials, sand, and clay.

Size	Any desired rectilinear size, based on 1.1 m (3.6 ft) long modules
Wall panels	16-gauge galvanized steel
Support frames	5 × 5 cm (2 × 2 in) welded structural steel, hot dip galvanized after fabrication
Hardware	Corrosion-resistant
Liners	A full range of membrane liner materials are available including HDPE, XR-5, PVC, and Hypalon

ModuTainer - Concept identical to ModuTank. Can be erected around existing storage tanks to contain potential spills.

Operational

Modutank/Econotank - A typical 23 × 23 m (75 × 75 ft), 757,000 L (200,000 gal) tank can be installed by six unskilled labourers in about eight hours, using only simple hand tools. Virtually any shape with right angles can be assembled from the modular components, which is useful for installations with special flow considerations or irregularly shaped sites. Floating covers are available made of the same material as the liners, for creating totally enclosed systems. Covers, buoyed up by flotation materials, ride the fluid's surface and sink to the bottom of the tank as it empties. Inlet and outlet pipes can be installed through the walls, over the top, or through the bottom. Bottom drains, sumps, and accessory fittings are available. A tank designed to meet Universal Building Code Seismic 4 conditions is available.

ModuStor - A 1,892,700 L (500,000 gal) ModuStor can be erected on a prepared site by six people in about eight to ten days. ModuStor tanks require a firm, level surface such as compacted earth, concrete, macadam, or rock. Floating membranes, conical roofs supported by rigid bar joists, as well as nets and screens are available to cover the ModuStor tanks. All conventional underdrains as well as input and discharge lines can be attached to the ModuStor tanks. Flange openings can be precut at the factory.

EconoStor/ComPakt/ChemStor - A 5,700 L (1,500 gal) ComPakt can be assembled by two unskilled people in less than three hours. The tank can be set up on slightly sloping or uneven surfaces. Virtually any shape with right angles is possible, so that the EconoStor, the ComPakt, and the ChemStor can be installed or set up in oddly shaped areas. Inlet and outlet pipes can be installed through the walls, over the top, or through the bottom. Bottom drains, sumps, and accessory fittings are available. The ChemStor tanks are engineered to store liquid chemicals.

QuickStor - QuickStor tanks are designed to be transported in a station wagon. A 21,200-L (5,000 gal) QuickStor can be prepared for operation in less than one hour by two people. While QuickStor tanks are designed primarily for over-the-wall loading and unloading, inlet and outlet pipes can be positioned and installed with optional through-the-wall bulkhead fittings. Aluminum QuickStors, which are available on special order, have been designed for airlifting to remote sites. They are available in capacities from 26,500 to 64,300 L (7,000 to 17,000 gal).

TerraStor - TerraStor steel containment systems are designed for rapid assembly on sites where berms or other retaining devices are not practical or cost-effective. A 2,020 m^2 (0.5 acre) TerraStor can be assembled in less than one day. Site-built earthen ramps allow easy truck access. Virtually any shape with right angles, such as a "T", "L", "+", can be assembled from the modular components, which is useful for irregularly shaped sites.

Power/Fuel Requirements

None.

Transportability

All parts can be comfortably hand-carried by one or two people and are packaged for space-efficient transportation.

PERFORMANCE

The manufacturer's literature cites many case histories of companies and government agencies successfully using ModuTank containment systems for various applications.

CONTACT INFORMATION

Manufacturer

ModuTank Inc.
41-04 35th Avenue
Long Island City, NY 11101
U.S.A.

Telephone (718) 392-1112
Facsimile (718) 786-1008

REFERENCES

(1) Manufacturer's Literature.

SKOLNIK STEEL OVERPACKS/SALVAGE DRUMS   No. 89

DESCRIPTION

Open-top, steel drums, used as overpack shipping containers for damaged or leaking drums, or as secondary containment and collective packaging for small quantities of containerized chemicals.

OPERATING PRINCIPLE

Salvage drums can fully enclose many sizes of drums or pails to contain leaks and facilitate transportation.

STATUS OF DEVELOPMENT AND USAGE

Commercial products developed to control leakage of chemicals from drums, or for co-packaging of small quantities of containerized chemicals (lab packs).

TECHNICAL SPECIFICATIONS

Physical - These drums range in capacity from 30 to 420 L (8 to 110 gal), weigh from 5 to 46 kg (12 to 100 lb), and are from 0.4 to 1 m (1.2 to 3.4 ft) high and 0.4 to 0.8 m (1.2 to 2.5 ft) in diameter.

Materials	16-gauge to 22-gauge steel
Styles	Full removable head (Open Head) and Closed Head (Tight Head)
Tight head	5 cm (2 in) and 1.9 cm (0.75 in) steel or plastic plugs in top, interior treated with a corrosive inhibitor, exterior painted black with a white top
Open head	Full removable cover, 16-gauge bolt ring closure, sponge rubber gasket, plated bolt and nut, exterior painted black with a white top
Fittings	Tri-Sure and Rieke style
Closures	18-gauge to 12-gauge bolt and leverlock ring.
Interior coatings	High-temperature, chemically resistant epoxy/phenolic, F.D.A. approved coatings, and rust inhibitors. Polyethylene inserts are also available.

Operational - Skolnik salvage drums meet the following specifications: the American National Standards (ANSI); the United States Department of Transport (DOT) Rule 40; the United States Code of Federal Regulations (CFR) 49, 173.3; the United Nations Specification 1A1, 1A2, 6HA1; and Nuclear Quality Assurance NQA-1. To respond to emergency needs for hazardous material containers, Skolnik Industries maintains a ready-to-ship inventory of most standard DOT/UN approved containers and are able to dispatch orders within 24 hours.

Power/Fuel Requirements and Transportability - No power or fuel are required. Salvage drums are designed for transporting chemicals. A drum hoist, hand truck, forklift and pallet, or other drum-handling equipment may be required when handling full drums. As Skolnik salvage drums are **not** nestable, they are bulky to transport when empty.

CONTACT INFORMATION

Manufacturer

Skolnik Industries Incorporated
4900 South Kilbourn Avenue
Chicago, Illinois 60632-4593, U.S.A.

Telephone (312) 735-0700
Facsimile (312) 735-7257

REFERENCES

(1) Manufacturer's Literature.

Section 3.3

Temporary Storage - Liners

GEOMEMBRANE LINERS - GENERAL LISTING   No. 90

DESCRIPTION

Geomembranes are membrane liners and barriers with low permeability that are used with any geotechnical engineering material to control fluid migrations. The term "liner" applies when a geomembrane is used as an interface or a surface revetment. The term "barrier" is usually applied when the geomembrane is used inside an earth mass. Geomembrane is a generic term that replaces terms such as polymeric membrane liners, flexible membrane liners, impermeable membranes, and impervious sheets. These terms are not appropriate for the following reasons: geomembranes are not always used as liners; the term "flexible membrane" is redundant; terms such as polymeric and synthetic are too restrictive; and no material is absolutely impermeable or impervious. Geomembranes should not be confused with geotextiles, which are permeable textiles used in engineering applications.

OPERATING PRINCIPLE

Geomembranes are used to line structural or earthen impoundments that contain recovered hazardous materials.

STATUS OF DEVELOPMENT AND USAGE

Commercially available products in general use.

TECHNICAL SPECIFICATIONS

Physical

Geomembranes are composed of materials with very low permeability, which may be reinforced with fabric. Such materials typically have a hydraulic conductivity or coefficient of permeability of 10^{-12} to 10^{-11} cm/s, such as compounds of asphalt and/or a polymer.

Asphalt is obtained either from natural deposits or as a by-product of oil distillation. Blown asphalt, which is often used to make geomembranes, is hardened by blowing air through the molten asphalt to raise its softening temperature and to decrease its tendency to flow.

Polymers are chemical compounds of high molecular weight. Only synthetic polymers are used to make geomembranes. The following are the most common types of polymers used as base products to manufacture geomembranes.

    Thermoplastics - Polyvinyl chloride (PVC); oil-resistant PVC (PVC-OR); thermoplastic nitrile-PVC (TN-PVC); and ethylene interpolymer alloy (EIA).

    Crystalline thermoplastics - Low-density polyethylene (LDPE); high-density polyethylene (HDPE); high-density polyethylene-alloy (HDPE-A); polypropylene; and elasticized polyolefin.

    Thermoplastic elastomers - Chlorinated polyethylene (CPE); chlorinated polyethylene-alloy (CPE-A); chlorosulfonated polyethylene (CSPE), also commonly referred to as "Hypalon"; and thermoplastic ethylene-propylene diene monomer (T-EPDM).

    Elastomers - Isoprene-isobutylene rubber (IIR), also commonly referred to as butyl rubber; ethylene-propylene diene monomer (EPDM); polychloroprene (CR), also commonly referred to as Neoprene; and epichlorohydrin rubber (CO). In addition to the asphalt and/or polymer base products, compounds in geomembranes generally include various additives. Additives typically compounded with asphalt include fillers, fibres, and elastomers. Additives typically compounded with polymers are fillers, fibres, processing aids, plasticizers, carbon black, stabilizers, antioxidants, and fungicides. Various types of fabric reinforcement are often used with both asphalt and polymer membranes, with the type depending on the manufacturing process of the geomembrane.

Geomembranes can be constructed on site, although they are usually manufactured in a plant using one of the following processes: extrusion, spread-coating, or calendering. In the extrusion process, a molten polymer is forced out and elongated into a non-reinforced sheet. When the sheet is still warm, it is laminated with a fabric to produce a reinforced membrane. The spread-coating process usually involves coating a fabric with a polymer or asphalt compound to reinforce the geomembrane. Non-reinforced geomembranes are made, using the spread-coating method, by spreading a polymer on a sheet of paper which is then removed and discarded.

Calendering is the most common manufacturing process. A calendered, non-reinforced geomembrane is usually a sheet of compound made by passing a heated polymeric compound through a series of heated rollers. Multi-laminate membranes, including reinforced geomembranes, are also manufactured using this process.

Geomembranes made on site are usually continuous (with no seams) and are made by spraying or otherwise placing a hot or cold viscous material onto a substrate. Geomembranes made by spraying are commonly referred to as "spray-applied geomembranes" or "spray-on membranes". The base product of the sprayed material is usually asphalt, an asphalt-elastomer compound, or a polymer such as polyurethane. The sprayed material forms a continuous flexible film with little or no tack after curing.

Important physical properties to consider when selecting an appropriate geomembrane for a specific application are chemical compatibility, permeability, tensile properties, modulus of elasticity, hardness, tear resistance, puncture resistance, vapour transmission, and absorption.

Operational

As all materials used as liners are permeable to fluids (liquids and gases), some leakage should be expected through liners. Defects such as cracks, holes, and faulty seams will cause additional leakage. A fluid permeates a membrane by osmosis, diffusion, and absorption. In most cases, this is expressed as a measure of unit volume per unit area per unit of time for the **vapour** phase of the fluid. These values, which can only be determined through actual lab test procedures, are supplied by most geomembrane manufacturers.

Some geomembrane manufacturers rate their products using the Darcy coefficient of permeability, "k" (cm/s). As the mechanisms by which fluids flow through geomembranes and soils are different, however, Darcy's equation can be used only if it is made clear that the coefficient of permeability used for the geomembrane depends on the flow conditions, hydraulic gradient, or hydraulic head. The resulting values are useful only to compare different liners or applications to give "apparent" hydraulic flow values and should be used as a guide only. Lined areas should be able to achieve an apparent permeability of at least 10^{-7} cm/s. At the present time, the best that could be expected is 10^{-10} cm/s. These values are greater than the actual permeability of the geomembranes themselves, as design and installation factors will both increase the permeability of the system.

The following are terms used to evaluate permeability.

Conductivity or Coefficient of Permeability (k) - Volume of fluid passing per unit area of geomembrane per unit hydraulic gradient per unit period of time.

Permittivity (Ψ) - Volume of fluid passing per unit area of geomembrane per unit hydraulic head per unit period of time.

Impedance (I) - Time necessary for a unit volume of fluid to pass through a unit area of geomembrane per unit hydraulic head.

Permeance (ω) - Mass of fluid passing per unit area of geomembrane per unit pressure per unit period of time.

Water Vapour Transmission Rate (WVT) - Mass of water vapour passing per unit area of geomembrane per unit period of time (under a specified pressure which may vary from one test to another).

When selecting a geomembrane, both site and operational conditions must be evaluated. These conditions include soil conditions, temperature ranges and fluctuations, exposure to sunlight or other weathering agents (sun-generated heat, ozone, fungi, and bacteria), installation and operating loads on liner, and anchoring/sealing requirements.

Power/Fuel Requirements

Special equipment may be required to install, seal, and test manufactured geomembranes and special applicators are required to construct geomembranes on site.

Transportability

The transportability of geomembrane liners varies greatly.

PERFORMANCE

Extensive research has been conducted on the ability of geomembranes to contain hazardous materials, especially in the context of hazardous waste landfills and industrial impoundment sites. In brief, such research has indicated that, if geomembranes are installed without flaws, they can provide a barrier that is much less permeable than an unlined impoundment. However, geomembranes are not resistant to complex mixtures of organic chemicals to which they may be exposed. Furthermore, certain chemicals may rapidly diffuse through such liners, even though these chemicals may not affect the physical properties of the geomembrane. No single liner material is effective against all the constituents in leachate from complex mixtures of chemicals.

3.3 Temporary Storage - Liners

While permeability is a major factor affecting the ability of geomembranes to contain hazardous materials, in a spill situation, liners are likely to be used only for short-term impoundment. Thus, when geomembranes are used in spill situations, permeability is not as important a factor.

CONTACT INFORMATION

Several representative manufacturers and their products and Canadian distributors are listed alphabetically here.

Manufacturer	Product(s)	Comments
Burke Environmental Products 2250 South Tenth Street San Jose, CA 95112 U.S.A. Telephone (408) 297-3500 Facsimile (408) 280-0938	Hypalon liners Ecoseal liners	Hypalon is a trade name of Dupont for chlorosulfonated polyethylene (CSPE) Ecoseal is an oil-resistant thermoplastic elastomer of synthetic rubber
CDF Corporation 77 Industrial Park Road Plymouth, MA 02360 U.S.A. Telephone (508) 747-5858 Toll-free (800) 443-1920 Facsimile (508) 747-6333	Polyethylene liners	See No. 91 in this section
Chemical Equipment Fabricators Ltd. 26 Riviere Drive Markham, Ontario L3R 5M1 Telephone (416) 438-9266 Facsimile (416) 438-0382	Fabrico environmental liners	A Canadian distributor of Fabrico environmental liners
Columbia Geosystems Limited 1415 - 28th Street, NE Calgary, Alberta T2A 2P6 Telephone (403) 273-5152 Facsimile (403) 235-6864	A full range of liners	
Donovan Enterprises Incorporated 2951 S.E. Dominica Terrace Stuart, FL 34997 U.S.A. Telephone (407) 286-3350 Toll-free (800) 327-8287 Facsimile (407) 287-0431	Liners made of: Polyvinyl chloride (PVC) Oil-resistant PVC (PVC-OR) Reinforced PVC Reinforced Hypalon	Non-UV-stabilized Resistant to most oils UV-stabilized Hypalon is a trade name of Dupont for chlorosulfonated polyethylene (CSPE)
Empire Buff Limited 1425 Tellier Street St. Vincent de Paul, Quebec H7C 2H1 Telephone (514) 664-1200 Facsimile (514) 664-1427	Fabrico environmental liners	A Canadian distributor of Fabrico environmental liners

Manufacturer	Product(s)	Comments
Environmental Protection Inc. 9939 US-131 South NE P.O. Box 333 Mancelona, MI 49659-0333 U.S.A. Telephone (616) 587-9108 Toll-free (800) 345-4637 Facsimile (616) 587-8020	UltraTech Polyvinyl Chloride (PVC) liners	Specifically designed for use when the liner will be exposed
Fabrico Environmental Liners Inc. 4222 S. Pulaski Road Chicago, IL 60632 U.S.A. Telephone (312) 890-5350 Toll-free (800) 621-8546 Facsimile (312) 890-4669	Liners made of: FH 618 Duralam RS 515 EPA 30+ 3134 AB 0322	 A thermoplastic polyester elastomer Similar to FH 618 but reinforced Urethane-coated polyester A composite liner of Dupont Dacron fibre and Elvaloy resin Oil-resistant PVC (PVC-OR) Polyester urethane
Fairplast Limited 40 King Street Montreal, Quebec H3C 2N9 Telephone (514) 871-8586 Facsimile (514) 871-8588	Fabrico environmental liners	A Canadian distributor of Fabrico environmental liners
Fred Cressman Sales Incorporated 264 Sunview Street Waterloo, Ontario N2L 3V9 Telephone (519) 884-3225 Facsimile (519) 884-1326	Fabrico environmental liners	A Canadian distributor of Fabrico environmental liners
Geogard Lining Incorporated 745 Main Street East Milton, Ontario L9T 3Z3 Telephone (416) 875-3200 Facsimile (416) 875-3211	Liners made of: Urethane Low density polyethylene (LDPE) High density polyethylene (HDPE) Hypalon Polychloroprene (CR) Chlorinated polyethylene (CPE) Polyvinyl chloride (PVC)	 Urethane liners are available in sheet form or as spray-on coatings Hypalon is a trade name of Dupont for chlorosulfonated polyethylene (CSPE) Polychloroprene (CR) is also commonly referred to as Neoprene
GSE Lining Technology Inc. 19103 Gundle Road Houston, TX 77073 U.S.A. Telephone (713) 350-1813 Toll-free (800) 435-2008 Facsimile (713) 875-6010	HyperFlex Gundline HD Gundline HDW Gundline VLT Gundseal UltraFlex	High-density polyethylene (HDPE) HDPE White-surfaced HDPE Very low density polyethylene liner HDPE/bentonite composite liner Unspecified material

3.3 Temporary Storage - Liners 179

Manufacturer	Product(s)	Comments
H.Q.N. Industrial Fabrics 760 Chester Street Sarnia, Ontario N7S 5N1 Telephone (519) 344-9050 Facsimile (519) 344-5511	Liners made of: High-density polyethylene (HDPE) Low-density polyethylene (LDPE) Polyvinyl chloride (PVC) Oil-resistant PVC (PVC-OR) Chlorinated polyethylene (CPE) Hypalon Ethylene-Propylene Diene Monomer (EPDM) Polychloroprene(CR) XR-5 and FuelTane Hytrel	See No. 94 in this section Hypalon is a trade name of Dupont for chlorosulfonated polyethylene (CSPE) Polychloroprene (CR) is also commonly referred to as Neoprene XR-5 and FuelTane are products of Seaman Corporation Hytrel is a Dupont hydrocarbon containment barrier
Harlock Industrial Incorporated 31 Passmore Avenue, Unit 9 Scarborough, Ontario M1V 4T8 Telephone (416) 291-4390 Facsimile (416) 291-4791	Fabrico environmental liners	A Canadian distributor of Fabrico environmental liners
HMT Canada 5614-A Burbank Road, SE Calgary, Alberta T2H 1Z4 Telephone (403) 259-3330 Toll-free (800) 461-3120 Facsimile (403) 259-2814	Fabrico environmental liners	A Canadian distributor of Fabrico environmental liners
Industrial Plastics Canada Limited P.O. Box 93, 625 Industrial Drive Fort Erie, Ontario L2A 5M6 Telephone (905) 871-0412 Facsimile (905) 871-6494	Fabrico environmental liners	A Canadian distributor of Fabrico environmental liners
MK Plastics 57 Queens Street Montreal, Quebec H3C 2N6 Telephone (514) 871-9999 Facsimile (514) 871-1753	Fabrico environmental liners	A Canadian distributor of Fabrico liners

Manufacturer	Product(s)	Comments
MPC Containment Systems Limited 4834 South Oakley Chicago, IL 60609 U.S.A. Telephone (312) 927-4120 Facsimile (312) 650-6028	Hytrel and Petrogard liners	Hytrel is a Dupont hydrocarbon containment barrier Petrogard liners are all reinforced and offer various chemical compatibility coatings
Montgomery Environmental 290 Mitchell Avenue Oshawa, Ontario L1H 2V9 Telephone (905) 432-2531 Toll-free (800) 668-7523 Facsimile (905) 404-0791	Permalon containment liners	Alloyed high-density polyethylene A Canadian distributor of Permalon containment liners for Reef Industries (listed below)
Packaging Research and Design Corp. P.O. Box 678 Madison, MS 39130-0678 U.S.A. Telephone (601) 856-9791 Toll-free (800) 833-9364 Facsimile (601) 853-1202	Heavy-duty polyethylene liners	See No. 92 in this section
Poly-Flex, Incorporated 2000 West Marshall Drive Grand Prairie, TX 75051 U.S.A. Telephone (214) 647-4374 Facsimile (214) 988-8331	Poly-Flex Dura-Flex	High-density polyethylene (HDPE) Very low-density polyethylene (VLDPE)
Reef Industries Incorporated P.O. Box 750245 Houston, TX 77275-0245 U.S.A. Telephone (713) 507-4200 Toll-free (800) 847-5616 Facsimile (713) 507-4295	Permalon containment liners	Alloyed high-density polyethylene (HDPE) Canadian distributor listed above See No. 93 in this section
RNG Equipment Inc. 1318 Ketch Court Coquitlam, British Columbia V3K 6W1 Telephone (604) 521-2088 Facsimile (604) 521-3498	MPC containment liners	Canadian distributor for MPC Containment Systems Limited

3.3 Temporary Storage - Liners

Manufacturer	Product(s)	Comments
RNG Equipment Inc. 8824 - 53 Avenue N.W. Edmonton, Alberta T6E 5G2 Telephone (403) 466-2171 Facsimile (403) 465-3064	MPC containment liners	Canadian distributor for MPC Containment Systems Limited
RNG Equipment Inc. 517 - 44th Street East Saskatoon, Saskatchewan S7K 0V9 Telephone (306) 655-0223 Facsimile (306) 665-1214	MPC containment liners	Canadian distributor for MPC Containment Systems Limited
RNG Equipment Inc. 187 - 189 Eagle Drive Winnipeg, Manitoba R2R 1V4 Telephone (204) 633-2502 Facsimile (204) 694-2375	MPC containment liners	Canadian distributor for MPC Containment Systems Limited
RNG Equipment Inc. 32 Stoffel Drive Rexdale, Ontario M9W 1A8 Telephone (905) 564-2422 Facsimile (905) 247-2973	MPC containment liners	Canadian distributor for MPC Containment Systems Limited
RNG Germain Inc. 305 Boul. Wilfrid Hamel Quebec City, Quebec G1M 2R7 Telephone (418) 683-2781 Facsimile (418) 683-8582	MPC containment liners	Canadian distributor for MPC Containment Systems Limited
RNG Germain Inc. 10220 L.H. Lafontaine Ville d'Anjou, Quebec H1J 2T3 Telephone (514) 355-3111 Facsimile (514) 355-3940	MPC containment liners	Canadian distributor for MPC Containment Systems Limited
RNG Equipment Inc. 101 Thornhill Drive, Suite 103 Dartmouth, Nova Scotia B3B 1S2 Telephone (902) 468-7342 Facsimile (902) 468-7341	MPC containment liners	Canadian distributor for MPC Containment Systems Limited

Manufacturer	Product(s)	Comments
RNG Equipment Inc. 48 Hathaway Crescent, Unit 4 P.O. Box 3068, Station B Saint John, New Brunswick E2M 4X7 Telephone (506) 635-1933 Facsimile (506) 635-8556	MPC containment liners	Canadian distributor for MPC Containment Systems Limited
RNG Equipment Inc. 97 Clyde Avenue Donovan's Industrial Park P.O. Box 156 Mount Pearl, Newfoundland A1N 4R9 Telephone (709) 747-0015 Facsimile (709) 747-0222	MPC containment liners	Canadian distributor for MPC Containment Systems Limited
SAI Engineering Sales Limited 2699 - 96 Street Edmonton, Alberta T6N 1C3 Telephone (403) 463-9000 Toll-free (800) 272-9615 Facsimile (403) 463-9402	Fabrico environmental liners	A Canadian distributor of Fabrico environmental liners
Seaman Corporation 9111 Cross Park Drive, Suite D-200 Knoxville, TN 37923 U.S.A. Telephone (615) 691-9476 Facsimile (615) 539-8294	8130 XR-5 3024 XR-5B 1732 and 1936 Polyurethane	8130 XR-5 is made with Dupont Dacron polyester fibres 3024 XR-5B is a nylon fabric-based membrane with a modified XR-5 coating Nylon fabric-based membranes with a urethane coating. 1936 is thicker and stronger than 1732
Sopers Engineered Fabric Products P.O. Box 277 144 Chatham Street Hamilton, Ontario L8P 2B6 Telephone (905) 528-7936 Facsimile (905) 528-8128	8130 XR-5 3024 XR-5B 1732 Polyurethane 1936 Polyurethane	A Canadian distributor for Seaman Corporation
Staff Industries, Incorporated 240 Chene Street Detroit, MI 48207-4494 U.S.A. Telephone (313) 259-1818 Toll-free (800) 526-1386 Facsimile (313) 259-0631	Polyvinyl chloride (PVC) Hypalon Chlorinated polyethylene (CPE) Elvaloy alloy	Hypalon is a trade name of Dupont for chlorosulfonated polyethylene (CSPE) Elvaloy alloy is also a trade name of Dupont. Elavaloy alloy is elasticized polyolefin

OTHER DATA

Consult local telephone directories or industrial trade registers for other manufacturers and local distributors.

REFERENCES

(1) Brown, K.W. and J.C. Thomas, "New Technology for Liners", in: *Proceedings of a Conference on the Prevention and Treatment of Groundwater and Soil Contamination in Petroleum Exploration and Production*, Info-Tech, Calgary, Alta. (May,1989).

(2) Giroud, J.P., "Geotextiles and Geomembranes, Definitions, Properties and Design", Industrial Fabrics Association International, St. Paul, MN (1984).

(3) Haxo, H.E., R.S. Haxo, N.A. Nelson, P.D. Haxo, R.M. White, S. Dakessian, and M.A. Fong, *Liner Materials for Hazardous and Toxic Wastes and Municipal Solid Waste Leachate*, Noyes Publications, Park Ridge, NJ (1985).

(4) McCallum, W.G., "Secondary Containment Liners - An Overview", in: *Proceedings of the Sixth Technical Seminar on Chemical Spills,* Environment Canada, Ottawa, Ont. (1989).

CDF DRUM LINERS

No. 91

DESCRIPTION

Disposable, low-density polyethylene liners for standard-sized pails and drums.

OPERATING PRINCIPLE

The liners are placed inside the pail or drum and reduce the risk of leakage, reduce washout costs, and increase container longevity.

STATUS OF DEVELOPMENT AND USAGE

Commercially available product.

TECHNICAL SPECIFICATIONS

Physical

Constructed of low-density polyethylene, the liners are made for standard-sized drums and pails. The liners are also available in many configurations to suit the specific needs of the users.

Operational

The liners can be easily installed by one person. No special tools are required.

Power/Fuel Requirements

None.

Transportability

Person-portable.

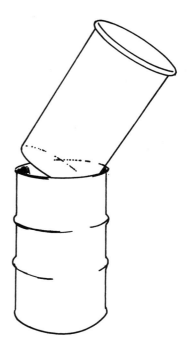

CONTACT INFORMATION

Manufacturer and Distributor

CDF Corporation
77 Industrial Park Road
Plymouth, MA 02360
U.S.A.

Telephone (508) 747-5858
Toll-free (800) 443-1920
Facsimile (508) 747-6333

REFERENCES

(1) Manufacturer's Literature.

HAZARDOUS WASTE CONTAINER BAG LINERS No. 92

DESCRIPTION

Disposable, heavy-duty, polyethylene liners for lugger buckets, roll-offs, dump trucks, and rail gondolas.

OPERATING PRINCIPLE

Liners are placed inside the waste container to reduce the risk of leakage, reduce washout costs, increase life of container, and ensure shorter turnaround time at the dump site.

STATUS OF DEVELOPMENT AND USAGE

Commercially available product.

TECHNICAL SPECIFICATIONS

Physical - Three bag liners are available: High Performance Bag Liners, lower cost "E" Bag Liners, and lowest cost "IA" Bag Liners. Liners are manufactured in 3.5, 4, 5, 6, 8, and 10 mil thicknesses. The bottom of each liner is 2.3 or 2.4 m (7.7 to 8 ft) wide and sides range from 1.7 to 3.7 m (5.5 to 12 ft) deep. The form-fit design ensures that liners fit squarely into the corners of the container to prevent tearing and provide triple-thickness leak protection at the bottom of the container's tailgate.

High Performance Bag Liners are made from a blend of prime virgin resins in full gauge 6, 8, and 10 mil. **"E" and "EB" Bag Liners** have the same design as the High Performance Bag Liners, but are made of a combination of virgin resins and recycled low-density polyethylene in thicknesses of 4, 6, 8, and 10 mil. "EB" Bag Liners are made the same as the "E" Bag Liners except they are black in colour. **"IA" Bag Liners** are made from the same material as the "E" Bag Liner, but are black in colour and run on "Industry Average" mil thickness standards, which means that the 4-mil liners average 3.5 mil, and the 6-mil liners average 5 mil.

Operational - The liners provide added security and safety when transporting or storing bulk solid and sludges. The liners can be easily installed by one person with no special tools required. The **High Performance Bag Liners** are extra heavy-duty polyethylene bag liners with the toughness, dart impact, and tensile strength to make them highly tear-resistant, puncture-resistant, and chemical-resistant. The **"E" and "EB" Bag Liners** provide about 85 to 90% of the strength of the High Performance Bag Liners. "EB" bag liners are the same as "E" except they are black in colour. **"IA" Bag Liners** have the same strength characteristics as the "E" Liners. The black colouring may result in heating of products stored within the liner. Large tubular bags, drum liners, bags for front loaders, small dump hopper bags, and tapered spout bags for use with asbestos vacuum systems are also available.

Transportability

Person-portable.

CONTACT INFORMATION

Manufacturer and Distributor

Packaging Research and Design Corporation
P.O. Box 678
Madison, MS 39130-0678, U.S.A.

Telephone (601) 856-9791
Toll-free (800) 833-9364
Facsimile (601) 853-1202

REFERENCES

(1) Manufacturer's Literature.

PERMALON DRUM LINERS

DESCRIPTION

Disposable, multi-layered, alloyed high-density polyethylene (HDPE) membrane liners for standard-sized drums.

OPERATING PRINCIPLE

The liners are placed inside the drum and reduce the risk of leakage, reduce washout costs, increase container longevity, and ensure shorter turnaround time at the dump site. Full liners are easily removed with a forklift for disposal.

STATUS OF DEVELOPMENT AND USAGE

Commercially available product.

TECHNICAL SPECIFICATIONS

Physical

Constructed of multi-layered, alloyed, high-density polyethylene membrane, these liners are made for standard-sized drums.

Operational

The liners are puncture- and abrasion-resistant with reinforced seams for added security and safety when transporting or storing liquids, solids, and sludges. The liners can be easily installed by one person with no special tools required.

Power/Fuel Requirements

None.

Transportability

Person-portable.

CONTACT INFORMATION

Manufacturer and Distributor

Reef Industries Incorporated
P.O. Box 750245
Houston, TX 77275-0245
U.S.A.

Telephone (713) 507-4200
Toll-free   (800) 847-5616
Facsimile (713) 507-4295

REFERENCES

(1) Manufacturer's Literature.

WASTE CONTAINER BAG LINERS

No. 94

DESCRIPTION

Disposable, woven, coated polyethylene liners for lugger buckets, roll-offs, dump trucks, and rail gondolas.

OPERATING PRINCIPLE

The liners are placed inside the waste container and reduce the risk of leakage as well as washout costs, increase the lifespan of the container, ensure shorter turnaround time at the dump site, and eliminate freeze-up problems.

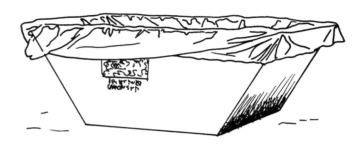

STATUS OF DEVELOPMENT AND USAGE

Commercially available product.

TECHNICAL SPECIFICATIONS

Physical

Constructed of woven, coated polyethylene, the liners are custom-fitted and available in stock sizes for most containers in use today. Each liner has a 6 mm (0.25 in) diameter poly rope drawstring incorporated into the material around the top edge.

Operational

The liners provide added security and safety when transporting or storing bulk solid and sludges. The 6 mm (0.25 in) diameter poly rope drawstring holds the liner in place when it is empty in the container, and then can be cinched up to seal off the top of the liner after the container is loaded. Liners are easily installed by one person, with no special tools required.

Power/Fuel Requirements

None.

Transportability

Person-portable.

PERFORMANCE

These liners have been used successfully for several years at waste sites by companies such as Laidlaw, BFI, WMX, Philip Environmental, and Sanifill.

CONTACT INFORMATION

Manufacturer and Distributor

H.Q.N. Industrial Fabrics
760 Chester Street
Sarnia, Ontario
N7S 5N1

Telephone (519) 344-9050
Facsimile   (519) 344-5511

REFERENCES

(1) Manufacturer's Literature.

Section 4.1

Transfer - Transfer of Liquids

4.1 Transfer - Transfer of Liquids

PUMPS - GENERAL LISTING   No. 95

DESCRIPTION

Pumps are machines used to raise and/or transfer liquids and sludges. See also the **General Listing on Vacuum Pumps for Chemicals** (No. 104) in this publication as well as the following entries in this section.

  Amflow-Lift Peristaltic Hose Pump (formerly Mastr-Pump) (No. 96)
  Flygt Submersible Pumps (No. 97)
  Gilkes Portable Emergency Turbo Pump (No. 98)
  Lutz Drum Pumps (No. 99)
  Megator Sliding-Shoe Pump (No. 100)
  Sala Roll Pump (No. 101)
  Solar-powered Pumping Systems (No. 102)
  Warren Rupp SandPIPER Pumps (No. 103)

OPERATING PRINCIPLE

Pumps are classified according to two basic operating principles: **centrifugal** and **positive displacement**. Centrifugal pumps use a rapidly spinning impeller to accelerate fluid through the pump. Positive displacement pumps use a device such as a diaphragm or a piston to impart mechanical energy directly to the fluid and literally "push" it along its way. For further details, see the figure, "Classification of Pumps by Operating Principle" in this section.

STATUS OF DEVELOPMENT AND USAGE

Commercially available products in general use. Except for electric motors, pumps are the most widely used machines in the world.

TECHNICAL SPECIFICATIONS

Physical

Pumps are available in nearly unlimited configurations. Variations include different operating principles, types of drives, power sources, hose connections, and construction materials. Physical specifications for individual pumps are available from the manufacturer. See also the specifications for selected pumps listed in the entries in this section.

Operational

The work that a pump must do to deliver fluid is expressed as the head and is generally measured in metres or feet, but may also be measured in kPa or psi. The following are other terms commonly used when discussing pumps or calculating head.

Suction lift - The lift required when the source of power supply is below the centre line of the pump.
Static suction lift - The vertical distance from the centre line of the pump to the free level of the liquid to be pumped.
Total suction lift - The static suction lift plus velocity head and all frictional losses in the suction line (friction head).
Suction head - The force exerted at the inlet to the pump when the source of supply is above the centre line of the pump, or when the liquid to be pumped flows to the pump under pressure.
Discharge head - The force exerted at the discharge side of the pump when the point of discharge or the free level of liquid at the point of discharge is above the centre line of the pump.
Static suction head - The vertical distance from the centre line of the pump to the free level of the liquid above the pump.
Static discharge head - The vertical distance from the centre line of the pump to the point of discharge, or to the free level of liquid at the point of discharge.
Friction head - The resistance to flow caused by the hose and fittings.
Total suction head - Static suction head less velocity head and all friction losses in the suction line.
Total discharge head - Static discharge head plus velocity head and all friction losses in the discharge line.
Total static head - The vertical distance between the free level of the source of power and the point of discharge, or to the level of the free surface of discharge liquid.
Total head - The total static head plus velocity head and all frictional losses (friction head).
Velocity head - The equivalent head in metres or feet through which the liquid would have to fall to acquire the same velocity.
Centrifugal pumps, which use a rapidly spinning impeller to accelerate fluid through the pump, are designed to be more efficient at "pushing" fluid than at "pulling" it. These pumps are therefore best used for applications with low suction lift.

Positive displacement pumps, which use a device to impart mechanical energy directly to the fluid to "push" it, generally have lower flow rates than centrifugal pumps, but are better for self-priming and for applications that require large suction lift, such as stripping a barge.

The intended use of the pump and its operational characteristics are important factors when selecting the right pump for a particular application. Features to consider include: suction lift capability; total discharge head; flow rates; flow pressures; capability to adjust or shut off flow while the pump is running; priming requirements; ability to run the pump dry without damage; temperature resistance; chemical compatibility; potential for vapour emissions from pump body/seals; drive type (some drive mechanisms cause the emulsification of products, which is often not desirable); the ability of the pump to handle viscous fluids, and/or trash and debris entrained within the fluid; and the power source of the pump (explosion-proof requirements).

Power/Fuel Requirements

Power sources for pumps include electricity, gasoline, diesel, compressed air, and high pressure water flow.

Transportability

The transportability of pumps varies greatly from unit to unit.

PERFORMANCE

Refer to No. 96 to 103 in this section or contact manufacturers for details on pump performance in chemical spill situations.

CONTACT INFORMATION

Because of the large number of manufacturers and suppliers of pumps, it is not possible to list all commercial contacts in this publication. Contact information for selected pumps is provided in the entries included in this section. Consult industrial registers and local telephone directories for other manufacturers and suppliers of pumps.

REFERENCES

(1) Manufacturer's Literature.

4.1 Transfer - Transfer of Liquids

Classification of Pumps by Operating Principle

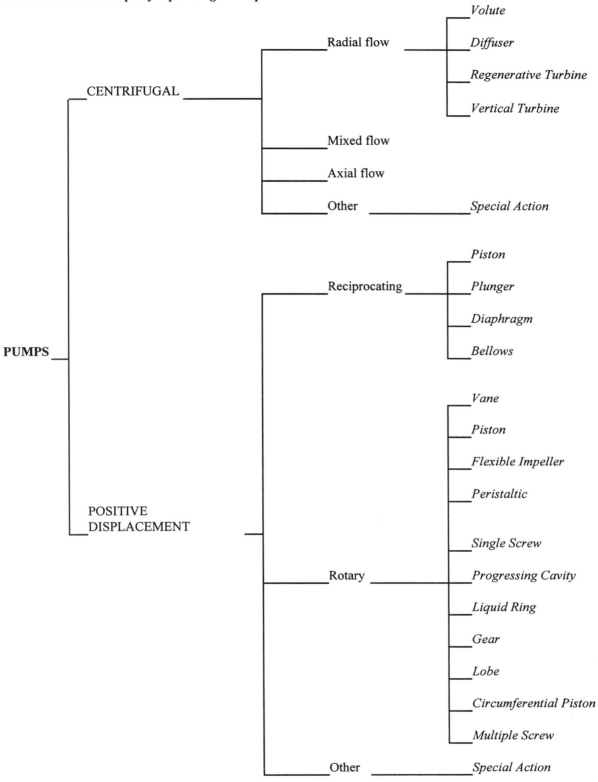

AMFLOW-LIFT
PERISTALTIC HOSE PUMP (formerly MASTR-PUMP)

No. 96

DESCRIPTION

A high vacuum peristaltic hose pump that transfers liquids, viscous slurries, and gaseous media. The term peristaltic refers to successive waves of contraction passing along the walls of a hollow structure which force the contents onwards. Material passes through the hose unrestricted and uncontaminated as it contacts only the inside of the hose.

OPERATING PRINCIPLE

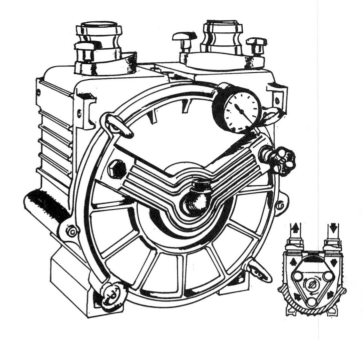

Though a true positive displacement design, the Amflow-Lift pump is difficult to fit into any of the general pump classifications. The pumping action is derived from three main components: a shaft-mounted rotor with three rollers, a stationary, internally toothed, rubber diaphragm/separator, and a smooth tube-type flexible hose, made of material compatible with the liquid to be pumped, which passes through the pump casing to connect directly to the inlet and outlet. The diaphragm/separator, which is positioned around and completely encloses the rollers, is screwed to the inside top of the pump casing and remains free to change shape in accordance with the rotor's motion. This movement ceates a vacuum on the suction side of the pump casing which remains totally sealed from the inner chamber of the pump. The pumped liquid, which is enclosed within the tube membrane, is positively displaced by the occlusive action of the rollers and passes straight through the pump without obstruction. The Amflow-Lift can be driven by a gasoline, diesel, electric, compressed air, or a hydraulic drive.

STATUS OF DEVELOPMENT AND USAGE

Commercially available product marketed for many applications including chemical spill response.

TECHNICAL SPECIFICATIONS

Physical

All pumps have 51 mm (2 in) hose connectors, which are available in aluminum, stainless steel, polypropylene, brass, bronze aluminum, and Teflon-coated aluminum. The peristaltic hose, housed inside the pump, is available in Polyurethane, Hypalon, natural rubber, Nitrile, Butyl, and Teflon. Several models of the Amflow-Lift pump are available with different drive types.

Model	Motor	Horsepower	Flow Rate	Weight
MBS	N/A	N/A	Up to 416 L/min (110 gpm)	35 kg (78 lb)
MG5	Honda Gasoline	4.1 kW (5.5 hp)	416 L/min (110 gpm)	59 kg (130 lb)
MG5	Briggs & Stratton Gasoline	3.7 kW (5 hp)	416 L/min (110 gpm)	59 kg (130 lb)
MA5	Gast Rotary Vane Air	3.7 kW (5 hp)	340 L/min (90 gpm)	54 kg (120 lb)
MD6	Yanmar Diesel	4.5 kW (6 hp)	416 L/min (110 gpm)	85 kg (190 lb)
MH5	Permco Hydraulic	3.7 kW (5 hp)	340 L/min (90 gpm)	54 kg (120 lb)
ME3-3 ODP (open, drip-proof)	Toshiba or Baldor Electric	2.2 kW (3 hp)	230 L/min (60 gpm)	57 kg (125 lb)
ME3-3 TEFC (totally enclosed fan)	Toshiba or Baldor Electric	2.2 kW (3 hp)	230 L/min (60 gpm)	61 kg (135 lb)
ME3-3 X-proof (explosion-proof)	Toshiba or Baldor Electric	2.2 kW (3 hp)	230 L/min (60 gpm)	68 kg (150 lb)

4.1 Transfer - Transfer of Liquids

Operational

This pump is an advance on the peristaltic principle and has been specially designed as a versatile, general purpose industrial pump that can be used with a wide variety of hazardous chemicals, solvents, corrosives, and otherwise toxic or chemically aggressive products. In fact, the range of liquids handled is limited only by the availability of compatible material for the flexible tube membrane, which is the only pump component in contact with the pumped media.

The Amflow-Lift has a gentle non-pulsating and non-emulsifying action, which is an important feature when delivering sludge to oil/water separators. It can pump up to 80% solids. Compressible solids can be handled up to the full bore diameter of the pump hose; noncompressible solids of up to approximately 12 mm (0.50 in) can also be handled. The pumps are dry, self-priming up to 8.8 m (29 ft) suction head and are able to run dry and perform well with intermittent air/liquid flow. With its dry running suction capability, the pump is particularly suitable for use with skimmers for removing floating oil and scum from the surface of tanks, ponds, lagoons, etc. Fitted with a suitable vacuum cleaner type of nozzle, these pumps have even been used to clear leaves, paper, and other litter from yards, loading bays, parking areas, and other similar areas. Maximum suction head is 9 m (30 ft) and maximum delivery head is 31 m (102 ft).

Power/Fuel Requirements

The Amflow-Lift is driven by gasoline, diesel, electric, compressed air drives, or a hydraulic motor. Electric drives are available as 115/230, 240/460 VAC, or 575/3/60 TEFC (totally enclosed fan) or explosion-proof model.

Transportability

Constructed almost entirely of aluminum, the pumps are extremely lightweight and easy to handle. All models are available as portable units with carrying handles.

PERFORMANCE

Maintenance is easy due to the pump's simple design without valves or seals. Most potential problems can be solved on site.

CONTACT INFORMATION

Manufacturer

AMFlow Master Environmental Inc.
9009 North Loop East
Suite 100
Houston, TX 77029
U.S.A.

Telephone (713) 675-5479
Facsimile (713) 675-6204

Canadian Distributors

R.M.S. Enviro Solv Incorporated
456 Alliance Avenue
Suite 102
Toronto, Ontario
M6N 2J2

Telephone (416) 766-7471
Facsimile (416) 766-7399

R.M.S. Enviro Solv Incorporated
2200 - 46th Avenue
Lachine, Quebec
H8T 2P3

Telephone (514) 631-3533
Facsimile (514) 631-8224

OTHER DATA

Skimmers, suction squeegees, and a two-way cleaning and suction nozzle designed for use with the Amflow-Lift are also available.

REFERENCES

(1) Manufacturer's Literature.

FLYGT SUBMERSIBLE PUMPS

No. 97

DESCRIPTION

A line of portable and permanent electric, dewatering, and process pumps for handling water containing corrosive chemicals and abrasives.

OPERATING PRINCIPLE

The Flygt pump is lowered into the material to be pumped and discharges to the surface through a discharge hose. No suction hose is required as the pump is submerged in the material to be pumped. This pump operates on the centrifugal principle, in which an impeller rotated inside the pump casing draws liquid into the centre (eye) of the impeller, and accelerates and displaces the flow outward along the impeller vanes. Tangential velocity head imparted to the fluid by the spinning impeller changes to a pressure head as the fluid flows along the inner casing (volute) and out the discharge port.

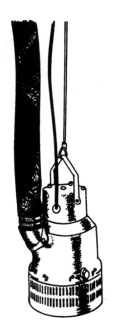

STATUS OF DEVELOPMENT AND USAGE

Commercially available for many pumping applications, including pumping water containing corrosive chemicals and abrasives.

TECHNICAL SPECIFICATIONS

Physical - A wide line of submersible pumps for various pumping applications, available in capacities up to 28,400 L/min (7,500 gpm) and with maximum heads of up to 95 m (310 ft). The pumps are available in discharge sizes from 3.8 to 25 cm (1.5 to 10 in) in diameter. The **B-, D-,** and **H-Pumps** are suitable for dewatering and drainage, the **C-Pumps** for pumping sewage, sludge, storm water, and process water, the **F-Pumps** for agricultural use, such as pumping manure, and the **P-Pumps** for handling large volumes of water in raw water supply, storm water, control of watercourse systems, and irrigation.

Special Performance Pumps are constructed of stainless steel or bronze and are used for pumping aggressive, corrosive, or warm liquids in industry. A modular system is also available in which certain standard cast iron parts can be used along with components in special materials depending upon the type of installation. Interchangeable parts include the separate cooling system, the volute, the motor cables, O-rings, shaft seals, and the impeller. These pumps include epoxy-coated models with sacrificial zinc anodes for cathodic protection and are available in B-, C-, and D-Pump configurations. Models B2060, C3060, C3126, D3060, and D3126 are made of stainless steel, models B2075, B2201, and D3080 are made of bronze, and models C3201 and C3300 are made of both stainless steel and bronze. The following are specifications for the Special Performance Pumps.

Model	Motor	Height	Width	Weight	Hose/PipeConnection
B2060	2.7 kW (3.6 hp) 3,300 rpm	56 cm (22 in)	24 cm (9 in)	32 kg (70 lb)	64 & 76 mm (2.5 & 3 in) hose
B2075	4.0 kW (5.5 hp) 3,400 rpm	56 cm (22 in)	42 cm (16 in)	75 kg (165 lb)	76 & 102 mm (3 & 4 in) hose
B2201	4.3 kW (5.8 hp) 3,500 rpm	120 cm (47 in)	62 cm (24 in)	415 kg (915 lb)	152 & 203 mm (6 & 8 in) hose
C3060	2.7 kW (3.6 hp) 3,300 rpm	not listed	not listed	41 kg (90 lb)	80 mm (3.1 in) flange
C3126	8.3 kW (11 hp) 1,750 rpm	82 cm (32 in)	36 cm (14 in)	195 kg (430 lb)	150 mm (6 in) flange
C3201	35 kW (47 hp) 1,750 rpm	132 cm (52 in)	105 cm (41 in)	590 kg (1,300 lb)	200 mm (8 in) flange
C3300	65 kW (87 hp) 1,770 rpm	165 cm (65 in)	105 cm (41 in)	1,130 kg (2,500 lb)	200 mm (8 in) flange
D3060	2.7 kW (3.6 hp) 3,300 rpm	60 cm (24 in)	27 cm (11 in)	41 kg (90 lb)	65 mm (2.6 in) flange
D3126	9.0 kW (12 hp) 3,450 rpm	not listed	not listed	165 kg (360 lb)	not listed
D3080	4.8 kW (6.5 hp) 1,700 rpm	74 cm (29 in)	33 cm (13 in)	110 kg (240 lb)	not listed

Operational - Flygt submersible pumps are portable, submersible, self-priming, frost-proof, silent, create no exhaust fumes, and can handle high percentages of solids. D-series pumps are available in explosion-proof configurations.

4.1 Transfer - Transfer of Liquids

The following are specifications for the **Special Performance Pumps**.

Model	Maximum Head	Maximum Flow
B2060	24 m (79 ft)	1,200 L/min (320 gpm)
B2075	48 m (157 ft)	1,440 L/min (380 gpm)
B2201	100 m (330 ft)	10,830 L/min (2,850 gpm)
C3060	28 m (92 ft)	960 L/min (250 gpm)
C3126	32 m (105 ft)	3,370 L/min (890 gpm)
C3201	50 m (160 ft)	13,500 L/min (3,560 gpm)
C3300	96 m (315 ft)	68,660 L/min (18,070 gpm)
D3060	14 m (46 ft)	600 L/min (160 gpm)
D3126	40 m (130 ft)	3,300 L/min (870 gpm)
D3080	45 m (150 ft)	1,690 L/min (440 gpm)

Power/Fuel Requirements and Transportability - The pumps are electric-powered with single- or three-phase motors rated from 2.7 to 9 kW (3.6 to 12 hp). Flygt pumps range from compact, lightweight, and person-portable units of 32 kg (70 lb) to pumps that weigh 1,130 kg (2,500 lb) and require lifting equipment to move.

PERFORMANCE

A chemical compatibility chart is available from the manufacturer for determining the right pump for a specific application.

CONTACT INFORMATION

Manufacturer

ITT Flygt, a Division of ITT Industries of Canada Ltd.
Pointe Claire, Quebec
H9R 4V5

Telephone (514) 695-0100
Facsimile (514) 697-0602

Canadian Distributors

ITT Flygt, A Division of ITT Industries of Canada Ltd.
St. John's, Newfoundland (709) 722-6717
Dartmouth, Nova Scotia (902) 450-1177
Moncton, New Brunswick (506) 857-2244
Pointe-Claire, Quebec (514) 695-0100
Beauport, Quebec (418) 667-1694
Val d'Or, Quebec (819) 825-0792

Etobicoke, Ontario (416) 675-3630
Nepean, Ontario (613) 225-9600
Sudbury, Ontario (705) 560-2141
Winnipeg, Manitoba (204) 235-0050
Saskatoon, Saskatchewan (306) 933-4849
Calgary, Alberta (403) 279-8371
Edmonton, Alberta (403) 489-1961
Coquitlam, British Columbia (604) 941-6664

REFERENCES

(1) Manufacturer's Literature.

GILKES PORTABLE EMERGENCY TURBO PUMP

DESCRIPTION

A lightweight, portable, flame-proof pump, powered by a turbo water motor, for handling volatile liquids.

OPERATING PRINCIPLE

The pump is powered by a reaction-type water motor (Francis principle) in which a circular impeller is rotated by the force of water entering the casing from an external pressurized water source, such as a fire truck or water main.

STATUS OF DEVELOPMENT AND USAGE

Commercially available product designed to safely and reliably transfer hazardous liquids in emergency situations as well as to meet the general pumping needs of the petro-chemical industry.

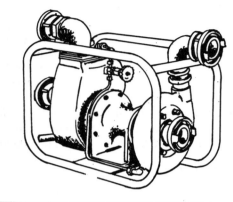

TECHNICAL SPECIFICATIONS

Physical - A self-priming, centrifugal pump built of cast iron or stainless steel, depending on liquid to be pumped. Both the pump inlet and discharge outlet have a 51 mm (2 in) flange. The motor is a 3 kW (4 hp) Francis water motor, with a bronze housing with stainless steel shaft and fittings for fresh or salt water operations. The water inlet is 32 mm (1.3 in) and the water outlet is 51 mm (2 in). The complete system is 51 cm (20 in) long, 27 cm (11 in) wide, 36 cm (14 in) high, and weighs 45 kg (100 lb).

Operational - This pump is ideal for pumping flammable liquids as no ignition source is needed. The pump is self-priming and can achieve a flow rate of up to 660 L/min (174 gpm) at 12 m (39 ft) head, when powered with a water supply of 348 L/min (92 gpm) at 965 kPa (140 psi). The pump's performance is affected by the water flow and pressure supplied to the pump's water drive.

Power/Fuel Requirements and Transportability - Pressurized water supplied at 400 to 1,170 kPa (60 to 170 psi) is required to run the motor. The pump is equipped with a carrying bar and is person-portable.

PERFORMANCE

In service for more than 10 years, this pump has been used successfully to recover liquid from road or rail accidents and spillage within containment berms, to remove aviation fuel from aircraft, and to safely transfer fire-fighting foam.

CONTACT INFORMATION

Manufacturer

Gilkes Incorporated
P.O. Box 628
Seabrook, TX 77586
U.S.A.

Telephone (713) 474-3016
Facsimile (713) 474 5304

Canadian Distributors

Co-Son Industries
3325 River Road
Gloucester, Ontario
K1G 3N3

Telephone (613) 521-0297
Facsimile (613) 526-5905

REFERENCES

(1) Manufacturer's Literature.

4.1 Transfer - Transfer of Liquids

LUTZ DRUM PUMPS

No. 99

DESCRIPTION

Centrifugal and progressive cavity pumps designed to transfer hazardous liquids from drums.

OPERATING PRINCIPLE

The pumps are powered by air or electric drives and lift liquids, by positive displacement or centrifugal force, through a long pump tube lowered into the liquid inside the drum. The pump motor is mounted on top of this tube and remains outside the drum.

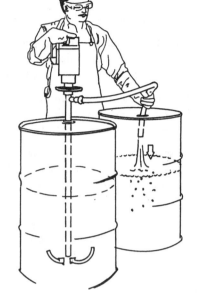

STATUS OF DEVELOPMENT AND USAGE

Commercially available product used for transferring corrosive and other hazardous liquids from drums, carboys, open tanks, and small reactors.

TECHNICAL SPECIFICATIONS

Physical - All components of Lutz drum pumps, such as motors and pump tubes, are completely interchangeable.

Motors

B55T - 115 VAC, 50/60 Hz, 1 amp, 510 Watts, totally enclosed, fan-cooled. Double insulated, thermal overload protected. Automatic reset. 5 m (16 ft) of three-wire power cord. Weighs 5.1 kg (11 lb).

B-40 - 115 VAC, 60 Hz, 1 amp, 620 Watts, U.L. listed Class I, Groups C&D. Class II, Group G, explosion-proof. Thermal overload protected, automatic reset. 5 m (16 ft) of three-wire power cord. Weighs 8 kg (18 lb).

4AL - 7,500 RPM @ 620 kPa (90 psi) and 8 L/s (17 cfm) pneumatic, 10 mm (3/8 in) air-line connection. Flame-proof muffler and filter. Suitable for hazardous duty. Weighs 1.8 kg (4 lb).

4-GT - 7,500 RPM @ 600 kPa (87 psi) and 6.6 L/s (14 cfm) pneumatic, 3 mm (1/8 in) air-line connection. Suitable for hazardous duty. Weighs 0.5 kg (1 lb).

Pump Tubes

MSL-PPH - Multipurpose, seal-less, polypropylene, centrifugal pump. Wetted parts include polypropylene, Hastelloy C shaft, polypropylene impeller and rotor, Teflon shaft sleeve, and high purity carbon stabilizer bushing with polypropylene bearing star. The pump tube is 38 mm (1.5 in) in diameter and the discharge hose connector is 19 mm (0.75 in). Maximum pumping temperature is 54°C (130°F).

MSL-SS - Multipurpose, seal-less, stainless steel, centrifugal pump. Wetted parts include stainless steel, stainless steel shaft, fibre-filled teflez impeller and rotor, Teflon shaft sleeve, high purity carbon stabilizer bushing with Teflon bearing star. The pump tube is 38 mm (1.5 in) in diameter and the discharge hose connector is 25 mm (1.0 in). Maximum pumping temperature is 93°C (200°F).

MSL-ALU - Multipurpose, seal-less, aluminum, centrifugal pump. Wetted parts include stainless steel, aluminum, stainless steel shaft, fibre-filled teflez impeller and rotor, Teflon shaft sleeve, high purity carbon stabilizer bushing with Teflon bearing star. The pump tube is 38 mm (1.5 in) in diameter and the discharge hose connector is 25 mm (1.0 in). Maximum pumping temperature is 88°C (190°F).

MMS-SS - Multipurpose, mechanical seal, stainless steel, centrifugal pump. Wetted parts include stainless steel, stainless steel shaft, fibre-filled tefzel impeller and rotor. The mechanical seal consists of a viton O-ring, ceramic seal, stainless steel spring. The pump tube is 38 mm (1.5 in) in diameter and the discharge hose connector is 25 mm (1.0 in). Maximum pumping temperature is 93°C (200°F).

MMS-PPH - Multipurpose, mechanical seal, polypropylene, centrifugal pump. Wetted parts include Hastelloy C shaft; polypropylene impeller and rotor. Mechanical seal consists of Viton O-rings, a ceramic seal, a Viton oil seal, Hastelloy C O-ring, and seal spring. Pump tube is 38 mm (1.5 in) and discharge hose connector is 25 mm (0.75 in). Maximum pumping temperature is 54°C (130°F).

MMS-ALU - Multipurpose, mechanical seal, aluminum, centrifugal pump. Wetted parts include stainless steel, aluminum, stainless steel shaft; fibre-filled teflez impeller and rotor. The mechanical seal consists of a Viton O-ring, ceramic seal, stainless steel spring. The pump tube is 38 mm (1.5 in) in diameter and the discharge hose connector is 25 mm (1.0 in). Maximum pumping temperature is 54°C (130°F).

Hastelloy C - Multipurpose, mechanical seal, centrifugal pump. Wetted parts include Hastelloy C and Teflon. The mechanical seal consists of a ceramic seal, Teflon O-ring, Viton oil seal, Hastelloy C seal spring. The pump tube is 38 mm (1.5 in) in diameter and the discharge hose connector is 19 mm (0.75 in). Maximum pumping temperature is 93°C (200°F).

SL-Kynar - Multipurpose, mechanical seal, centrifugal pump. Wetted parts include Kynar, Hastelloy C, fibre-filled tefzel impeller, Teflon shaft sleeve, and high purity carbon stabilizer bushing. The pump tube is 38 mm (1.5 in) in diameter and the discharge hose connector is 19 mm (0.75 in). Maximum pumping temperature is 82°C (180°F).

B-70 - A progressive cavity, positive displacement pump designed to transfer highly viscous fluids. Constructed of stainless steel with a Teflon stator and Buna-N stators also available. Available with either a mechanical seal or a stuffing box. The pump tube is 50 mm (2 in) in diameter and the discharge hose connector is 32 mm (1.25 in). Immersion depth is 1 m (39 in).

RE-88 - A stainless steel, centrifugal pump. A special pump foot prevents liquid in the pump from running back into the drum. All pump tubes have several tapered ridges on the hose connection to prevent slippage and leakage during operation and available in diameters of 69 cm (27 in) for carboys, 86 cm (34 in) for 208-L (55-gal) drums, and 119 cm (47 in) for vats.

Operational

Motors

Model	Maximum Flow Rate	Maximum Head	Maximum Viscosity	Maximum Specific Gravity
B55T	182 L/min (48 gpm)	16 m (55 ft)	750 cP	1.8
B-40	190 L/min (50 gpm)	18 m (59 ft)	750 cps	1.8
4AL	182 L/min (48 gpm)	16 m (55 ft)	320 cps	1.4
4-GT	133 L/min (35 gpm)	12 m (38 ft)	150 cps	1.4

Pump Tubes - When used with a 4AL motor, the B-70 transfers materials with viscosities up to 60,000 cP and is capable of flow rates of up to 21 L/min (5.5 gpm) with discharge pressures to 550 kPa (80 psi). The RE-88 is capable of removing all but 1.3 L (0.3 gal) from a standard upright 208-L (55-gal) drum, as opposed to the standard pump which leaves 4.7 L (1.2 gal). If the drum is tilted, almost 100% removal is possible, leaving less than 100 mL (3.4 oz) in the standard drum.

Power/Fuel Requirements and Transportability - 115 VAC, 50/60 Hz electric power or compressed air supplied at 620 kPa (90 psi) and 8 L/s (17 cfm) pneumatic, 10 mm (3/8 in) air-line connection, or 600 kPa (87 psi) and 6.6 L/s (14 cfm) pneumatic, 3 mm (1/8 in) air-line connection. The pumps are all lightweight and person-portable.

PERFORMANCE

A chemical compatibility chart is available from the manufacturer to select the right pump and tube for a specific use. Other accessories are available, such as mixing tubes, strainers, and barrel adaptors, to fix pumps within barrel bungs.

Manufacturer

Lutz Pumps, Incorporated
1160 Beaver Ruin Road
Norcross, GA 30093, U.S.A.

Telephone (404) 925-1222
Toll-free   (800) 843-3901
Facsimile (404) 923-0334

Canadian Distributors

James Morton Ltd.
180B Sheldon Drive
Cambridge, Ontario, N1R 5V5

Telephone (519) 621-7240
Facsimile   (519) 621-3442

REFERENCES

(1) Manufacturer's Literature.

MEGATOR SLIDING-SHOE PUMP

No. 100

DESCRIPTION

A positive displacement, "sliding shoe" pump capable of passing particulate and solids up to 5 mm (.2 in) in diameter.

OPERATING PRINCIPLE

The pumping mechanism is comprised of three or more "shoes", or plastic displacement chambers, each of which is reciprocated vertically by an eccentric rotor disk inside. Parts in the face of each shoe register alternately with the suction and discharge ports in the port plate, thus inducing a pumping action. Sliding shoe pumps are protected by Megator patents in the United States and elsewhere.

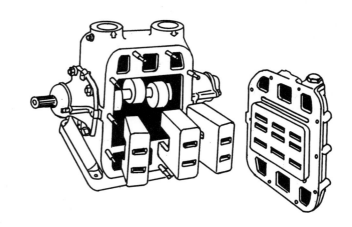

STATUS OF DEVELOPMENT AND USAGE

Originally designed for mine dewatering, the sliding shoe pump is now also marketed as part of the "Megator Recovery System" designed for use in all types of industry to recover clean or dirty oils of high or low viscosity, any mixture of oil and water, and any amount of air, without loss of prime.

TECHNICAL SPECIFICATIONS

Physical

Available in models with total heads up to 76 m (250 ft) and suction lifts up to 8.3 m (27 ft). Drive selection includes electric, gasoline, or diesel with V-belt or direct coupled arrangements. All models are 111 × 56 × 58 cm (44 × 22 × 23 in) in size, with cast iron bodies and stainless steel rotors and port plates. Standard shoes are made of phenolic material lined with synthetic rubber or optional urethane shoes. Pumps and drive units are mounted on a tubular frame with a flat steel base, belt guard, and two 30 cm (12 in) semi-pneumatic wheels.

Model	Hose Connections	Weight
L-100	25 mm (1 in)	107 kg (235 lb)
L-125	32 mm (1.2 in)	116 kg (255 lb)
L-150	38 mm (1.5 in)	150 kg (330 lb)

Operational

Megator sliding shoe pumps have high self-priming and suction powers and are able to run without damage under dry suction conditions. They are designed to handle clean or dirty oils of high or low viscosity and any mixture of oil and water, with any amount of air, without loss of prime. They are capable of passing particulate and solids up to 5 mm (.2 in) in diameter. Access through a single cover allows easy maintenance. Megator sliding shoe pumps are available with capacities ranging from 23 to 980 L/min (6 to 260 gpm) with total heads up to 76 m (250 ft) and suction lifts up to 8.3 m (27 ft).

Power/Fuel Requirements

The pumps are powered by electric, gasoline, or diesel drives. Electric drives require a 230/460 VAC, 60 Hz, 3-phase electrical supply.

Transportability

Models L-100, L-125, and L-150, which are supplied as part of the "Megator Recovery System", are all mounted on tubular frames with two 30 cm (12 in) semi-pneumatic wheels. These models weigh up to 150 kg (330 lb) and require more than one person to transport. The larger pumps are mounted on stationary bases and require lifting equipment to transport.

PERFORMANCE

Megator sliding shoe pumps are suitable for spill response because of their suction lift, cold tolerances, and ease of maintenance and handling. A chemical compatibility chart is available from the manufacturer to assist in determining the right pump for a specific application.

CONTACT INFORMATION

Manufacturer

Megator Corporation
562 Alpha Drive
Pittsburgh, PA 15238
U.S.A.

Telephone (412) 963-9200
Facsimile   (412) 963-9214

Canadian Distributors

Geostructure Instruments, Inc.
1410 Taschereau Blvd., B-200
La Prairie, Quebec
J5R 4E8

Telephone (514) 444-8420
Facsimile   (514) 444-8422

McIntyre Industries
12 Petersburg Circle, R.R. # 2
Port Colborne, Ontario
L3K 5V4

Telephone (905) 835-6761
Facsimile   (905) 835-6790

REFERENCES

(1) Manufacturer's Literature.

SALA ROLL PUMP

No. 101

DESCRIPTION

A wheel-mounted peristaltic hose pump that can be transported like a wheelbarrow. The term peristaltic refers to successive waves of contraction passing along the walls of a hollow structure which force the contents onwards. Material passes through a hose unrestricted and uncontaminated as it contacts only the inside of the hose. This pump is powered by a separate diesel-driven hydraulic power unit that can be remotely positioned if fire or explosion hazards are a concern.

OPERATING PRINCIPLE

The pump is driven by a hydraulic motor. Pumping and suction are developed by a rotating squeeze ring, which acts upon the main hose passing through the pump body. As the hose is squeezed, it pushes fluid ahead of the squeeze ring to the discharge and, as the squeeze ring passes a point on the hose, the hose rebounds to its former shape, creating a vacuum within that portion of hose behind the squeeze ring. This vacuum creates the suction action of the pump.

STATUS OF DEVELOPMENT AND USAGE

Developed primarily for use in cleaning up oil or chemical spills.

TECHNICAL SPECIFICATIONS

Physical - The pump is 0.9 m (36 in) high, 0.8 m (2.7 ft) wide, 1.9 m long (6 ft), and weighs 170 kg (375 lb). The hose is made of Nitrile/Chloroprene and has an outer diameter of 9.1 cm (3.6 in), an inner diameter of 5 cm (2 in), and is 3 m (9.8 ft) long.

Operational - The Sala Roll Pump can handle very high viscosity liquids while at the same time passing solids up to 30 mm (1.2 in) in size. The pump has no valves to obstruct the free flow of particles through the pump. The main hose is of thick double-ply construction with an outer cover of chloroprene and inner surface of nitrile rubber. Nitrile compounds are excellent for use with most oils and chemicals. The thick wall hose is capable of developing up to 95 kPa (14 psi) suction. If the pump is blocked, the drive can be reversed and the pump quickly cleared.

Power/Fuel Requirements and Transportability - The remote power supply unit is diesel-powered. The pump can be transported like a wheelbarrow over most terrain. In demanding terrain, a self-propelled, tracked vehicle is available to transport the pump and power pack, with the power unit towed behind on a sleigh.

PERFORMANCE

The Roll Pump was developed under guidelines from the TOBOS 85 Programme (Technique for Oil Combatting in Beach Zone and Off-Shore, managed by the Swedish Association of Local Authorities) and supported by the Swedish Board for Technical Development. It has been tested and approved for use with chemicals.

Manufacturer

Faltech AB
P.O. Box 66
S-79622 Alvdalen, Sweden

Telephone +46 251 511 95
Facsimile +46 251 511 75

REFERENCES

(1) Manufacturer's Literature.

SOLAR-POWERED PUMPING SYSTEMS   No. 102

DESCRIPTION

Solar-powered pumping systems that provide reliable and economical pumping at remote locations.

OPERATING PRINCIPLE

Photovoltaic panels convert sunlight into DC electricity that operates the solar pumps directly or through batteries charged by the solar panels.

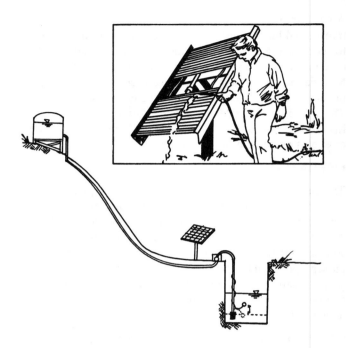

STATUS OF DEVELOPMENT AND USAGE

Commercially available product that provides reliable, clean water for many uses, including pumping water to livestock in vast grazing areas, to wildlife, farmsteads, remote homes, and for irrigation.

TECHNICAL SPECIFICATIONS

Physical - Custom pumping systems can be built for a specific application. The solar pump systems are complete, ready for field installation and operation. Pumps are available from 100 to 3,000 Watts in configurations ranging from high-volume, low-lift systems capable of delivering up to 3,785 L/min (1,000 gpm) to high-lift models capable of overcoming 150 m (500 ft) total head. The company also manufactures small 112-volt pumps suitable for groundwater sampling.

Operational - The solar-powered pumps are designed to operate unattended for several years and are useful for long-term pumping at remote sites. The only routine maintenance required is occasional removal of dust from the solar panels. The pump's components are easily replaced in the field and the solar panels have a typical lifespan of 20 years. The complete pumping system is designed individually for each project to ensure that all site variables are considered. Factors considered during design are the average daily pumping requirement, the total head, and the available hours of sunlight. A computer program with worldwide records of solar radiation data is used to design systems for specific locations.

Power/Fuel Requirements and Transportability - Solar-powered, with or without batteries, for storage of accumulated DC charge. The entire system is compact and lightweight and can be transported by highway or by air.

CONTACT INFORMATION

Manufacturer

CAP International Inc.
5037 - 50th Street
Suite 105
Olds, Alberta
T4H 1R8

Telephone (403) 556-8779
Facsimile (403) 556-7799

REFERENCES

(1) Jensen, E., "Photovoltaic Pumping: Environmentally-friendly Applications", *SOL Magazine*, Ottawa, Ont. (May-June, 1992).
(2) Manufacturer's Literature.

4.1 Transfer - Transfer of Liquids

WARREN RUPP SANDPIPER PUMPS   No. 103

DESCRIPTION

A line of air-powered, positive displacement, double diaphragm pumps that can handle highly abrasive and corrosive fluids, as well as high viscosity fluids.

OPERATING PRINCIPLE

Flexible diaphragms mounted in chambers on each side of the pump divide/seal the pumping section from the air-driving section. Compressed air introduced into the non-wetted side of the chamber moves the diaphragm outward on a discharge/pumping stroke, forcing the liquid out of the wetted side of the chamber. As one diaphragm pushes the fluid out, a rod connecting the two chambers pulls the other diaphragm on a suction stroke, filling the second chamber with fluid. At the end of each stroke, an air distribution valve automatically shifts, reversing the entire sequence, so that filling and pumping occur in alternate chambers. Suction and discharge check valves control the flow of liquid through the chambers and out the discharge port.

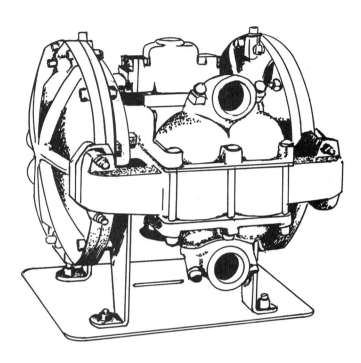

STATUS OF DEVELOPMENT AND USAGE

Commercially available for many pumping applications, including hazardous environments in which the use of air as a power source eliminates the risk of explosion and fire.

TECHNICAL SPECIFICATIONS

Physical - Four basic SandPIPER configurations are offered for different pumping needs. **S Series** pumps are available with ball or flap check valves. Bottom-discharge porting uses the natural settling effect of gravity to reduce or eliminate the buildup of solids in the product chambers and avoid damage to the diaphragm-connecting rod. These pumps are available with hose connector sizes of 25, 38, 51, 76, and 102 mm (1, 1.5, 2, 3, and 4 in). **E Series** are general purpose units with ball check valves without the costlier porting and valve options of the S Series units. This fixed porting design makes this a more competitively priced pump for handling fluids that may be viscous, contain entrained air or vapours, and small solids. These pumps are available with hose connector sizes of 13, 25, 38, 51, 76, and 102 mm (0.5, 1, 1.5, 2, 3, and 4 in). **P Series** pumps have ball check valves and are made of chemical-resistant polypropylene, with some models also available in nylon, PVDF, and PFA which offer increased structural strength and lightweight construction. These pumps are available with hose connector sizes of 13, 19, 25, 32, and 51 mm (0.5, 0.8, 1, 1.3, and 2 in).

Spill Containment Series for Hazardous/Toxic Liquids features pumps with spill containment chambers and additional diaphragms. If a pumping diaphragm fails, accidental spills are contained in the spill chamber and the pump can often complete its operation before repairs are done. Pumps and spill chambers are available in aluminum, cast iron, and stainless steel, with spill chambers also available in Teflon. Mechanical, visual, or electronic leak detection devices are available to alert the pump user if a pump diaphragm fails. Hose connectors are available in sizes from 25 to 75 mm (1 to 3 in).

Operational - SandPIPER pumps handle heavy abrasives and the low internal velocity is ideal for handling slurries. They can also handle high viscosities. Pumps equipped with flap valves can handle line-sized solids. As the pumps have no mechanical seals or packing, they are ideal for handling toxic, corrosive, and abrasive fluids. The pumps are self-priming and can be run dry without damage. Regulating the air supply provides variable flows from 0 to 985 L/min (0 to 260 gpm). Discharge pressure on all pumps is 860 kPa (125 psi). Closing the discharge stops the pump without damage. The pumps are explosion-proof as air-driven diaphragms eliminate sparks associated with rotating or electrical equipment. Models SA2-A and SA3-A have been tested with oils by Environment Canada (Environment Canada, 1978; 1979). A chemical compatibility chart is available from the manufacturer to assist in determining the right pump for a specific application.

The following are specifications for the **Spill Containment Series** of pumps.

Model	Inlet Size	Discharge Size	Flow Capacity
ET11/2-SM	38 mm (1.5 in)	38 mm (1.5 in)	0 to 246 L/min (0 to 65 gpm)
ET2-M	50 mm (2 in)	50 mm (2 in)	0 to 511 L/min (0 to 135 gpm)
ET3-SM	75 mm (3 in)	75 mm (3 in)	0 to 772 L/min (0 to 204 gpm)
ET3-M	75 mm (3 in)	75 mm (3 in)	0 to 984 L/min (0 to 260 gpm)
ET1-M	32 mm (1.3 in)	25 mm (1 in)	0 to 204 L/min (0 to 54 gpm)
ET11/2-M	50 mm (2 in)	38 mm (1.5 in)	0 to 466 L/min (0 to 123 gpm)
ST1-A	25 mm (1 in)	25 mm (1 in)	0 to 159 L/min (0 to 42 gpm)
ST11/2-A	38 mm (1.5 in)	38 mm (1.5 in)	0 to 340 L/min (0 to 90 gpm)

Power/Fuel Requirements and Transportability - The pumps are driven by compressed air. A **Hydraulic Powered Ball-Valve SandPIPER (Model SB11/2-H)** is designed for use when an adequate air supply is not available. Sandpiper pumps are compact, lightweight, and person-portable.

CONTACT INFORMATION

Manufacturer

Warren Rupp Incorporated
800 North Main Street, P.O. Box 1568
Mansfield, OH 44901
U.S.A.

Telephone (419) 524-8388
Facsimile (419) 522-7867

Canadian Distributors

Windsor Pump Company
3057 Marentette Avenue
Windsor, Ontario
N8X 4G1

Telephone (519) 969-2190
Facsimile (519) 969-2047

OTHER DATA

Warren Rupp also produces **Submersible Pumps,** and four models of **Waste Treatment Pumps**, as well as **Electronic Leak Detectors, Modular Air Preparation Systems, Drum Pump Adapter Kits**, and **Surge Suppressors** for the SandPIPER pumps. Free job-site demonstrations of SandPIPER pumps can be arranged. Authorized distributors are located worldwide. For the name of a local distributor, check the Yellow Pages of the telephone directory under "Warren Rupp SandPIPER Pumps", or contact the factory at (419) 524-8388.

Similar pumps are produced by Wilden Pump & Engineering Company.

Manufacturer

Wilden Pump & Engineering Company
22069 Van Buren Street
Colton, CA 92324
U.S.A.

Telephone (714) 422-1730
Facsimile (714) 783-3440

Canadian Distributors

Affiliated - Dynesco, a Division of Wajax Industries Limited
6644 Abrams
Ville-St-Laurent, Quebec
H4S 1Y1

Telephone (514) 333-9064
Facsimile (514) 333-9077

A complete line of non-metallic diaphragm pumps for corrosive fluid applications is also manufactured by Osmonics Incorporated. The pumps, manufactured under the name "American Pump Line", are made of pure virgin PTFE Teflon or polypropylene.

Manufacturer

Osmonics, Incorporated
5951 Clearwater Drive
Minnetonka, MN 55343, U.S.A.

Telephone (612) 933-2277
Facsimile (612) 933-0141

REFERENCES

(1) Environment Canada, "Pumps for Oil Spill Cleanup", EPS 4-EC-78-3, Ottawa, Ont. (1978).
(2) Environment Canada, "Evaluation of Pumps and Separators for Arctic Oil Spill Cleanup", Environmental Protection Series EPS 4-EC-79-3, Ottawa, Ont. (1979).
(3) Manufacturer's Literature.

Section 4.2

Transfer - Transfer of Gases

VACUUM PUMPS FOR CHEMICALS - GENERAL LISTING   No. 104

DESCRIPTION

Vacuum pumps are used to capture and/or transfer gases. These pumps often operate on the same principles as pumps used to transfer liquids (see Section 4.1, Pumps - General Listing, No. 95). The Busch Vacuum Pump (No. 105) and the Gast Vacuum Pump (No. 106) are examples of vacuum pumps listed in this survey. Some pumps used to transfer liquids are also used to transfer gases, such as the Amflow-Lift Peristaltic Hose Pump (No. 96) and the Sala Roll Pump (No. 101), listed in Section 4.1. In addition, vacuum pumps are used with the following soil vapour extraction units (listed in Section 2.4) in soil remediation projects to draw the contaminants from the soil. These units can also be used solely to transfer recovered gases.

Carbtrol Regenerative and Multi-Phase Extraction Systems	(No. 50)
EG&G Rotron Regenerative Blowers for Environmental Remediation	(No. 51)
Global Vacuum Extraction and Vapour Liquid Separator Modules	(No. 52)
Lamson Centrifugal Exhausters for Environmental Remediation	(No. 55)
M-D Rotary Positive Displacement Blowers	(No. 56)
NEPPCO SoilPurge Soil Vapour Treatment System	(No. 57)
Pego Soil Vapour Extraction Units	(No. 58)
RETOX (Regenerative Thermal Oxidizer)	(No. 59)
Roots Blowers and Vacuum Pumps for Environmental Remediation	(No. 60)

OPERATING PRINCIPLE

Pumps, including vacuum pumps, are classified according to two basic operating principles: **centrifugal** and **positive displacement**. Centrifugal pumps use a rapidly spinning impeller to accelerate material through the pump. Positive displacement pumps use a device such as a diaphragm or a piston to impart mechanical energy directly to the fluid and literally "push" it along. See also the figure "Classification of Pumps by Operating Principle" in Section 4.1.

STATUS OF DEVELOPMENT AND USAGE

Commercially available products in general use.

TECHNICAL SPECIFICATIONS

Physical - Vacuum pumps are available in many configurations. Variations include different operating principles, drive types, power sources, hose connections, and construction materials. Specifications for individual pumps are available from the manufacturer. See also the specifications for selected pumps listed in Entries No.105 and 106 of this section.

Operational - Refer to entries on specific pumps or contact the manufacturer for details on operating specifications.

Power/Fuel Requirements - Power sources for pumps include electricity, gasoline, diesel, compressed air, and high pressure water flow.

Transportability - The transportability of pumps varies from unit to unit.

PERFORMANCE

Refer to entries on specific pumps or contact the manufacturer for details on pump performance in chemical spill situations.

CONTACT INFORMATION

Because of the large number of manufacturers and suppliers of pumps, it is not possible to list all commercial contacts within this publication. Contact information is provided in the separate entries for selected vacuum pumps. Consult local telephone directories and industrial registers for other manufacturers and suppliers of pumps.

REFERENCES

(1) Manufacturer's Literature.

BUSCH VACUUM PUMPS FOR CHEMICALS

No. 105

DESCRIPTION

Cobra Dry Screw Pump - A pump with a unique screw rotor design for use in the chemical and pharmaceutical processing industries. The Cobra has a nitrogen-purging system with flow control devices for dilution of toxic, corrosive, and condensable process gases. The pump is single-stage, direct-coupled, and has no intercooler that requires extensive external piping, which minimizes downtime due to maintenance.

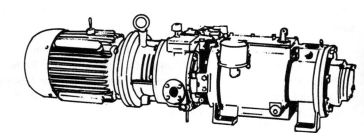

Rotary Vane Centrifugal Pump - A "once-through", oil-lubricated pump for critical industrial use and for pumping vapours and solvents in the chemical industry.

OPERATING PRINCIPLE

Cobra Dry Screw Pump - Two parallel screw rotors rotate in opposite directions in the cylinder at a speed of 3,600 rpm. Gases are trapped between the screw rotors and housing, transported by the rotation to the gas outlet, and discharged. As the distance between the outlet and the discharge is short and straight, agitation of the gas is minimal.

Rotary Vane Centrifugal Pump - This pump operates on the centrifugal principle using a rotary vane to create the vacuum suction.

STATUS OF DEVELOPMENT AND USAGE

Commercially available pumps used in a variety of applications requiring the transport of vapours. Specific uses for the Cobra pump include recovering solvents from drying, filtration, crystallization, and distillation processes, degasifying liquid streams, recycling process gases without adding contamination, vacuum separation of azeotropic mixtures, central vacuum systems for laboratories and pilot plants, evacuating gas cylinders before filling with high purity gas, any vacuum process requiring oil-free operation, and any other applications that require a contamination-free vacuum source.

TECHNICAL SPECIFICATIONS

Physical

Cobra Dry Screw Pump - Standard equipment includes explosion-proof motor, stage-cooling air inlet filters, cooling water inlet solenoid valve, cooling water flow control system, cooling water temperature switch and gauge, exhaust gas temperature switch, exhaust check valve, and steel base for pump and motor. Optional accessories are inlet particulate filter, combination knock-out pot/inlet filter, automatic start/stop package, automatic solvent flushing package, inert gas cooling system, exhaust silencer, junction box control panel, and aftercondenser. The following are specifications for the various models of Cobra pump.

Model	Height	Width	Length	Weight (with motor & base)
C 100	56 cm (22 in)	43 cm (17 in)	118 cm (47 in)	360 kg (88 lb)
C 200	65 cm (25 in)	76 cm (30 in)	127 cm (50 in)	540 kg (1,200 lb)
C 400	73 cm (29 in)	86 cm (34 in)	160 cm (63 in)	770 kg (1,700 lb)
C 800	81 cm (32 in)	69 cm (27 in)	175 cm (69 in)	1,088 kg (2,400 lb)
C 2500	87 cm (34 in)	210 cm (83 in)	177 cm (70 in)	1,800 kg (4,000 lb)

Rotary Vane Centrifugal Pump - Available in three models: the single-stage Monovac, the two-stage Huckepack, and the three-stage Lotos. These three models are available in a wide range of sizes and configurations to meet varying needs. Standard equipment includes explosion-proof motor, class 1; intrinsically safe instrumentation; cast iron, water-jacket

4.2 Transfer - Transfer of Gases

design; coolant temperature control system; exhaust muffler; non-metallic vanes; and an inlet flange on Huckepacks. The following are specifications for the Monovac and Huckepack models.

Model	Height	Width	Length	Shipping Weight
Monovac				
216	88 cm (35 in)	109 cm (43 in)	127 cm (50 in)	390 kg (860 lb)
225	94 cm (37 in)	109 cm (43 in)	130 cm (51 in)	410 kg (900 lb)
240	106 cm (42 in)	109 cm (43 in)	142 cm (56 in)	540 kg (1,200 lb)
263	112 cm (44 in)	109 cm (43 in)	147 cm (58 in)	710 kg (1,575 lb)
Huckepack				
429	79 cm (31 in)	69 cm (27 in)	125 cm (49 in)	430 kg (950 lb)
433	84 cm (33 in)	79 cm (31 in)	124 cm (48 in)	470 kg (1,000 lb)
437	106 cm (42 in)	84 cm (33 in)	125 cm (49 in)	635 kg (1,400 lb)
441	106 cm (42 in)	97 cm (38 in)	132 cm (52 in)	950 kg (2,100 lb)
445	145 cm (57 in)	119 cm (46 in)	157 cm (62 in)	1,680 kg (3,700 lb)

Operational

Cobra Dry Screw Pump - This innovative pump is built with state-of-the-art manufacturing techniques. Due to its simple design, it has fewer moving parts than most chemical-duty vacuum pumps which results in less maintenance and lower operating costs. In addition to being extremely durable, this pump is oil-free and features single-stage operation which requires no intercoolers. Other advantages of this pump include the high volumetric efficiency of the rotors, low vibration and noise levels, Teflon shaft seals that isolate bearings from process gas, and anti-corrosion coating on all parts in contact with gases. A nitrogen gas purging system for dilution of toxic, corrosive, and condensable process gas is also available. The following are operating specifications.

Model	Displacement	Vacuum	Power	Sound level
C100	27 L/s (57 cfm)	40 Pa (0.16 in H_2O)	3.8 kW (5 hp)	82 dBa
C200	57 L/s (120 cfm)	13 Pa (0.05 in H_2O)	5.6 kW (7.5 hp)	84 dBa
C400	120 L/s (260 cfm)	1 Pa (0.004 in H_2O)	15 kW (20 hp)	85 dBa
C800	230 L/s (490 cfm)	0.7 Pa (0.003 in H_2O)	23 kW (30 hp)	85 dBa
C2500	755 L/s (1,600 cfm)	10 Pa (0.04 in H_2O)	75 kW (100 hp)	88 dBa

Rotary Vane Centrifugal Pump - These pumps are built around Busch "Once-Through-Oiling (OTO)" vacuum pumps. With OTO, internal parts are continuously flushed with small amounts of fresh, clean oil. Oil passes through the pumps only once, eliminating the contaminated oil sump problems inherent with recirculating oil systems. OTO also keeps corrosive vapours away from pump surfaces so the pump can be built of cast iron and carbon steel instead of expensive alloys. Other process-compatible fluids may be used instead of oil, provided that they meet the required vapour pressure and lubrication properties for the application. Any gases handled by the pump are collected at the exhaust. The water-cooled Huckepack and Monovac use a water-jacket design that provides excellent temperature control while preventing cooling water from contacting process gases. This eliminates both the problems and cost of effluent treatment. As the operating temperature of the pump is accurately controlled, all corrosive vapours remain as gas during the pumping cycle and the pump will not corrode internally. The following are operating specifications for the Monovac and Huckepack models.

Model	Displacement	Vacuum	Power	Sound Level
Monovac				
216	57 L/s (120 cfm)	5 kPa (20 in H_2O)	7.5 kW (10 hp)	80 dB
225	90 L/s (190 cfm)	5 kPa (20 in H_2O)	11 kW (15 hp)	80 dB
240	146 L/s (310 cfm)	5 kPa (20 in H_2O)	15 kW (20 hp)	82 dB
263	210 L/s (440 cfm)	5 kPa (20 in H_2O)	19 kW (25 hp)	83 dB
Huckepack				
429	57 L/s (120 cfm)	50 Pa (0.02 in H_2O)	7.5 kW (10 hp)	81 dB
433	90 L/s (190 cfm)	50 Pa (0.02 in H_2O)	11 kW (15 hp)	81 dB
437	146 L/s (310 cfm)	50 Pa (0.02 in H_2O)	15 kW (20 hp)	82 dB
441	210 L/s (440 cfm)	50 Pa (0.02 in H_2O)	19 kW (25 hp)	84 dB
445	370 L/s (780 cfm)	50 Pa (0.02 in H_2O)	30 kW (40 hp)	85 dB

Power/Fuel Requirements

Busch vacuum pumps are powered by TEFC high efficiency electric motors.

Transportability

The pumps are heavy, ranging from 360 to 1,800 kg (800 to 4,000 lb), and require lifting equipment. Pumps are equipped with lifting lugs and mounted on a skid-type base.

PERFORMANCE

Performance in a chemical spill cleanup has not been documented.

CONTACT INFORMATION

Manufacturer and Canadian Distributor

Busch Vacuum Technics Inc. (Head Office)
1740 Lionel Bertrand
Boisbriand, Quebec
J7H 1N7

Telephone (514) 435-6899
Facsimile (514) 430-5132
Toll-free   (800) 363-6360

Busch Vacuum Technics Inc. (Regional Office)
10582 - 169th Street
Surrey, British Columbia
V4N 3H7

Telephone (604) 583-1927
Facsimile (604) 377-4473

Busch Vacuum Technics Inc. (Regional Office)
5004 Timberlea Blvd., Unit 5
Mississauga, Ontario
L4W 5C5

Telephone (905) 238-6090
Facsimile (905) 238-9195

Busch Vacuum Technics Inc. (Regional Office)
R.R. # 1
Souris, Prince Edward Island
C0A 2B0

Telephone (902) 969-3100
Facsimile (902) 969-3822

REFERENCES

(1) Manufacturer's Literature

GAST VACUUM PUMPS

No. 106

DESCRIPTION

A complete line of rotary vane pumps, reciprocating pumps, and regenerative blowers for transferring gases.

OPERATING PRINCIPLE

A variety of pumps featuring different operating principles. Centrifugal pumps include rotary vane pumps and regenerative blowers. Positive displacement pumps include diaphragm, piston, and rocking piston pumps.

STATUS OF DEVELOPMENT AND USAGE

Commercially available pumps used in a variety of applications requiring the transport of vapours.

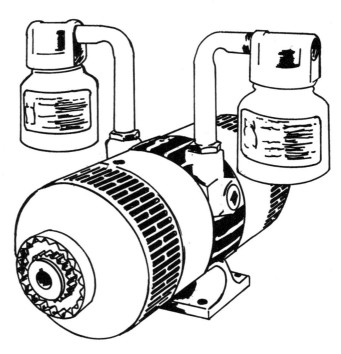

TECHNICAL SPECIFICATIONS

Physical

Available in a complete range of operating principles, drive types, and construction materials.

Operational

Gast Manufacturing offers a complete line of fractional horsepower vacuum pumps. Each type of vacuum pump and compressor has its own performance range, maintenance requirements, and resistance to ambient conditions. The following is a sampling of specifications for the various types of Gast vacuum pumps.

Type	Possible Applications	Size	Max Pressure/Vacuum Range	Displacements
Regenerative Blower	Supplied breathing air Pond and lake reclamation Soil reclamation Aerobic sewage treatment	370 W to 11 kW (0.5 to 15 hp)	40/30 kPa (6 psi/10 in Hg)	10 to 230 L/s (22 to 480 cfm)
Rotary Vane	Supplied breathing air Pond and lake reclamation Aerobic sewage treatment Portable air sampler Waste oil heater	2.2 W to 3.7 kW (0.03 to 5 hp)	170/100 kPa (25 psi/30 in Hg)	.28 to 26 L/s (0.6 to 55 cfm)
Piston	Pond and lake reclamation Deep well testing Sewage pump station Stack gas sampler	130 W to 1.5 kW (0.2 to 2 hp)	860/100 kPa (125 psi/29 in Hg)	.47 to 5 L/s (1 to 11 cfm)
Diaphragm	Air sampler Stack gas sampler	130 to 370 W (0.2 to 0.5 hp)	410/100 kPa (60 psi/29 in Hg)	.24 to 1.4 L/s (0.5 to 3 cfm)

Power/Fuel Requirements

Gast vacuum pumps are powered by electric motors.

Transportability

The transportability of pumps varies from unit to unit.

PERFORMANCE

Performance in a chemical spill cleanup has not been documented.

CONTACT INFORMATION

Manufacturer
Gast Manufacturing Corporation
P.O. Box 97
Benton Harbour, MI 49023-0097
U.S.A.

Telephone (616) 926-0006
Facsimile (616) 925- 8288

Canadian Distributors
Wainbee Limited

Calgary, Alberta	(403) 236-1133
Edmonton, Alberta	(403) 434-9528
Dartmouth, Nova Scotia	(902) 468-1787
Kitchener, Ontario	(519) 748-5391
London, Ontario	(519) 681-6266
Montreal, Quebec	(514) 697-8810
North Bay, Ontario	(800) 461-9534
Ottawa, Ontario	(613) 744-1720
Quebec, Quebec	(418) 683-1956
Rexdale, Ontario	(416) 243-1900
Richmond, B.C.	(604) 278-4288
Winnipeg, Manitoba	(204) 632-4558

REFERENCES

(1) Manufacturer's Literature.

Section 4.3

Transfer - Transfer Hoses

ACID, CHEMICAL TRANSFER HOSES - GENERAL LISTING No. 107

DESCRIPTION

Acid, chemical transfer hoses are made of natural or synthetic materials and are designed to withstand specific chemicals and acids, including petroleum products, dry material, and vapours. More transfer hoses suitable for use with chemicals are available than are listed in this section. This general listing simply provides contact information on some Canadian manufacturers. See the following entries in this section for information on specific products.

Flexaust Hose and Duct	(No. 108)
Gates Industrial, Acid-Chemical Hoses	(No. 109)
Goodall Chemical Hoses	(No. 110)

OPERATING PRINCIPLE

Hoses are used in conjunction with pumps either to pick up spilled materials or to transfer substances already collected.

STATUS OF DEVELOPMENT AND USAGE

Commercially available products. Made to customer specifications or sold as standard hardware for a variety of applications.

TECHNICAL SPECIFICATIONS

Physical

Important physical characteristics to consider when selecting hoses for transferring chemicals are the type of construction material and its flexibility, the internal diameter, length, and weight of the hose, appropriate fittings, i.e., connectors, reinforcement, application (including whether the hose is designed specifically for suction or discharge), and chemical compatibility. Hoses are generally made of Teflon, Hypalon (chlorosulfonated polyethylene), Viton, polyvinyl chloride (PVC), pure gum, rubber, metal, and other custom plastics. Individual manufacturers should be consulted for specifications on their products.

Operational

Individual manufacturers should be consulted for operational specifications. Usage should be specified to ensure that the hose is compatible with the material being transferred, ambient temperature, and the temperature of the substance of concern. Weight per unit length, flexibility, operating pressure, and chemical versatility are other operational concerns.

Power/Fuel Requirements

The only hose that requires power or fuel is heated hose used to carry a product that could be damaged by exposure to a drop in temperature. An example of such an application in a spill scenario is transporting highly viscous fluid in a cold environment when the material must be kept at a pre-heated temperature to keep the viscosity low and allow pumping and transfer.

Transportability

Most hoses are supplied in lengths that can easily be rolled up and handled by one person. Ease of handling and transportation varies with hose diameter, reinforcement, and construction materials.

PERFORMANCE

Performance for chemical spills has not been documented. Transfer hoses have been used for chemical spills, however, as well as for transferring chemicals in-plant and other movement of various chemical substances.

CONTACT INFORMATION

The following lists manufacturers and distributors of hoses that can be used with acids and chemicals.

Manufacturer

ABCO Industries Incorporated
280 Arran
St. Lambert, Quebec
J4R 2T5

Telephone (514) 671-1160
Facsimile (514) 671-4339

Associated/I.R.P. Limited
280 Matheson Blvd. East
Mississauga, Ontario
L4Z 1P5

Telephone (905) 890-2960
Toll-free   (800) 387-9537
Facsimile (905) 890-6354

City Industrial Sales Corporation
5937 Dixie Road
Mississauga, Ontario
L4W 1E8

Telephone (905) 670-1360
Facsimile (905) 670-1974

Deetag Limited
63 Clarke Road
London, Ontario
N5W 5Y2

Telephone (519) 659-4673
Toll-free   (800) 461-4673
Facsimile   (519) 659-4677

Downs JJ Industrial Plastics Incorporated
243 Bering Avenue
Toronto, Ontario
M8Z 3A5

Telephone (416) 236-1884
Facsimile (416) 236-3726

Flexaust Company Inc.
P.O. Box 4275
Warsaw, IN 46381
U.S.A.

Telephone (219) 267-7909
Facsimile (219) 267-4665

(See No. 108 in this section)

Distributor

Flex-Pression Limited
6590 Abrans
St. Laurent, Quebec
H4S 1Y2

Telephone (514) 334-9888
Facsimile (514) 334-1044

Gates Canada Incorporated
300 Henry Street
Brantford, Ontario
N3S 7R5

Telephone (519) 759-4141
Facsimile (519) 759-0944   (See No. 109 in this section)

Goodall Rubber Company of Canada Limited
264 Tecumseh Street
Sarnia, Ontario
N7T 2K9

Telephone (519) 336-9394
Facsimile   (519) 336-1277   (See No. 110 in this section)

Hydro-Flex Incorporated
1354 Sandhill Drive
Ancaster, Ontario
L9G 4V5

Telephone (905) 648-9999
Facsimile (905) 648-7038

Industrial Plastics Canada Limited
625 Industrial Drive
Fort Erie, Ontario
L2A 5M6

Telephone (905) 871-0412
Facsimile (905) 871-6494

Linatex Canada Incorporated
5235 Henri Bourassa West
Montreal, Quebec
H4R 1B8

Telephone (514) 334-0252
Facsimile   (514) 334-4588

4.3 Transfer - Transfer Hoses

Manufacturer

N.R. Murphy Limited
430 Franklin Boulevard
Cambridge, Ontario
N1R 8G6

Telephone (519) 621-6210
Facsimile (519) 621-2841

Northflex Manufacturing
428 Millan Road, Unit # 5
Stoney Creek, Ontario
L8E 2N9

Telephone (905) 664-1063
Facsimile (905) 664-1644

Omniflex Industrial Sales Limited
1177 Franklin Boulevard
Cambridge, Ontario
N1R 7W4

Telephone (519) 622-0261
Facsimile (519) 622-0525

Ontario Hose Specialties Limited
6295 Kestrel Road
Mississauga, Ontario
L5T 1Z4

Telephone (905) 670-0113
Facsimile (905) 670-4958

Distributor

Ontario Rubber, A Division of ROBCO Incorporated
281 Ambassador Drive
Mississagua, Ontario
L5T 2J3

Telephone (905) 564-6890
Facsimile (905) 564-6901

RPM Mechanical Incorporated
2290 Industrial Street
Burlington, Ontario
L7P 1A1

Telephone (905) 335-5523
Facsimile (905) 335-5843

Taylor Fluid Systems Incorporated
Box 781, 81 Griffith Road
Stratford, Ontario
N5A 6W1

Telephone (519) 273-2811
Facsimile (519) 273-2051

Utility & Industrial Supply Limited
645 Keddco Street
Sarnia, Ontario
N7T 7K6

Telephone (519) 344-3614
Facsimile (519) 344-3181

Versatile Wales Limited
9 Roddis Road
Elliot Lake, Ontario
P5A 2T1

Telephone (705) 848-8666
Facsimile (705) 848-6445

REFERENCES

(1) Manufacturer's Literature.
(2) *Frasers Canadian Trade Directory*, Toronto, Ont. (1992).

FLEXAUST HOSE AND DUCT

No. 108

DESCRIPTION

Lightweight hoses for air movement, dust collection, fume control, and handling lightweight materials.

OPERATING PRINCIPLE

Hollow tubing used for the initial pickup or transfer of materials.

STATUS OF DEVELOPMENT AND USAGE

Commercially available products used for air movement, dust collection, fume control, and handling lightweight materials in a variety of applications.

TECHNICAL SPECIFICATIONS

Physical

The following are specifications for two of the more than 30 types of hose/duct available. **FSP Flexible Duct** is single- or double-ply PVC/fabric ducting, reinforced by a spring-steel wire helix and protected by a tough, orange vinyl wearstrip. Available from 8 to 30 cm (3 to 12 in) in diameter and in standard lengths of 7.6 m (25 ft). **Flx-Thane I** is an extremely flexible, medium-wall polyurethane hose, reinforced with a spring-steel wire helix, designed for chemical resistance, abrasive dust, or handling lightweight materials and fume removal applications. Available from 5 to 30 cm (2 to 12 in) in diameter and in standard lengths of 7.6 m (25 ft).

Operational

Single-ply **FSP-I Flexible Duct** is lightweight and extremely flexible and is suitable for most air and fume applications. The double-ply **FSP-II** is suitable for most applications including conveying air, fumes, dust, and other lightweight materials. Both feature chemical resistance, low friction loss, mildew and moisture resistance, flame retardance, and an abrasion-resistant "safety orange" wearstrip. All FSP duct has an operating temperature range of -34° to +66°C (-30° to +150°F). **Flx-Thane I** is an extremely flexible, medium-wall polyurethane hose with a temperature range of -32° to +66°C (-25° to +150°F). The following are specifications for FSP-I and FSP-II.

Inside Diameter	Working Pressure	Burst Pressure	Suction	Minimum Bend Radius	Weight
FSP-I					
10 cm (4 in)	<60 kPa (<8 psi)	<170 kPa (<24 psi)	33 kPa (10 in Hg)	8 cm (3.3 in)	0.6 kg/m (0.5 lb/ft)
15 cm (6 in)	<60 kPa (<8 psi)	<170 kPa (<24 psi)	20 kPa (6 in Hg)	13 cm (5.0 in)	0.9 kg/m (0.7 lb/ft)
20 cm (8 in)	<60 kPa (<8 psi)	<170 kPa (<24 psi)	17 kPa (5 in Hg)	17 cm (6.6 in)	1.6 kg/m (1.1 lb/ft)
25 cm (10 in)	<60 kPa (<8 psi)	<170 kPa (<24 psi)	10 kPa (3 in Hg)	18 cm (7.1 in)	1.9 kg/m (1.3 lb/ft)
30 cm (12 in)	<60 kPa (<8 psi)	<170 kPa (<24 psi)	10 kPa (3 in Hg)	22 cm (8.5 in)	2.2 kg/m (1.5 lb/ft)
FSP-II					
7 cm (3 in)	120 kPa (18 psi)	380 kPa (55 psi)	47 kPa (14 in Hg)	7 cm (3.0 in)	0.5 kg/m (0.4 lb/ft)
10 cm (4 in)	110 kPa (16 psi)	340 kPa (50 psi)	44 kPa (13 in Hg)	10 cm (4.1 in)	0.8 kg/m (0.6 lb/ft)
15 cm (6 in)	100 kPa (14 psi)	290 kPa (42 psi)	27 kPa (8 in Hg)	14 cm (5.3 in)	1.2 kg/m (0.8 lb/ft)
20 cm (8 in)	80 kPa (12 psi)	250 kPa (36 psi)	24 kPa (7 in Hg)	17 cm (6.8 in)	1.8 kg/m (1.3 lb/ft)
25 cm (10 in)	70 kPa (10 psi)	210 kPa (30 psi)	13 kPa (4 in Hg)	20 cm (7.8 in)	2.3 kg/m (1.6 lb/ft)
30 cm (12 in)	60 kPa (8 psi)	170 kPa (25 psi)	13 kPa (4 in Hg)	20 cm (8.0 in)	2.7 kg/m (1.8 lb/ft)

4.3 Transfer - Transfer Hoses

Power/Fuel Requirements

None.

Transportability

Most hoses are supplied in lengths that can be easily rolled up and handled by one person. Ease of handling and transportation varies with hose diameter, type of reinforcement, and construction materials.

PERFORMANCE

Performance in a chemical spill cleanup has not been documented. A chemical compatibility chart is available from the manufacturer to assist in determining the correct hose for a specific application.

CONTACT INFORMATION

Manufacturer	*Canadian Distributors*
Flexaust Company Inc. P.O. Box 4275 Warsaw, IN 46381 U.S.A.	Associated/I.R.P. Limited 280 Matheson Blvd. East Mississauga, Ontario L4Z 1P5
Telephone (219) 267-7909 Facsimile (219) 267-4665	Telephone (905) 890-2960 Toll-free (800) 387-9537 Facsimile (905) 890-6354

REFERENCES

(1) Manufacturer's Literature.

GATES INDUSTRIAL, ACID-CHEMICAL HOSES No. 109

DESCRIPTION
A line of hoses constructed for acid-chemical transfer.

OPERATING PRINCIPLE
Hollow tubing for the initial pickup or transfer of materials.

STATUS OF DEVELOPMENT AND USAGE
Commercially available products for transferring acids and chemicals.

TECHNICAL SPECIFICATIONS

Physical

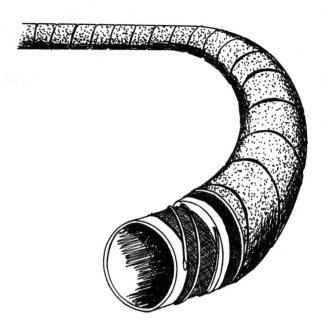

Stock Hoses - The **45HW (Modified XLPE), 42HW (FKM), 37HW (EPDM)**, and **Masterflex 112CL (PVC)** hoses are available with an inside diameter of 25 to 102 mm (1 to 4 in). The **45HW hose** is suited for suction, discharge, or gravity feed transfer service for a wide range of chemicals and solvents. It bends to a tight-bend radius and still maintains a uniform cross-section at full suction. The tube is made of "Type K" Gatron material, which is specially compounded cross-linked polyethylene with excellent resistance to most corrosive chemicals. The hose is reinforced with synthetic, high-tensile, textile fabric with a wire helix and covered with black Neoprene, which is resistant to abrasion, weather, and sunlight. A static wire is available on special order. Also available are the **41HW (Hypalon)** hose with an inside diameter of 25 to 76 mm (1 to 3 in), the **49SB (Butyl)** hose with inside diameters of 25, 76, and 102 mm (2, 3, and 4 in), and the **Masterflex 300 (EPDM)** with an inside diameter of 25 to 152 mm (1 to 6 in).

Made-to-order Hoses - The **401SB (CSM, FKM, XLPE)** and **400SB (Natural Rubber)** are available with inside diameters of 2.5 to 25 cm (1 to 10 in) and the **49SB (Butyl)** with inside diameters of 2.5 to 27 cm (1 to 11 in).

Operational

Stock Hoses - The **45HW (Modified XLPE)** has the best chemical coverage of all Gates hoses and is used to convey most strong acids, strong bases, inorganic salts, alcohols, aldehydes, ketones, bleaching agents, paints, hydrocarbons, and oils. The operating temperature range is -40° to +66°F (-40° to +150°F). Standard length is 30 m (100 ft) and maximum length is 61 m (200 ft). The following are specifications for this hose.

Inside Diameter	Outside Diameter	Maximum Working Pressure	Suction	Minimum Bend Radius	Weight
25 mm (1 in)	40 mm (1.6 in)	1,380 kPa (200 psi)	100 kPa (30 in Hg)	13 cm (5 in)	1.1 kg/m (0.75 lb/ft)
32 mm (1.3 in)	46 mm (1.8 in)	1,380 kPa (200 psi)	100 kPa (30 in Hg)	15 cm (6 in)	1.3 kg/m (0.88 lb/ft)
38 mm (1.5 in)	53 mm (2.1 in)	1,380 kPa (200 psi)	100 kPa (30 in Hg)	20 cm (8 in)	1.6 kg/m (1.1 lb/ft)
51 mm (2 in)	66 mm (2.6 in)	1,380 kPa (200 psi)	100 kPa (30 in Hg)	23 cm (9 in)	2.1 kg/m (1.4 lb/ft)
64 mm (2.5 in)	80 mm (3.1 in)	860 kPa (125 psi)	100 kPa (30 in Hg)	30 cm (12 in)	2.6 kg/m (1.8 lb/ft)
76 mm (3 in)	93 mm (3.7 in)	860 kPa (125 psi)	100 kPa (30 in Hg)	46 cm (18 in)	3.3 kg/m (2.2 lb/ft)
102 mm (4 in)	118 mm (4.7 in)	860 kPa (125 psi)	100 kPa (30 in Hg)	61 cm (24 in)	4.2 kg/m (2.9 lb/ft)

The **42HW (FKM)** hose provides the best resistance to hot acid of all Gates hoses and is used to transfer strong acids, oxidizing agents, dilute alkaline solutions, aromatic solvents, certain chlorinated solvents, hydrocarbons, and aircraft and automotive fuels and lubricants. The **41HW (Hypalon)** hose is used to transfer inorganic acids, bases, and salts, alcohols, aldehydes, aliphatic hydrocarbons, and hydraulic fluids.

4.3 Transfer - Transfer Hoses

The **49SB (Butyl)** hose is used to convey some strong acids, strong bases, inorganic salts, alcohols, aldehydes, ketones, as well as detergents and bleaches and the **37HW (EPDM)** hose is used to convey dilute solutions of acids and bases, as well as ketones, aldehydes, glycols, and ethers.

Power/Fuel Requirements

None.

Transportability

Most hoses are supplied in lengths that can be easily rolled up and handled by one person. Ease of handling and transporting varies with hose diameter, reinforcement, and construction materials. A 30-m (100-ft) length of 51-mm (2-in) hose weighs 45 kg (140 lb).

PERFORMANCE

Performance in a chemical spill cleanup has not been documented. A chemical compatibility chart is available from the manufacturer to assist in determining the correct hose for a specific application.

CONTACT INFORMATION

Manufacturer

Gates Canada Incorporated
300 Henry Street
Brantford, Ontario
N3S 7R5

Telephone (519) 759-4141
Facsimile (519) 759-0944

REFERENCES

(1) Manufacturer's Literature.

GOODALL CHEMICAL HOSES No. 110

DESCRIPTION

A line of hoses for transferring acid-chemicals.

OPERATING PRINCIPLE

Hollow tubing used in conjunction with pumps for the initial pickup of materials or their subsequent transfer.

STATUS OF DEVELOPMENT AND USAGE

Commercially available products used in the chemical process industries to transfer a wide range of chemicals.

TECHNICAL SPECIFICATIONS

Physical

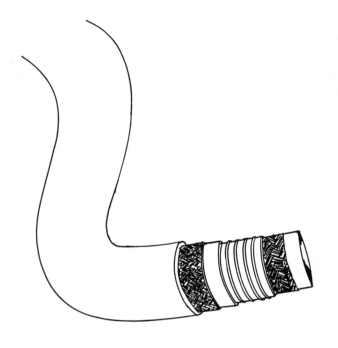

Corru-Chem N-2639 - An extremely flexible chemical transfer hose for suction and discharge applications. Constructed with a modified cross-linked tube and a specially corrugated Neoprene cover, reinforced with textile braid and helix wire, this hose can handle most chemicals at a temperature range of -40° to 66°C (-40° to 150°F). Available from 25 to 76 mm (1 to 3 in) in diameter.

Kem Flex N-2494 - A chemical and solvent transfer hose constructed of clear cross-linked polyethylene tube with a Hypalon cover, reinforced with polyester braid and helix wire. This hose can handle 90% of common industrial chemicals, including esters and ketones at up to 66°C (150°F). Available from 19 to 76 mm (0.8 to 3 in) in diameter and in up to 30 m (100 ft) lengths.

Chemall –2546 - Clear cross-linked polyethylene tube with an EPDM cover which protects the hose from severe weather conditions, and reinforced with polyester braid and helix wire. This hose can handle temperatures up to 66°C (150°F). Available from 25 to 76 mm (1 to 3 in) in diameter and in 30 m (100 ft) lengths.

Super Chem II N-2442 - Clear Teflon tube with a Hypalon cover, reinforced with textile braid and helix wire. This hose can be used to transfer virtually any chemical or acid at temperatures up to 120°C (250°F). Available from 25 to 76 mm (1 to 3 in) in diameter and in 15 m (50 ft) lengths.

Acid King N-2493 - A hose constructed specifically for transferring acids. Hypalon tube with Hypalon cover reinforced with polyester braid and helix wire. Available from 25 to 102 mm (1 to 4 in) in diameter and in up to 30 m (100 ft) lengths.

Flexlon C N-8020 - A Teflon-lined chemical transfer hose with a tough abrasion-resistant Neoprene cover, reinforced with a textile braid, and able to handle temperatures up to 120°C (250°F). Available from 13 to 25 mm (0.5 to 1 in) in diameter.

Chloraflex - A chlorine transfer hose manufactured of corrugated Monel 400, with Monel 400 reinforcement, and a stainless steel interlock metal cover. This hose meets the Chlorine Institute's recommended specifications. Standard size is 25 mm (1 in) in diameter and 15 m (50 ft) lengths.

Teleflex - This hose is the only alternative to Monel hoses for transferring chlorine/bromine. It is manufactured of convoluted, white, non-conductive Teflon, with Kynar monofilament reinforcement and a pin-pricked, CPE rubber jacket cover. Standard size is 25 mm (1 in) in diameter and 15 m (50 ft) lengths.

4.3 Transfer - Transfer Hoses

Operational

Inside Diameter	Outside Diameter	Minimum Bend	Max. Working Pressure	Weight
Corru-Chem N-2639				
25 mm (1 in)	39 mm (1.5 in)	13 cm (5 in)	1,030 kPa (150 psi)	1.1 kg/m (0.73 lb/ft)
38 mm (1.5 in)	52 mm (2.1 in)	20 cm (8 in)	same	1.6 kg/m (1.1 lb/ft)
51 mm (2 in)	67 mm (2.6 in)	23 cm (9 in)	same	2.1 kg/m (1.4 lb/ft)
76 mm (3 in)	94 mm (3.7 in)	46 cm (18 in)	same	3.3 kg/m (2.2 lb/ft)
Kem Flex N-2494				
19 mm (0.8 in)	34 mm (1.3 in)	10 cm (4 in)	1,380 kPa (200 psi)	0.9 kg/m (0.6 lb/ft)
25 mm (1 in)	39 mm (1.5 in)	13 cm (5 in)	same	1.2 kg/m (0.8 lb/ft)
32 mm (1.3 in)	46 mm (1.8 in)	15 cm (6 in)	same	1.3 kg/m (0.9 lb/ft)
38 mm (1.5 in)	53 mm (2.1 in)	20 cm (8 in)	same	1.6 kg/m (1.1 lb/ft)
51 mm (2 in)	66 mm (2.6 in)	30 cm (12 in)	same	1.9 kg/m (1.3 lb/ft)
76 mm (3 in)	91 mm (3.6 in)	46 cm (18 in)	same	3.2 kg/m (2.2 lb/ft)
Chemall –2546				
25 mm (1 in)	38 mm (1.5 in)	15 cm (6 in)	1,030 kPa (150 psi)	1 kg/m (0.7 lb/ft)
38 mm (1.5 in)	52 mm (2.1 in)	25 cm (10 in)	same	1.5 kg/m (1 lb/ft)
51 mm (2 in)	65 mm (2.6 in)	36 cm (14 in)	same	1.7 kg/m (1.2 lb/ft)
76 mm (3 in)	90 mm (3.6 in)	51 cm (20 in)	same	3 kg/m (2.1 lb/ft)
Super Chem II N-2442				
25 mm (1 in)	39 mm (1.5 in)	30 cm (12 in)	1,030 kPa (150 psi)	1.2 kg/m (0.8 lb/ft)
32 mm (1.3 in)	46 mm (1.8 in)	38 cm (15 in)	same	1.5 kg/m (1 lb/ft)
38 mm (1.5 in)	53 mm (2.1 in)	46 cm (18 in)	same	1.7 kg/m (1.2 lb/ft)
51 mm (2 in)	66 mm (2.6 in)	61 cm (24 in)	same	2.3 kg/m (1.6 lb/ft)
76 mm (3 in)	94 mm (3.7 in)	107 cm (42 in)	same	3.7 kg/m (2.5 lb/ft)
Acid King N-2493				
25 mm (1 in)	41 mm (1.6 in)	8 cm (3 in)	1,030 kPa (150 psi)	1.2 kg/m (0.8 lb/ft)
32 mm (1.3 in)	48 mm (1.9 in)	10 cm (4 in)	same	1.5 kg/m (1 lb/ft)
38 mm (1.5 in)	54 mm (2.1 in)	13 cm (5 in)	same	1.7 kg/m (1.1 lb/ft)
51 mm (2 in)	66 mm (2.6 in)	20 cm (8 in)	same	2.1 kg/m (1.4 lb/ft)
76 mm (3 in)	92 mm (3.6 in)	30 cm (12 in)	same	3.4 kg/m (2.3 lb/ft)
102 mm (4 in)	119 mm (4.7 in)	38 cm (15 in)	same	4.9 kg/m (3.3 lb/ft)
Flexlon C N-8020				
13 mm (0.5 in)	17 mm (0.7 in)	15 cm (6 in)	2,070 kPa (300 psi)	0.4 kg/m (0.3 lb/ft)
19 mm (0.8 in)	30 mm (1.2 in)	20 cm (8 in)	2,070 kPa (300 psi)	0.6 kg/m (0.4 lb/ft)
25 mm (1 in)	38 mm (1.5 in)	30 cm (12 in)	1,720 kPa (250 psi)	0.9 kg/m (0.6 lb/ft)
Chloraflex				
25 mm (1 in)	49 mm (1.9 in)	30 cm (12 in)	2,590 kPa (375 psi)	3.2 kg/m (1.8 lb/ft)
Teleflex				
25 mm (1 in)	51 mm (2 in)	15 cm (6 in)	2,590 kPa (375 psi)	2.1 kg/m (1.2 lb/ft)

Power/Fuel Requirements

None.

Transportability

Most hoses are supplied in lengths that can be easily rolled up and handled by one person. Ease of handling and transporting varies with hose diameter, reinforcement, and construction materials.

PERFORMANCE

Performance in a chemical spill cleanup has not been documented. A chemical compatibility chart is available from the manufacturer to assist in determining the right hose for a specific application.

CONTACT INFORMATION

Manufacturer

Goodall Rubber Company of Canada Limited
264 Tecumseh Street
Sarnia, Ontario
N7T 2K9

Telephone (519) 336-9394
Facsimile (519) 336-1277

Canadian Distributors

Winnipeg, Manitoba	(204) 633-5553
Edmonton, Alberta	(403) 465-0301
Montreal, Quebec	(514) 336-5610
Sudbury, Ontario	(705) 682-0663
Saskatoon, Saskatchewan	(306) 931-7300
Vancouver, British Columbia	(604) 273-0091

REFERENCES

(1) Manufacturer's Literature.

Section 5.1

Treatment/Disposal - Spill Treating Agents

ANSUL SPILL TREATMENT KITS AND APPLICATORS

No. 111

DESCRIPTION

A line of spill control agents, available in kits for cleaning up smaller spills of many mineral and organic acids, caustics, and organic solvents, as well as applicators for applying agents to larger spills. SPILL-X spill control agents are available in either a **Multipurpose Spill Treatment Kit** or individual **Spill Treatment Kits** for acid, caustic, or solvent spills.

OPERATING PRINCIPLE

SPILL-X-A, SPILL-X-C, and **SPILL X-S** are available in either Multipurpose or Individual Kits. The Multipurpose Kit contains six polypropylene containers - two of each type of SPILL-X spill control agent. The containers are easy to handle and are designed for either a pouring-type or a shaker-type application, depending on the nature of the spilled chemical. None of these agents produces toxic by-products. **SPILL-X-FP** agent shakers are available in a complete formaldehyde/solvent Spill Treatment Kit which contains three SPILL-X-FP shakers and three SPILL-X-S shakers, as well as safety and cleanup equipment.

Spill Gun SC-30 - Similar to a dry chemical fire extinquisher, this unit consists of an agent canister, an expellant gas cartridge, a discharge hose, and a nozzle, which when activated, fluidizes the control agent, pressurizes the container, and discharges the spill control agent through the hose. The **Spill Gun SC-30** contains enough agent to treat up to 9.3 m^2 (100 ft^2) of spill area per discharge.

Wheeled Spill Control Applicator - Similar to a large dry chemical fire extinquisher, this applicator consists of an agent canister, an expellant gas cartridge, a discharge hose, and a nozzle. It can be operated by one person and is available in sizes to treat and solidify a spill of up to 133 L (35 gal) or 98 m^2 (1,050 ft^2) per discharge.

STATUS OF DEVELOPMENT AND USAGE

Commercially available products in general use.

TECHNICAL SPECIFICATIONS

Physical

SPILL-X-A Acid Neutralizer/Solidifier - A red, magnesium oxide-based agent used to treat small spills of many mineral and organic acids that can both neutralize and solidify the spilled material when properly applied.

SPILL-X-C Caustic Neutralizer/Solidifier - A tan, citric acid-based agent containing chemical additives for treating small spills of many caustic and organic bases that can both neutralize and/or solidify the spilled material when properly applied.

SPILL-X-S Solvent Adsorbent - A black, activated carbon-based agent for treating small spills of many organic liquid solvents and fuels, other than organic peroxides or hydrazine compounds. It can both rapidly adsorb the spilled material and elevate the flashpoint above 60°C (140°F) when properly applied.

SPILL-X-FP Formaldehyde Polymerizer - A fluorescent green, urea-based agent effective on spills of a variety of formaldehyde solutions of different strengths. Proper application yields a solid residue for formaldehyde solutions of between 15 and 37% (weight/weight) in strength. A cloudy appearance to the aqueous formaldehyde 1 to 5 minutes after application indicates that the treatment reaction has begun. The reaction is complete when a plastic-like solid (polynoxylin) has formed. Properly applied, SPILL-X-FP agent can help reduce contamination of the workplace caused by formaldehyde spills and their vapours. Polymerized spill residue must be disposed of in accordance with federal, state, and local regulations as well as a company's own waste-handling guidelines.

Spill Gun SC-30 - Corrosion-resistant construction designed for industrial environments. Each applicator contains 5 kg (11 lb) of spill-treating agent.

Wheeled Spill Control Applicator - These applicators are available in a large and smaller model and can be used with three spill treatment agents produced by Ansul. The unit is made of welded steel with noncorrosive materials and standard corrosion-resistant coating. Wide wheels or rubber tires are available as options for both models.

	Wheeled Spill Control Applicator 150	**Wheeled Spill Control Applicator 350**
Capacity	68 kg (150 lb) SPILL-X-A agent	160 kg (350 lb) SPILL-X-A agent
	52 kg (116 lb) SPILL-X-C agent	130 kg (280 lb) SPILL-X-C agent
	20 kg (48 lb) SPILL-X-S agent	60 kg (120 lb) SPILL-X-S agent
Height	1.3 m (4.3 ft)	1.5 m (5 ft)
Width	0.7 m (2.4 ft)	1.5 m (5 ft)
Diameter	1 m (3.5 ft)	0.9 m (3 ft)
Charged weight	220 kg (480 lb) SPILL-X-A agent	400 kg (880 lb) SPILL-X-A agent
	200 kg (450 lb) SPILL-X-C agent	370 kg (820 lb) SPILL-X-C agent
	170 kg (380 lb) SPILL-X-S agent	300 kg (660 lb) SPILL-X-S agent

Operational

SPILL-X Agents - Contact the manufacturer for details on the types of chemicals that can be treated and the amounts required.

Spill Gun SC-30 - This applicator can be used with SPILL-X-A agent, SPILL-X-C agent, and SPILL-X-S agent. It applies the agent from a distance of 3 to 3.7 m (10 to 12 ft) and treats a spill of up to 11 L (3 gal) [equivalent to approximately 9.3 m^2 (100 ft^2)]. Spill Guns can be quickly recharged on site without special tools.

Wheeled Spill Control Applicators - These can be used with the three spill treatment agents described above. The wheeled applicators apply the agent from a distance of 4.6 to 6.1 m (15 to 20 ft). The 150-lb capacity unit treats a spill of up to 57 L (15 gal) [equivalent to about 42 m^2 (450 ft^2)] and the 350-lb capacity unit treats a spill of up to 130 L (35 gal) [equivalent to about 98 m^2 (1,050 ft^2)]. Both models are suitable for operation in extreme temperatures [-54 to 49°C (-65 to 120°F)]. The following are typical spill coverages for the 68-kg (150-lb) capacity unit for some common products.

Wheeled Spill Control Applicator 150

Acid Spills SPILL-X-A Agent		Caustic Spills SPILL-X-C Agent		Solvent or Fuel Spills SPILL-X-S Agent	
Acid	Coverage	Caustic	Coverage	Solvent/Fuel	Coverage
Sulphuric (93%)	43 L (11 gal)	Sodium hydroxide (50%)	32 L (8 gal)	Methyl ethyl ketone	35 L (9 gal)
Hydrochloric (36%)	47 L (12 gal)	Morpholine	20 L (5 gal)	Acrylonitrile	27 K (7 gal)
		Diethylenetriamine	20 L (5 gal)	Petroleum	36 L (10 gal)
Nitric (70%)	47 L (12 gal)	Diethylamine	20 L (5 gal)	Toluene	22 L (6 gal)
Phosphoric (85%)	47 L (12 gal)	Aniline	17 L (4 gal)	Xylene	22 L (6 gal)
		Potassium hydroxide (45%)	50 L (13 gal)	Jet A-1	22 L (6 gal)
Formic (90%)	47 L (12 gal)	Ammonium hydroxide (29%)	76 L (20 gal)	Gasoline	30 L (8 gal)

5.1 Treatment/Disposal - Spill Treating Agents

Power/Fuel Requirements

Spill Gun SC-30 and Wheeled Spill Control Applicator - The agent and expellant gas cartridge must be replaced after each discharge.

Transportability

Spill Guns are hand-portable and both models of the Wheeled Applicator can be wheeled by one person. The wide wheels are recommended if the applicator is to be wheeled over sandy, loose, or rough terrain or flooring. Rubber wheels are available if sparks from the metal wheels could cause problems.

CONTACT INFORMATION

Manufacturer	*Canadian Representative*
Ansul Incorporated	Levitt Safety Limited
One Stanton Street	2872 Bristol Circle
Marinette, WI 54143-2542	Oakville, Ontario
U.S.A.	L6H 5T5
Telephone (715) 735-7411	Telephone (416) 829-3299
Facsimile (715) 732-3471	Facsimile (416) 829-2929

OTHER DATA

Contact the manufacturer for complete details on the full line of SPILL-X Spill Control Products.

REFERENCES

(1) Manufacturer's Literature.

CAPSUR®, MEXTRACT®, AND PENTAGONE® SPILL AGENTS

No. 112

DESCRIPTION

CAPSUR® is an aqueous-based solvent with emulsifiers for cleaning up PCB spills on solid surfaces. **MEXTRACT®** is an aqueous system developed specifically for cleaning up heavy metals from concrete, asphalt, and metal surfaces. **PENTAGONE®** is an aqueous-based solvent decontamination product for cleaning up pentachlorophenol (PCP), creosote, petroleum hydrocarbons, chlorinated hydrocarbons, polynuclear aromatic hydrocarbons (PAHs), and select pesticide and herbicide spills on concrete, asphalt, and metal surfaces.

OPERATING PRINCIPLE

These products interact chemically with the hazardous materials, allowing their extraction from surfaces and their suspension in water for easy removal. These products can be applied as a foam blanket which allows easy application on overhead, vertical, and horizontal surfaces. This foaming ability provides increased contact time with the surface and thus improves extraction efficiency, while reducing the volume of material destined for disposal.

STATUS OF DEVELOPMENT AND USAGE

Commercially available products marketed for spill response and decontamination involving PCBs, heavy metals, PCPs, creosote, petroleum hydrocarbons, chlorinated hydrocarbons, PAHs, and select pesticides and herbicides.

TECHNICAL SPECIFICATIONS

Physical

Each of these products is a patented liquid formulation available in 19-L and 208-L (5-gal and 55-gal) steel containers.

Operational

These products can be applied as a foam blanket allowing application to overhead, vertical, and horizontal surfaces. The Model T Jr. Foamer is the only foam applicator endorsed and sold for use with these products. Either as a liquid or a foam, they are applied in a manner similar to laboratory extraction procedures. The contaminated area is foamed, agitated with a stiff broom, and left for five minutes. The residues are vacuumed up and the surface is lightly rinsed with water and then re-vacuumed. Application is repeated, this time omitting the agitation step, and it is again left for five minutes. The area is then vacuumed, triple-rinsed with water, and vacuumed again. Application coverage rates vary depending on the porosity of the surface. These products can also be applied without a foamer. After the product is diluted, mixing of the solution generates foam. The solution is then applied and the surface is vigorously agitated for a full five minutes.

To use CAPSUR as an example, with either application technique, emulsified PCBs are suspended in water and vacuumed up and out of the surface. In areas with initial spill concentrations of less than 200 μg/100 cm^2, one application has produced results within regulatory standards. Two cycles are required in areas with concentrations ranging from 200 to 800 μg/100 cm^2. Concentrations greater than 800 μg/100 cm^2 require three or more cycles. The extraction capacity seems to be a function of the extent of PCB contamination and the capacity of the solvent. CAPSUR has successfully cleaned surfaces, such as concrete, asphalt, and metals contaminated with varying PCB concentrations, to government standards, and has been successful on new and old spills.

These products are most effective when applied to a dry surface that is free of excess grease, dirt, or oil buildup and at a surface temperature of between 7 and 32°C (45 and 90°F).

Power/Fuel Requirements

No power or fuel is required to directly apply or use any of these products. A supply of fresh water is required to dilute the agent and to rinse between applications. An industrial wet vacuum with an appropriate power supply is required to remove and recover contaminated product and rinse water. Compressed air is required to power the foamer.

Transportability

All of these products are supplied in 19-L and 208-L (5-gal and 55-gal) steel containers which are readily transportable.

PERFORMANCE

Some performance data are provided here for **CAPSUR**, which has demonstrated more than 85% extraction efficiencies in laboratory tests and customer use. In one application, CAPSUR was applied to a spill of 380 L (100 gal) of oil containing >500 ppm of PCBs covering a 14 m^2 (150 ft^2) area. Initial concentrations ranged from 7 to >107 µg/100 cm^2. After treatment, 50% of sample points were below 10 µg/100 cm^2, while the remaining sample points were below 100 µg/100 cm^2. In another example, CAPSUR was applied to a one-year-old spill with initial PCB concentrations of 50 to 500 ppm. Final PCB concentrations were 0 ppm after one application. CAPSUR has also proven effective on older spills. It was applied to a 30- to 50-year-old spill with initial PCB concentrations of less than 500 µg/100 cm^2. After one application, plus some repeat applications to hot spots, the final PCB concentration was less than 10 µg/100 cm^2.

CONTACT INFORMATION

Manufacturer

Integrated Chemistries Incorporated
1970 Oakcrest Avenue, Suite 215
St. Paul, MN 55113
U.S.A.

Telephone (612) 636-2380
Facsimile (612) 636-3106

OTHER DATA

Integrated Chemistries also manufactures and distributes industrial products, many of which have been identified as environmentally sensitive by the U.S. Defence General Supply Centre and can be used to replace traditional products containing hazardous constituents such as chlorinated solvents.

REFERENCES

(1) Manufacturer's Literature.

CARTIER SPILL RESPONSE KITS   No. 113

DESCRIPTION

A series of spill response kits which contain VYTAC® neutralizers and are designed for total response to spills of hazardous material. The kits combine neutralizers, absorbents, containment materials, and personnel safety equipment.

OPERATING PRINCIPLE

These kits contain VYTAC® neutralizers which treat contained spills and render them harmless.

STATUS OF DEVELOPMENT AND USAGE

Commercially available products marketed to eliminate accidental spills of hazardous materials at industrial, institutional, and laboratory facilities.

TECHNICAL SPECIFICATIONS

Physical

Spill Response Kits

Containers that can be used to customize spill response kits include 23 L (6 gal) pails, 55 L (15 gal) or larger tote boxes, 115 L (30 gal) or 208 L (55 gal) drums, 295 L (78 gal) lockers with or without casters, 380 L (100 gal) overpack drums, or wall-mounted cabinets for labs. Extra large mobile response units are available for larger spills. The following are examples of kit components.

Acid Spill Response Kit - This kit contains VYTAC® ACX powder/liquid, acid neutralizer with colour indicator; VYTAC® ATF liquid, acid neutralizer - no foam, no gas generation, with colour indicator; three sizes of 3M Mini booms; 3M pads; 3M - CFL 550DD folded sorbent; 3M Acid gas and vapour respirator; 3M - P300 pillows; DrumRoll - Teflon-coated CO_2 inflatable belt for perforated 208-L (55-gal) drums; HD-8 Neoprene drain cover; spill identifier strips; Tyvek™ Saranex chemical-resistant, disposable coverall; Chemrel™ C101 encapsulating suit for strong acids; chemical-resistant gloves; goggles; and H.D. disposal bags.

Other available kits are **Alkalis Spill Response Kits**, which include VYTAC®CS powder/liquid - alkali neutralizer; **Hazardous Chemicals Mini-spill Response Kits for Laboratories**, which include VYTAC®VSP vapour suppressant and adsorbent for volatile fuels; OCLANSORB™ peat moss absorbent; VYTAC® ACX powder/liquid, acid neutralizer; VYTAC® ATF liquid, acid neutralizer; VYTAC®CS powder/liquid, caustic neutralizer with colour indicator; VYTAC®FAS formaldehyde absorbent and immobilizer; VYTAC®FNC formaldehyde neutralizer; and VYTAC®MIS mercury cleanup system; **Petroleum Spill Response Kits; Formaldehyde Spill Response Kits**; and **Mercury Spill Response Kits**.

VYTAC Neutralizers

These agents quickly and efficiently neutralize concentrated spills. The residue is nontoxic and easily disposed of. These agents contain an indicator system that shows when the spill is completely neutralized. Suitable protective equipment must be used when applying these agents.

For Acids - VYTAC ACX Powder and Liquid, ATF Liquid, and AGA Liquid - These products are applied to the spill by spraying, flooding, or pouring, working from the outside to the centre. The mixture turns permanently pink, purple, or red when sufficient ACX, ATF, or AGA has been applied to completely neutralize the spill. The residue can be removed with absorbents or flushed into the sewer system, in accordance with local regulations.

For Alkalis - VYTAC CS Powder and Liquid - CS Powder is for direct use on dilute alkaline solutions and CS Liquid is for spray and manual application. If possible, concentrated spills should be diluted with water before applying CS to minimize heat buildup. CS should be sprayed or otherwise applied to the spill while mixing in, working from the outside to the centre of the spill. The mixture becomes pink at first but the pink vanishes when sufficient CS has been applied to completely neutralize the spill. Residues can be removed with absorbents or flushed into the sewer system, in accordance with local regulations.

For Formaldehyde - VYTAC FNC LCX, FNC HCS Liquid, and FAS Powder - FNC LCX is a concentrated neutralizer for low concentration solutions of formaldehyde or glutaraldehyde. The proprietary blend of low toxicity

5.1 Treatment/Disposal - Spill Treating Agents

oxygenated sodium salts and activators is effective in removing formaldehyde down to levels of less than 100 ppm. It can be applied either in a batch process or continuous process and the neutralization process takes 2 to 30 minutes to complete depending on temperature, mixing rate, and other chemicals present. FNC HCS is optimized for use and enhanced storage stability in hot climates, is phosphate-free, and very safe to use. It is applied by spray or other means directly onto the spill, starting from the outside and working toward the centre. Concentrated solutions are transformed into an orange solid within 15 to 30 minutes (possibly longer at lower temperatures). VYTAC FAS is a high performance absorbent which is compatible with the FNC neutralization system. An effective odour control agent combined with super absorbency controls formaldehyde vapour emissions.

For Mercury - VYTAC MIS Powder and MVS Powder - MIS is a unique formulation that transforms liquid mercury into an immobile solid amalgam, which almost completely suppresses the emission of mercury vapours. It works on contact and the solid granules can be easily and safely removed with pick up tools or vacuum devices. MIS is also effective on spills in drains and can be used to make mercury waste safer for transport. MVS is specially developed to deal with highly dispersed mercury. The chemically activated particles bind to mercury surfaces and absorb mercury vapour.

For Other Hazardous Materials - VYTAC GNF Liquid - Institutional and process neutralizer for glutaraldehyde.
VYTAC VSP Powder - Vapour suppressant for highly flammable liquids, which raises flash point.
VYTAC HRX Powder - High performance absorbent for heavy water and radioactive liquid.
VYTAC JCN Liquid - Quick, safe and efficient neutralizer for sodium hypochlorite solutions.
VYTAC PX Liquid and HPBA Powder - Safe, efficient neutralizer and absorbent/neutralizer for hydrogen peroxide solutions.

Power/Fuel Requirements

No power or fuel is required to directly apply or use any of these products. A supply of fresh water is required to dilute the agent and to rinse after application. Compressed air may be required to power a foamer.

Transportability

Spill response kits are provided in many sizes of containers, ranging from a 23 L (6 gal) pail to extra large mobile response units. The VYTAC agents are supplied in person-portable containers.

CONTACT INFORMATION

Manufacturer

Cartier Chemical Ltd.
Environmental Products Division
445-21E Avenue
Lachine, Quebec
H8S 3T8

Telephone (514) 637-4631
Facsimile (514) 637-8804

OTHER DATA

Cartier provides specialized consultation to determine an appropriate environmental or emergency response system.

REFERENCES

(1) Manufacturer's Literature.

EPOLEON ODOUR-NEUTRALIZING AGENTS No. 114

DESCRIPTION

Concentrated water-based products comprised of all-natural, nontoxic, and biodegradable compounds. Instead of masking odours, each formula is designed to chemically react with and neutralize the odour associated with particular types of gases.

OPERATING PRINCIPLE

Through chemical reaction, counteraction, and absorption, Epoleon formulas neutralize odorous gases upon contact. The following are examples of some reactions.

(1) Ammonia (NH_3)
  $-CH_2COOH + NH_3 \rightarrow -CH_2COONH_4$
(2) Trimethylamine $(CH_3)_3N$
  $-CH_2COOH + (CH_3)_3N \rightarrow -CH_2COONH(CH_3)_3$
(3) Hydrogen Sulphide (H_2S)
  $-CH_2COONa + H_2S \rightarrow NaHS + -CH_2COOH$
(4) Methyl Mercaptan (CH_3SH)
  $-CH_2COONa + CH_3SH \rightarrow CH_3SNa + -CH_2COOH$

STATUS OF DEVELOPMENT AND USAGE

Commercially available product marketed for use in removing odours from scrubber systems, wastewater and sludge treatment systems, landfill applications, and spill response applications.

TECHNICAL SPECIFICATIONS

Physical

Epoleon –100 - Formulation consists of organic and salt of organic acids, amine compounds, betaine compounds, and water. The following are its physical properties.

pH	5.0 to 6.7 at 25°C (77°F)
Boiling point	100°C (212°F)
Freezing point	0°C (32°F)
Vapour pressure	Same as water
Solubility in water	Completely
Specific gravity	1.17 ± 0.05 at 25°C (77°F)
Molecular weight	50 to 800
Appearance	Transparent, slightly yellow
Odour	None

Epoleon –7C - Formulation consists of glycine betaine, organic and salt of organic acids, amine compounds, essential oils, and water. The following are its physical properties.

pH	5.0 to 6.7 at 25°C (77°F)
Boiling point	100°C (212°F)
Freezing point	0°C (32°F)
Vapour pressure	Same as water
Solubility in water	Completely
Specific gravity	1.06 ± 0.05 at 25°C (77°F)
Molecular weight	50 to 800
Appearance	Transparent, slightly yellow
Odour	Floral scent

5.1 Treatment/Disposal - Spill Treating Agents

Operational

Epoleon can be applied directly by drip application to a process stream or with a spray fogger to large surface areas. Depending on the application, Epoleon can be applied at full concentration or diluted down to a 1:100 ratio. The following table indicates the amount of some gases that can be deodorized with 1 g of Epoleon.

Type of Epoleon	Ammonia	Hydrogen Sulphide	Trimethylamine	Methyl Mercaptan
N-100	24 mg	43 mg	84 mg	22 mg
N-7C	6.1 mg	20 mg	21 mg	2.6 mg
NnZ	12 mg	14 mg	41 mg	5.5 mg

Power/Fuel Requirements

A fogger or mister system may be required to apply or use Epoleon agents.

Transportability

Epoleon is supplied in person-portable containers.

PERFORMANCE

Head space deodorization tests were conducted in a laboratory. One mL of deodorizer and 19 mL of solution, which included odorous gas and water (total 20 mL), were mixed and put into a 500-mL flask. The flask was sealed with a rubber stopper and stirred for ten minutes. The concentration of odorous gas in the head space of the upper portion of the flask was measured by a gas analyzer. The following are the results of this testing.

Rate of Elimination for N-100

Odorous Gas Contaminant	Original Concentration	Post-treatment Concentration
Ammonia	700 ppm	20 ppm
Hydrogen Sulphide	3,000 ppm	40 ppm
Methyl Mercaptan	10 ppm	< 1 ppm
Trimethylamine	120 ppm	trace

Rate of Elimination for N-7C

Odorous Gas Contaminant	Original Concentration	Post-treatment Concentration
Ammonia	700 ppm	100 ppm
Hydrogen Sulphide	3,000 ppm	120 ppm
Methyl Mercaptan	10 ppm	3 ppm
Trimethylamine	120 ppm	5 ppm

CONTACT INFORMATION

Manufacturer

Epoleon Corporation of America
19160 S. Van Ness Ave.
Torrance, CA 90501
U.S.A.

Telephone (310) 782-0190
Facsimile (310) 782-0191

REFERENCES

(1) Manufacturer's Literature.

FOUL-UP HAZARDOUS WET-WASTE SOLIDIFYING AGENT — No. 115

DESCRIPTION

A water-insoluble, dry, granular adsorbent that adsorbs, solidifies, and deodorizes hazardous wet wastes.

OPERATING PRINCIPLE

Water-based liquids and slurries are physically adsorbed and converted into granulated solid waste so that they can be removed with a vacuum or a shovel.

STATUS OF DEVELOPMENT AND USAGE

Commercially available products marketed for cleanup and decontamination of sewage, unknown odorous or odourless water-based spills or seepage, water-based insecticides, rodenticides, fungicides and fumigants, bacteriological/pharmaceutical, and other wet wastes.

TECHNICAL SPECIFICATIONS

Physical

Polyacrylic acid metallic salt and calcium aluminum silicate. Available in 19-L (5-gal) pails, weighing 14 kg (30 lb).

Operational

Foul-Up can be applied directly to a spill and scooped up into the pail provided. Larger masses can be recovered with an appropriate vacuum system and dealt with as hazardous solid wastes rather than hazardous wet wastes.

Power/Fuel Requirements

No power or fuel is required for the application or use of Foul-Up.

Transportability

Foul-Up is supplied in 19-L (5-gal) pails.

PERFORMANCE

No performance data were received, but according to the manufacturer, Foul-Up's use in property loss insurance claims is well documented.

CONTACT INFORMATION

Manufacturer

Mateson Chemical Corporation
1025 Montgomery Avenue
Philadelphia, PA 19125
U.S.A.

Telephone (215) 423-3200
Facsimile (215) 423-1164

REFERENCES

(1) Manufacturer's Literature.

MERCON™ SPILL-TREATING KITS   No. 116

DESCRIPTION

A line of spill kits for acids, caustics, and solvents, as well as for suppressing mercury vapour.

OPERATING PRINCIPLE

These products work immediately to neutralize and adsorb spills of acids, caustics, and solvents. The mercury spill kits form a shell over spilled mercury. A chemical change then converts vapour-producing elemental mercury into the more stable mercury salts (mercuric sulphide) which occur naturally in the environment. Mercury vapour and dust are also absorbed from the air which stops the methylation of mercury in water.

STATUS OF DEVELOPMENT AND USAGE

Commercially available products marketed for acid, caustic, solvent, and mercury spill response and decontamination at industrial, institutional, and laboratory facilities.

TECHNICAL SPECIFICATIONS

Physical - **Acid Spill Kit, Caustic Spill Kit, and Solvent Spill Kit** - The Acid Spill Kit comes with 2.3 kg (5 lb) of neutralizer/adsorbent, the Caustic Kit with 1 kg (2.2 lb) of neutralizer and 290 g (10 oz) of adsorbent, and the Solvent Kit with 1 kg (2.2 lb) of adsorbent. All these kits come with 2 sealable disposable bags and labels, 1 pair of safety goggles, 1 pair of nitrile gloves, 1 scoop, 1 squeegee and sponge, and an instruction booklet. Bulk reagents are also available.

Multi-agent Package Spill Kit - This kit contains one each of the Mercon™ Acid, Caustic, and Solvent Kits.

Mercury Spill Kit - The Mercon product line includes kits in which the agent is available as a powder, liquid, gel, pump spray, or impregnated into towelettes. For example, **MerconDrum** is a self-contained industrial spill kit in which the shipping container is actually a mercury containment drum that holds everything needed for spills up to 9 kg (20 lb). The **MerconVap** Industrial kit is a 22-L (5.8-gal) drum of a solution specially formulated to decontaminate large areas.

Operational - In the **Acid Spill Kit** and the **Caustic Spill Kit**, a built-in indicator changes from pink to dark blue and from blue to pink, respectively, when neutralization is complete. In the **Solvent Spill Kit**, a carbon-based adsorbent reduces vapours and can raise the flash point to above 60°C (140°F). Mercon™ mercury spill-treating products control the vapours from a mercury spill while encapsulating the mercury and converting it to a more stable state. They are biodegradable, non-corrosive, and odourless and allow spilled and recovered mercury to be recycled.

Power/Fuel Requirements

No power or fuel is required to apply or use Mercon™ products.

Transportability

Mercon products are supplied in containers that are person-portable.

CONTACT INFORMATION

Manufacturer

EPS Chemicals Incorporated
Unit 8 - 7551 Vantage Way
Delta, British Columbia
V4G 1C9

Telephone (604) 940-0975
Toll-free   (800) 663-8303
Facsimile (604) 946-3663

REFERENCES

(1) Manufacturer's Literature.

NOCHAR SOLIDIFYING AGENTS No. 117

DESCRIPTION

Liquid solidification agents designed for solidifying acidic spills and liquids in general.

OPERATING PRINCIPLE

The dry granular materials immobilize spills by solidifying the liquid. The liquid is bonded together to produce a solid mass from which liquid cannot be squeezed out.

STATUS OF DEVELOPMENT AND USAGE

Commercially available products marketed specifically for cleaning up liquid spills.

TECHNICAL SPECIFICATIONS

Physical

A660 Acid Bond

Weight by volume	1 kg = 1.3 L (1 lb = 35 in^3)
Specific gravity	0.80
pH	6.5 to 7
Packaging	1.8-kg (4-lb) shaker or 18-kg (40-lb) drum

A630R Retardant Liquid Bond

Weight by volume	1 kg = 4.1 L (1 lb = 114 in^3)
Specific gravity	0.91
pH	6.5 to 7
Packaging	Assorted sizes and containers are available. Standard package is 4.5-kg (10-lb) drum.

Operational

A660 Acid Bond - A nontoxic, nonhazardous, noncorrosive acid solution solidifying agent, which has been tested on phosphoric, sulphuric, hydrofluoric, hydrochloric, and acetic acids as well as hydrazine and chromium trioxide.

Pickup ratio	Varies with the acid
Volatility before use	Zero
Volatility after use	Varies with flammability of liquid bonded

A630R Retardant Liquid Bond - A nontoxic, nonhazardous, fire-retarded, all-purpose liquid bonding agent. A630R bonds water, or water contaminated with oils, fuels, solvents, or a mixture of these. A630R has passed the U.S. Toxic Characteristics Leachate Procedure (TCLP) testing.

Pickup ratio	1:25, 1 kg of product recovers up to 25 kg of liquid
Volatility before use	Zero
Volatility after use	Varies with flammability of liquid bonded

Power/Fuel Requirements and Transportability

No power or fuel is required to directly apply or use the Nochar agents. Nochar's products are transportable as a solid according to local regulations.

Manufacturer

Nochar, Inc.
10333 North Meridian, Suite 215
Indianapolis, IN 46290

Telephone (317) 573-4860
Facsimile (317) 573-4865

REFERENCES

(1) Manufacturer's Literature.

Section 5.2

Treatment/Disposal - Liquid-Solid Separation Systems

ALFA-LAVAL CENTRIFUGES   No. 118

DESCRIPTION

A complete line of centrifuges that can be used for dewatering, filtering, drying, or washing solids, or for separating different phase liquids.

OPERATING PRINCIPLE

Centrifugal force is used to promote solid/liquid separation. The feed slurry is introduced into a rotating vessel and solids/contaminants are pulled out of the liquid by centrifugal force and collected.

STATUS OF DEVELOPMENT AND USAGE

Commercially available products marketed for reclaiming cutting and lubricating oils, recovering valuable particulates from industrial processes, and treating waste.

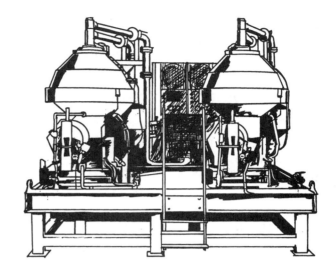

TECHNICAL SPECIFICATIONS

Physical

Alfa-Laval makes more than 200 models of centrifuges, including solids-retaining separators, solids-ejecting separators, nozzle bowl separators, and decanter types.

Operational

Solids-retaining Separators - This is the original type of centrifuge separator, developed to meet the requirements of the chemical industry. They are used for liquid/liquid or liquid/solids separation when the solids content in the feed is very low, i.e., up to a maximum of 1% by volume. Alfa-Laval has a range of solids-retaining separators with hydraulic capacities of up to 1,000 L/min (260 gpm) and solids-handling capacities up to 64 kg (17 gal). They feature high-grade stainless steel gaskets in resistant rubber materials and explosion-protected equipment such as flame-proof motors, gas-tight and nitrogen-purged frames, and spark-proof machine parts.

Solids-ejecting Separators - These separators are normally used when a continuous separation of a process liquid is desired with discontinuous discharge of a solids-phase. It is especially applicable if the required clarity of the separated liquid is very high; the solids are sub-micron particles; the solid particles are sticky, hazardous, or otherwise difficult to deal with; and the feed has a high solids volume and/or the solids volume varies and cannot be predicted. These separators are available for flow rates from 3.8 to 1,665 L/min (1 to 440 gpm) containing solids from 1 ppm to approximately 15% v/v. They are used for liquid/solid separation as well as for liquid/liquid/solid separation.

Nozzle Bowl Separators - These separators are primarily suited to applications with high flow volumes containing large amounts of fine solids. They can process flow rates from 3.8 to 3,330 L/min (1 to 880 gpm) and solids concentrations ranging from 5 to 30% v/v. Alfa-Laval nozzle bowl separators are suitable for classification of particles down to the 0.1 micron range.

Decanter Centrifuges - These centrifuges provide a means for the continuous dewatering of crystalline materials and plastics, dewatering fibrous and amorphous materials, solids concentration, liquid clarification, and wet classification. The decanter centrifuge is designed to handle a wide range of solid particle sizes from 5 mm diameter to a few microns. Unit capacities vary from tenths of gallons per minute to hundreds of gallons per minute, with slurry concentrations having solids contents from 0.5 to 50% v/v. Solids are continuously separated from the liquid phase by centrifugal force ranging to well beyond 3,000 times the force of gravity. Decanter centrifuges are available in either horizontal or vertical configurations.

Power/Fuel Requirements

The systems operate on electric power. Drive motors are available in all voltages and frequencies.

Transportability

As Alfa-Laval centrifuges are available in more than 200 models, their transportability varies considerably. Most systems, however, can be manufactured in a mobile or transportable configuration, including skid-mounted and containerized units. Alfa-Laval also builds ultralight and compact centrifugal separators for use on offshore drilling rigs.

PERFORMANCE

Performance varies from model to model. Refer to the operating specifications provided here, or contact the manufacturer for specific performance data.

CONTACT INFORMATION

Manufacturer

Alfa-Laval Separation Inc.
955 Mearnes Road
Warminster, PA 18974-0556
U.S.A.

Telephone (215) 443-4019
Facsimile (215) 443-4112

Canadian Representative

Alfa-Laval Separation Incorporated
101 Milner Avenue
Scarborough, Ontario
M1S 4S6

Telephone (416) 297-6345
Facsimile (416) 299-5864

OTHER DATA

Alfa-Laval will evaluate customer samples/products to assist in selecting an appropriate separator. This evaluation can range from the analysis of a 1.9 L (0.5 gal) sample to an on-site, full-scale equipment test.

REFERENCES

(1) Manufacturer's Literature.

ANDRITZ MOBILE DEWATERING SYSTEMS

No. 119

DESCRIPTION

Complete, trailer-mounted, mobile dewatering systems for dewatering sludges.

OPERATING PRINCIPLE

The mobile systems use feed pumps, chemical conditioning, and belt-filter press technology to feed and dewater sludge, and a discharge conveyor to discharge the dewatered material.

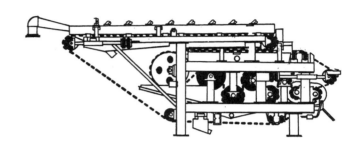

STATUS OF DEVELOPMENT AND USAGE

Commercially available products marketed primarily for use in dewatering industrial lagoon sludges before further treatment or disposal.

TECHNICAL SPECIFICATIONS

Physical

The filter press belt has a 2-m (6.6-ft) wide processing surface. The full production unit is mounted on a trailer. The mobile systems can be customized to suit specific application needs.

Operational

The system can be towed to the site for processing of the sludge on site. The system uses belt-filter press technology to physically dewater sludges. The trailer features removable side panels to allow access to all equipment.

Power/Fuel Requirements

The power/fuel requirements vary according to the unit configuration. Specific requirements are available from the manufacturer.

Transportability

The system is designed to be transported on the highway. A transport truck is required to tow the mobile trailer.

PERFORMANCE

Performance varies from model to model. Specific performance data are available from the manufacturer.

CONTACT INFORMATION

Manufacturer

Andritz Ruthner, Inc.
1010 Commercial Boulevard South
Arlington, TX 76017
U.S.A.

Telephone (817) 465-5611
Facsimile (817) 472-8589

Canadian Representative

Envirocan Wastewater Treatment Equipment Company Limited
26 McCauley Drive
Bolton, Ontario, L7E 5R8

Telephone (416) 880-2418
Facsimile (416) 880-2327

REFERENCES

(1) Manufacturer's Literature.

DENVER SAND-SCRUBBER™ SYSTEM

No. 120

DESCRIPTION

A separator system for removing oils from sand.

OPERATING PRINCIPLE

Oil-contaminated sand is fed into the system in batches and allowed to cycle within the unit. Clean sand and recyclable oil are discharged.

STATUS OF DEVELOPMENT AND USAGE

Commercially available products marketed for on-site removal of oil from sands raised to the surface in offshore oil exploration and production.

TECHNICAL SPECIFICATIONS

Physical - The system measures 2.7 × 3.4 × 4 m (9 × 11 × 13 ft).

Operational - The batch system cycles two barrels (1 production barrel = 42 U.S. gal and 160 L) of produced sand at a time. Cycle time varies from 10 to 45 minutes. The U.S. Minerals Management Service (U.S. MMS) has authorized the discharge of the clean

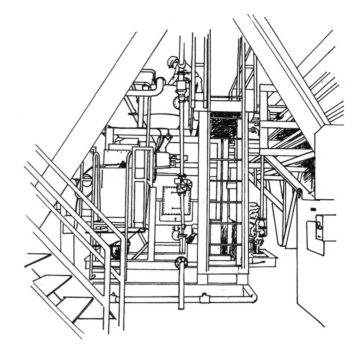

processed sand back into the waters of the Gulf of Mexico. It is not known how this system will work on soils with a greater particle size distribution than sand.

Power/Fuel Requirements - The system operates on electric power.

Transportability - The system is mounted on skids and can be transported by truck.

PERFORMANCE

In June 1991, oil and grease content and particle size distribution analyses were performed by two independent test labs on an offshore oil production platform in the Gulf of Mexico where the Denver sand-scrubber system was in use. The testing was witnessed by the U.S. MMS. The content of oil entrained with the solids feed was 72,000 mg/L and the treated sand contained 27 mg/L at discharge. When the discharged sand was subjected to a sheen test, in which it was immersed in water, no oil or sheen was found on the water surface.

CONTACT INFORMATION

Manufacturer

Svedala Pumps & Process
621 South Sierra Madre
Colorado Springs, CO 80903
U.S.A.

Telephone (719) 471-3443
Facsimile (719) 471-4469

REFERENCES

(1) Manufacturer's Literature.

FLO TREND HYDROCLONE CENTRIFUGAL SEPARATOR, VIBRATORY SCREEN, AND CONTAINER FILTER

No. 121

DESCRIPTION

The **Hydroclone Centrifugal Separator** is a fixed wall device that separates particles according to density and mass and is designed for separating solids from liquids. The **Vibratory Screen** is a mobile, transition screen for separating liquid/solids and for dewatering slurries. The **Container Filter** is a patented, one-step receptacle for separating, dewatering, and transferring sludge. It can be used to separate, clarify, and recover either the liquid or solid phase or to separate one from the other for disposal.

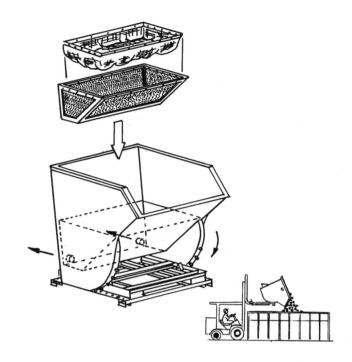

OPERATING PRINCIPLE

Unlike in other centrifuges, in the **Hydroclone Centrifugal Separator**, the liquid mixture rotates within a fixed, conically shaped chamber. Liquid entering the hydroclone forms a primary vortex along the cylindrical and conical wall forcing the heavier phase against the wall and downward toward the solids outlet orifice on the bottom. The orifice also permits air to enter the cyclone creating a secondary, rising, low-pressure vortex within the cleaned liquid. The secondary vortex has a boundary separate from the air core so that the primary vortex, the secondary vortex, and the air core all exist simultaneously, producing thorough separation of a wide range of liquid/solids. The simultaneously acting forces move the liquid phase and lighter solids inward toward the vertical axis and upward where they are discharged through the top of the cyclone.

The **Vibratory Screen** has an inclined dewatering screen and two horizontal dewatering screens. The horizontal screens are spring-mounted over a collection pan and vibrated by attached vibratory motors. The slurry enters the system through a feed or mud box and then flows over an adjustable weir onto the inclined screen. The high energy impact against the screen surface rapidly separates the solids from the liquid fraction, with most of the liquid passing through the inclined screen. As the partially dewatered material is introduced to the middle screen, the slurry goes through a transition from peak acceleration onto the horizontal screen. This change in acceleration imparts G forces and allows the remainder of the material to be dewatered and dried.

A liquid suspension is pumped or conveyed into the **Container Filter** for dewatering. The solids are retained by filter media in the container, while filtered liquid leaves the container through ports located below the filter support basket. The dewatering process can be accelerated by using a vacuum pump, such as the Amflow-Lift (No. 96), or an air diaphragm pump. When the slurry level rises above the perforated zone in the filter support basket, a vacuum chamber is formed between the basket and the inside container wall, enhancing the separation process.

STATUS OF DEVELOPMENT AND USAGE

Commercially available products designed specifically for liquid/solid separation in waste management applications.

TECHNICAL SPECIFICATIONS

Physical - **Hydroclone Centrifugal Separators** are typically mounted in series on a common manifold. Individual hydroclones are available in 5, 7.6, 10, or 13 cm (2, 3, 4, or 5 in) sizes. Flo Trend will fabricate unitized hydroclone units with other liquid/solid separation systems such as the Vibratory Screen separator. These systems, including the hydroclones, manifold vibratory screen, and collection reservoir, are completely self-contained on a single skid.

The **Vibratory Screen** system is completely self-contained on a single skid. The system is 3.3 m (11 ft) long, 2 m (6.5 ft) wide, 1.4 m (4.7 ft) high, and weighs 2,300 kg (5,100 lb). The screen deck with the attached vibratory motors is suspended by four coil springs, mounted on a base frame. The frame has a feed box with an adjustable weir and collection pan positioned in front of the entire screen frame. There are three screen panels with a total screen area of 3.3 m² (35 ft²) and a screen area per panel of 1.1 m² (12 ft²). The deck angle adjustment is -1° to +6°. The system has two 1.5 kW (2 hp) motors, with an rpm of 1,728 and a vibration amplitude of 9 mm (0.4 in).

The **Container Filter** is available in hoppers, luggers, dumpsters, roll-off boxes, and sealed vessels of various sizes. Custom units can be fabricated to specifications. Standard units are constructed of carbon steel or stainless steel. A wide range of filter fabrics is available for selected micron size filtration and product compatibility. The filter media or material can be either reusable or disposable.

Operational - The **Hydroclone Centrifugal Separator** can separate solids in the range of 38 to 80 microns (1.5 to 1.7 mil) from liquids. The solids discharge is in a wet state and may contain up to 80% solids by weight. The **Vibratory Screen** was developed for high flow rates, removal of fine solids, and a drip-free cake. The adjustable weir provides an even and uniform feed onto the screens. Speed can be adjusted with dual motor controls. The system can separate solids in the range of 6.4 mm to 44 microns (0.3 in to 1.7 mil), at flow rates from 950 to 2,840 L/min (250 to 750 gpm). All electrical components are explosion-proof. For the **Container Filter**, the characteristics of suspended particles in a sludge determine the rate of separation. Some applications require only gravity for drainage, while others require filter aids, flocculents, and a liquid vacuum pump to increase the separation rate. The pump can also be used to transfer the liquid as it is being filtered. The smaller units feature a self-dumping mechanism, which tips the hopper forward, empties it, returns the hopper to the upright position, and locks it.

Power/Fuel Requirements and Transportability - No power/fuel supply is required by the Hydroclone or the Container Filter. However, the feed liquid must be provided at an appropriate rate and pressure, and in the case of the Container Filter, the effluent must be removed at an appropriate rate. The **Vibratory Screen** requires an electric power supply of 380/420 VAC, 3-phase, 50 Hz or 230/460 VAC, 3-phase, 60 Hz. All systems come mounted on skids for portability.

CONTACT INFORMATION

Manufacturer

Flo Trend Systems, Incorporated
707 Lehman
Houston, TX 77018
U.S.A.

Telephone (713) 699-0152
Facsimile (713) 699-8054

Canadian Distributors

Stady Oil Tools
100, 635 - 6th Avenue
Calgary, Alberta
T2P 0T5

Telephone (403) 262-8025
Facsimile (403) 240-3357

PAP Process Engineering Services
34 Jasmine Road
Weston, Ontario
M9M 2P9

Telephone (416) 743-9601
Facsimile (416) 745-3655

OTHER DATA

Flo Trend manufactures a complete line of liquid/solid separation equipment that can be used individually for pre-treatment separation or in a series for complete liquid/solid separation. Other systems manufactured by Flo Trend include the **Decanting Centrifuges** for liquid/solids separation from 6.4 mm to 44 microns (0.3 in to 1.7 mil) and for oil recovery; **Filter Presses** for dewatering and clarifying with a separation range from coarse to the submicron level; **Cartridge, Bag, and Automatic Backflush Filters** for separation from 0.13 to 1 micron (0.005 to 0.04 mil); **Oil/Water Separators** for liquid/liquid separation from 11 to 1,140 L/min (3 to 300 gpm); the **Jet Shear** for mixing, blending, and shearing; and a **Soil-washing System** which incorporates many of the above systems as required for the specific sites.

REFERENCES

(1) Manufacturer's Literature.

5.2 Treatment/Disposal - Liquid-Solid Separation Systems

KASON DRUM SIFTER   No. 122

DESCRIPTION

A compact, portable batch sifter for drum-filling or bag-dumping operations that require intermittent, dependable, and fast separation of dry materials.

OPERATING PRINCIPLE

The Drum Sifter incorporates a solids hopper, mounted on a spring assembly, which is designed to be placed on top of open-top drums. A vibratory motor shakes the unit, separating solids according to the screen mesh size placed in the sifter. Fines pass through the sifter into the drum while oversize material remains in the sifter.

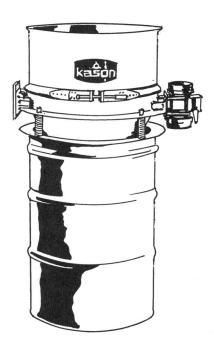

STATUS OF DEVELOPMENT AND USAGE

Commercially available product marketed for drum-filling operations requiring separation of dry materials.

TECHNICAL SPECIFICATIONS
Physical
0.5, 0.6, and 0.8 m (1.5, 2, and 2.5 ft) diameter units are standard. All units can be equipped with Kason screens with mesh sizes ranging from 0.035 to 51 mm (0.0014 to 2 in). The unit can be provided with a stand and casters for ease of movement. Dust-tight versions are also available.

Weight	0.5 m (1.5 ft) diameter unit - 30 kg (65 lb)
	0.6 m (2 ft) diameter unit - 39 kg (85 lb)
	0.8 m (2.5 ft) diameter unit - 59 kg (130 lb)
Frame materials	Carbon steel and 304 stainless steel are standard construction materials. Other materials are available for specific requirements.
Screen materials	304 stainless steel and synthetic are standard materials for the screens. Other materials are available for specific requirements.
Motor	0.08 or 0.06 kW (0.1 or 0.08 hp) with variable-speed controls

Operational

Production rates for the Drum Sifter depend on the material being screened. Examples for a 0.6 m (2 ft) unit include 2,540 kg/h (5,600 lb/hr) of evaporated salt with 28-mesh, tensile bolting cloth screen and 2,270 kg/h (5,000 lb/h) of hard wheat flour with a 20-mesh, tensile bolting screen. Anti-binding devices are available for use with troublesome materials. Hinges can also be added so that the Drum Sifter can be attached to blenders and other process equipment to ensure that only on-size product enters the product stream.

Power/Fuel Requirements

The systems operate on 115-VAC, 1-phase, 60 Hz electric power.

Transportability

The units are compact and can be handled by one person.

CONTACT INFORMATION

Manufacturer

Kason Corporation
67-71 East Willow Street
Millborn, NJ 07041
U.S.A.

Telephone (201) 467-8140
Facsimile (201) 258-9533

Canadian Distributors

Separator Engineering Limited
2220 Midland Avenue, Unit 85
Scarborough, Ontario
M1P 3E6

Telephone (416) 292-8822
Facsimile (416) 292-3382

Separator Engineering Limited
810 Ellingham Street
Pointe Claire, Quebec
H9R 3S4

Telephone (514) 694-4440
Facsimile (514) 694-2074

OTHER DATA

Kason Corporation manufactures a broad line of separation equipment for the chemical, pharmaceutical, petrochemical, plastics, food, mineral, pulp and paper, and other process industries. This equipment includes **Flo-Thru**, **Vibroscreen®** and **Blo-Thru Circular Screen Separators, Centri-Sifter® Centrifugal Sifters**, and **Cross-Flo Sieves**.

REFERENCES

(1) Manufacturer's Literature.

LAVIN CENTRIFUGES

No. 123

DESCRIPTION

Liquid/liquid and liquid/solid centrifugal separators for dewatering, filtering, drying, or washing solids, or for separating different phase liquids.

OPERATING PRINCIPLE

The solid-wall basket centrifuge uses up to 2,200 gravities of centrifugal force to promote solid/liquid separation. The feed slurry is introduced into the rotating basket assembly and is accelerated to the basket speed. Solids/contaminants are pulled out of the liquid and collect on the basket wall. The clarified liquid flows up and out of the basket, is collected by the centrifuge housing, and flows out by gravity. Liquid clarification continues until the basket can no longer retain additional solids and must be emptied, which can be done automatically or semi-automatically.

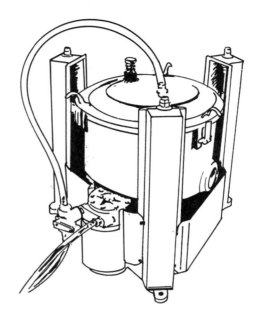

STATUS OF DEVELOPMENT AND USAGE

Commercially available products marketed for use in reclaiming cutting and lubricating oils, recovering valuable particulates from industrial processes, and treating waste.

TECHNICAL SPECIFICATIONS

Physical

Standard Lavin centrifuges are constructed with aluminum bodies and lids and stainless steel baskets. Optional materials include 316 stainless steel for the body, lid and baskets, and a wide variety of epoxy and fluorocarbon coatings for the body and lid. Also available are motors in all voltages and frequencies, variable-speed drives with dynamic braking, an anti-wave device to stabilize the flow of low-viscosity fluids, explosion-proof motor controls, and portable assemblies mounted on hand trucks, platform trucks or skids, with or without tanks and pumps. Models 12-157, 12-413, and 12-413V are 0.6 m wide × 0.6 m high × 0.6 m long (2 × 2 × 2 ft) and Model 20-1160-V is 0.9 m wide × 1.3 m high × 0.6 m long (3 × 4.2 × 2 ft).

	Model 12-157	Model 12-413	Model 12-413V	Model 20-1160V
Drive power	1.1 kW (1.5 hp)	1.1 kW (1.5 hp)	1.1 kW (1.5 hp)	1.5 to 3.7 kW (2 to 5 hp)
Basket sludge capacity	2.6 L (157 in^3)	6.8 L (413 in^3)	6.8 L (413 in^3)	19.0 L (1,160 in^3)
Basket height	6.4 cm (2.5 in)	15 cm (6 in)	15 cm (6 in)	23 cm (9 in)
Basket diameter	31 cm (12 in)	31 cm (12 in)	31 cm (12 in)	51 cm (20 in)
System weight	27 kg (60 lb)	57 kg (125 lb)	61 kg (135 lb)	136 kg (300 lb)

Operational

All Lavin Centrifuges are easy to install and simple to operate. The triangulated, cushioned mounting method eliminates vibration, improves separation, and extends the life of all moving parts. The following are general operating specifications for some Lavin Centrifuges.

	Model 12-157	Model 12-413	Model 12-413V	Model 20-1160V
Maximum flow rate	760 L/h (200 gph)	1,600 L/h (420 gph)	2,700 L/h (720 gph)	11,000 L/h (3,000 gph)
Centrifugal force	2,200 g	2,200 g	2,200 g	1,800 g

Soft or gelatinous sludges can be automatically or semi-automatically removed while the centrifuge rotates. A fully automatic sludge discharge uses a field-programmable, solid-state timer to periodically empty the basket while the centrifuge operates. At the selected time interval, a self-contained hydraulic pump is actuated. The hydraulic pressure closes the inlet valve to momentarily shut off the inlet flow, and simultaneously actuates the automatic discharge mechanism. Sludge is discharged through a reinforced hose to a collection drum. After about 30 seconds, the pump stops, the inlet valve opens, the discharge mechanism retracts, and operation continues. The semi-automatic sludge discharge allows periodic cleaning of the centrifuge basket with a hand-operated discharge mechanism. To empty the basket, the operator shuts off the inlet flow and manually rotates a lever to discharge the sludge through a reinforced hose to a collection drum. After about 30 seconds, the handle is released, inlet flow is resumed, and operation continues.

A diffuser is recommended for materials that are difficult to separate. By uniformly diffusing the feed material into the basket, rapid acceleration and separation are achieved with minimal turbulence. Uniform distribution eliminates preferential buildup of the solids and ensures vibration-free operation.

By using a perforated basket instead of the standard solid-wall centrifuge basket, the centrifuge can be used for filtering, drying, and washing. Entering liquid must then escape through the walls of the basket. The liquid can be made to pass through one or more barriers before being discharged. For filtering, the inside surface of the perforated basket can be lined with any filter paper or cloth in pre-cut strip or bag form. A wide variety of materials and micron ratings is available. For drying, liquids-solvents, oils, or water are extracted by centrifugal force and discharged through the basket perforations. For washing, spray jets or nozzles thoroughly wash the contents of the basket. Automatic controls provide desired washing/drying sequences. An optionally available sensor indicates when the basket is clogged or full.

Power/Fuel Requirements

The systems operate on electric power. Drive motors are available in all voltages and frequencies.

Transportability

The units are compact, with the smallest centrifuge requiring less than .28 m^3 (3 ft^2) of floor space. Since no special anchoring is necessary, they are readily portable. The centrifuges are also available as systems mounted on hand trucks, platform truck or skids, with or without tanks and pumps.

PERFORMANCE

The efficiency of separation attained is determined by the difference in the specific gravity of the two materials to be separated. A minimum difference in specific gravity is required to achieve a successful separation. Large specific gravity differentials and low flow rates make it possible to remove 99% (and more) of contaminants. Typically, solids down to 1 micron are removed, although submicron separations can be achieved under ideal conditions.

CONTACT INFORMATION

Manufacturer

A.M.L. Industries Incorporated
3500 Davisville Road
Hatboro, PA 19040
U.S.A.

Telephone (215) 674-2424
Facsimile (215) 674-3252

OTHER DATA

A.M.L. Industries offers professional reconditioning and original parts replacement for most basket type centrifuges and separators. To determine whether a material can be treated with a centrifuge, for a small fee, A.M.L. Industries will process a 38 L (10 gal) sample in its testing laboratory in Hatboro, Pennsylvania and send out a report on the results, as well as the clarified liquid and sludge. A.M.L. also manufactures the **Lavin Miniature Lab Centrifuge** for bench-top use.

REFERENCES

(1) Manufacturer's Literature.

Section 5.3

Treatment/Disposal - Fundamental Water Treatment Processes

FUNDAMENTAL WATER TREATMENT PROCESSES – GENERAL LISTING   No. 124

DESCRIPTION

Fundamental water treatment process technologies are summarized in this section with an emphasis on their application to chemical spill response. Water treatment technologies use biological, chemical, and/or physical processes to destroy, remove, or concentrate contaminants. Although in theory, physical and chemical processes are separate, in actual practice it is difficult to distinguish between them or to determine which is predominant. Water treatment processes have therefore been broken down into two categories: **Biological** and **Physical/Chemical.**

Biological - Biodegradation and Biosorption

Physical/Chemical - Adsorption, Air Stripping and Steam Stripping, Chelation, Chemical Reduction/Oxidation, Clarification/Flotation, Filtration, Flocculation/Precipitation, Ion Exchange, Neutralization, and Photochemical Oxidation/Reduction

BIOLOGICAL WATER TREATMENT

Biodegradation

Biodegradation removes organic matter from water through microbial oxidation. Dissolved organic matter is biologically oxidized to simple organic or inorganic end products by microorganisms growing in either aerobic (with oxygen) or anaerobic (without oxygen) environments. These microorganisms are naturally occurring or genetically engineered. Aerobic oxidation is most common as it is easier to use and the end products are more acceptable than those from anaerobic processes, e.g., water, carbon dioxide, and sulphates versus methane, mercaptons, and hydrogen sulphide.

In the most conventional process, activated sludge flows into an aeration basin where it is mixed with active acclimated microorganisms and is aerated for several hours. During aeration, the biodegradable portion of the waste is destroyed biologically. When the aeration cycle ceases, tank contents are allowed to settle. Solids go to the bottom of the tank, while the clarified effluent is drawn off and discharged to the receiving stream for disposal. The solids, including micro-organisms, are retained in the tank for use with the next batch of waste, except for that portion which is periodically withdrawn for dewatering and disposal.

Many organic compounds are considered to be amenable to biological treatment, although the relative ease of biodegradation varies widely. Biological treatment, whether aerobic or anaerobic, is not suitable for treating high-strength organic waste or aqueous waste streams with suspended solids concentrations of more than 1% (10,000 ppm) and high oil and grease content. They are also not suitable for treating waste streams with high concentrations of chlorinated compounds, or inhibitory concentrations of other organics and inorganics or pH extremes (less than 6 or greater than 9). Biological processes are also subject to failure from "shock loads".

Biological treatment processes are used in many water treatment technologies, as well as in-situ treatment of contaminated groundwaters and soils. The following entries in Section 5.4 are examples of this type of treatment: **Bioacceleration Treatment System** (No. 130) and the **PACT Batch Wastewater Treatment System** (No. 150).

Biosorption

Biosorption is based on the strong natural affinity of biological materials, such as the cell walls of plants and microorganisms, for heavy metal ions. Biological materials, both living and nonviable biomass, can be incorporated into particulate forms such as gels or beads, to produce a "biological" ion exchange resin which can be used to effectively remove dissolved metal ions from water. The biological ion exchange material can be packed into columns through which waters containing heavy metal ions are flushed. The heavy metal ions are adsorbed to the biomass so that the water leaving the column is almost metal-free and can be reused or discharged.

See the entry on the **Algasorb Process** (No. 128, Section 5.4) for an example of biosorption.

PHYSICAL/CHEMICAL WATER TREATMENT

Adsorption

Adsorption is a surface phenomenon by which soluble substances in a solution are adsorbed, or collected, on a suitable surface. Although silica and clays have served as adsorbents, granular activated carbon (GAC) or powdered activated carbon (PAC) are the most common adsorbents. Activated carbon is produced from cellulosic materials, such as wood, coal, peat, and lignin, which have been dehydrated and carbonized. Pore openings are then enlarged by activation to increase the surface area and, subsequently, the adsorptive capacity.

Two common configurations of adsorption systems are batch and continuous flow. In a batch system, incoming water is introduced to a reactor, where it is mixed with the adsorbent, allowed to settle, and withdrawn after the contaminants have been adsorbed. In continuous-flow systems, incoming water flows continuously through a porous bed containing the adsorbent material.

Continuous-flow systems are often used in spill response applications. The following entries in Section 5.4 are examples of this form of treatment: **Calgon Activated Carbon Adsorption Treatment Systems** (No. 133), **Carbtrol L-1, L-4, and L-5 Activated Carbon Adsorption Canisters** (No. 134), **Liquid-Miser Activated Carbon Adsorption Canisters** (No. 146), **PACT Batch Wastewater Treatment System** (No. 150), and **Water Scrub Units - Activated Carbon Adsorption Canisters** (No. 153).

The effectiveness of either batch or continuous-flow systems is based on accurate determination of the amount of adsorbent required to provide active adsorption (mass transfer zone) and the minimum amount of carbon that can be used without allowing the constituents being removed to flow through the adsorbent material (breakthrough). Because mixtures of pollutants react differently than single-pollutant streams, pilot-scale studies for multiple-pollutant streams are usually required to determine the mass transfer zone and breakthrough characteristics of the waste stream. Once the carbon's removal capacity has been exhausted, the carbon is referred to as "spent" and must be replaced, disposed of, or regenerated for reuse. Thermal regeneration, which is the most common method used, can be done in the field through the use of mobile systems.

When using carbon adsorption methods for spill response, the following factors should be considered.

- One to two days lead time may be required when ordering carbon for field use, as well as an additional 24 hours to wet the carbon before use.
- As wetting carbon results in more efficient adsorption, a source of clean water must be available on site.
- Pre-filtration of suspended solids from the influent may be required to prevent fouling of the adsorption system and waste of carbon.
- Other forms of carbon, such as "tea bags", powdered, and other packaging concepts, have also been tried for in-situ treatment of spilled materials. In general, however, these forms have a lower capacity than engineered systems.

Air Stripping and Steam Stripping

Both air stripping and steam stripping are separation processes that use the differences in thermodynamic properties of liquids and gases to separate VOCs from wastewaters. In the air stripping process, contaminated water is mixed with air by cascading it down through a packed column countercurrent to a flow of air from a blower at the bottom, or by flowing it over aeration trays. This flow allows the air to come into contact with the contaminated water which results in the VOCs being transferred from the water phase to the air. The VOCs then exit the column with the air stream. Generally, the off-gas from the air stripper has to be further treated, which is usually done by adding a carbon adsorption unit to the air vent system.

Air stripping is a popular technology as it is well suited to unmanned, on-site treatment. The following entries in Section 5.4 are examples of this technology: **Air Plastics - Air Stripping Towers** (No. 127), **Breeze Air Stripping System** (No. 131), **Carbtrol Diffused Air Strippers** (No. 135), **ESI Cascade LP500 Series Low Profile Air Strippers** (No. 140), **Hazelton Maxi-Strip Systems** (No. 142), **Hydro-Stripper Air Stripper** (No. 144), **ORS LO-PRO Air Strippers** (No. 147), and **ShallowTray Low Profile Air Strippers** (No. 151).

Steam stripping uses the same operating principle as air stripping, but is a more flexible technology as it is capable of removing a wider range of compounds, including alcohols, phenols, aldehydes, and amines. In the steam stripping process, the steam is fed into the bottom of the column and the feed stream enters the middle or top of the column. When the liquid feed stream comes into contact with the steam in the column, the volatile and some semi-volatile compounds are transferred from the liquid to the vapour phase. The outlet vapour stream (tops) is condensed and sent to a decanter where any sufficiently concentrated organics are separated from the aqueous stream. The treated liquid outlet (bottoms) are either re-injected into the groundwater well, or if the levels are above discharge limits, treated further. If this stream is very low in organics and de-ionized, it can be used to feed the boiler and close the process cycle. **Environment Canada's Steam Stripping Units** (No. 138) are examples of steam stripping.

Chelation

Chelating agents are compounds or ligands (usually organic) that bind a metal ion to more than one position on the chelating agent. This binding prevents the metal from reacting chemically, thus reducing its toxicity. Chelating agents may be useful for detoxifying spills containing heavy metals, but the resultant ligand-metal complexes must ultimately be removed by other water treatment techniques.

The two groups of chelating agents are sequestrants, which bind the metal ion but remain soluble in water, and precipitants, which form insoluble complexes. EDTA (ethylenediaminetetraacetic acid) has been identified as a promising sequestering chelating agent. It forms complexes with copper, nickel, zinc, cadmium, iron, and other heavy metals. EDTA is less effective with silver, antimony, and titanium, and compounds such as potassium dichromate, potassium permanganate, chromic anhydride, and sodium ferrocyanide. Oxine (8-hydrocyquinoline), a precipitant, performs like EDTA, but is limited in application due to its low solubility. EDTA is widely available and is lower in toxicity than oxine.

Chemical Reduction/Oxidation

Reduction/Oxidation ("Redox") reactions constitute a transfer of electrons from one species to another. The species **accepting electrons**, i.e., becoming more negatively charged, are **reduced** and the species **losing electrons**, i.e., becoming more positively charged, are **oxidized**. For example:

$$2MnO_4^- + CN^- + 2OH^- \rightarrow 2MnO_4^{2-} + CNO^- + H_2O$$
Permanganate Cyanide → Manganate Cyanate

Cyanide acts as a reducing agent, transferring one electron to permanganate and reducing it to manganate. Simultaneously, cyanide is oxidized to cyanate which has one less electron than cyanide. Therefore, oxidizing agents themselves are reduced when inducing oxidation and reducing agents are similarly oxidized when reducing another species. Redox reactions can be used to change hazardous species into less harmful forms. They are generally most efficient when controlled in an engineered wastewater treatment system.

Oxidation - Oxidizing agents include fluorine, hydrogen peroxide (H_2O_2), permanganate (MnO_4^-), chlorine, dichromate ($Cr_2O_7^{2-}$), and air. Fluorine is the strongest oxidizer, but is rarely used for industrial applications because of the hazards of handling and controlling it. Air is the safest oxidizer and offers other benefits, such as precipitating ferrous iron, oxidation of sulphites and hydrogen sulphide, and stripping of VOCs from water. As air is only slightly soluble in water, however, oxidation by aeration requires equipment to deliver large volumes of air. Oxidizing agents may react violently in the presence of significant quantities of readily oxidizable organics. These agents should, therefore, be well mixed and added slowly to avoid momentary excesses and violent reactions. Oxidation reaction times occur within seconds to minutes.

Waste solutions to be treated by chemical oxidation must be pH-adjusted to ensure efficient oxidation. For example, phenol oxidizes at pH 5 to 6, while cyanides require more alkaline conditions (pH 8.5 to 10). This can limit the applicability of oxidation as a response technique in a spill situation where the need for immediate response and limited control make it difficult to prepare the spill site.

The most common products of chemical oxidation are carbon dioxide, water, partially oxidized organics, i.e., short chain organics from long chain compounds, and insoluble inorganics. The latter two products may require further treatment. Oxidizing agents, such as oxygen and hydrogen peroxide, have the advantage of not introducing additional foreign ions to the mixture.

Reduction - Commonly used reducing agents include sulphate, sulphites, sulphur dioxide, and waste iron. These agents lower the oxidation state of a substance, reducing its toxicity or solubility, or transforming it into more easily handled forms. Reducing agents have been used primarily for treating inorganics, notably chromium. The end products of reduction are similar to those of oxidation reactions.

Clarification/Flotation

In most cases, the separation process uses gravity. Solids either float to the surface or settle on the bottom of the tank, depending on the difference in density of the liquid and solids. For systems treating solids that tend to float to the top of the tank, fine air bubbles (dissolved air flotation) may be introduced into the liquid. These air bubbles cause solids that are heavier than water to rise, or speed up the rise of lighter solids to the surface. Bubbles attach to the solids and the buoyant force of the combined particle-bubble causes it to rise. At the surface, the captured solids can be removed for disposal, reuse, or further treatment.

Clarifiers and centrifuges are examples of devices that separate solids by gravity. Coalescing oil/water separators use coalescence and flotation to separate oil and water. The following entries in Section 5.4 are examples of these devices: **GLE Slant Rib Coalescing (SRC) Oil/Water Separator** (No. 141), **Hudson CS Series Coalescing Oil/Water Separators** (No. 143), and **PACE Oil/Water Separators** (No. 149).

Filtration

Filtration is a method for removing solids from a liquid by passing the liquid through porous media or a semi-permeable membrane. The driving force of filtration is either gravity or applied pressure. Most municipal wastewater filters are operated by gravity, whereas pressure filters are common in industrial applications.

Granular media, membranes, or diatomaceous earth are used as filtration media. Sand and anthracite coal are the most common granular media used. Straining is the predominant mechanism for removing suspended solids. As large particles are removed from the liquid, smaller particles penetrate the bed and are removed within the bed. Other mechanisms, including sedimentation within the bed and chemical adsorption, are also involved in the granular media filtration process.

Membrane filtration technology has expanded dramatically as more selective techniques are required to deal with the ever-increasing complexity of water treatment. The technique now includes reverse osmosis (RO), also called hyperfiltration, nanofiltration (NF), ultrafiltration (UF), and microfiltration (MF). Membrane technology can be used to remove heavy metals, organic compounds, salts, and large molecules, such as dissolved oils.

Crossflow membrane filtration is a self-cleaning process suitable for many spill applications. The components of a fluid are separated with semi-permeable membranes through the application of pressure and flow. In crossflow, the influent or "feed" stream is separated into two effluent streams: the "permeate", which passes through the membrane, and the "concentrate", which retains the dissolved and suspended solids rejected by the membrane. The process is termed "crossflow" because the feed and the concentrate flow parallel to the membrane instead of perpendicular to it. The pressure required to drive the process is determined by the specific nature of the feed solution and the pore size of the membrane. If the solution contains significant levels of ionic solutes, the operating pressure must first overcome the osmotic pressure, which is the natural energy potential between the more concentrated and the less concentrated streams.

In crossflow filtration, the solutes and solids are swept away with the concentration stream, eliminating the need to frequently change the filter media or to regenerate resin. Crossflow membranes produce precise separations at the ionic, molecular, and macromolecule levels. Systems can be designed to concentrate all feed stream materials, purify solvents, or selectively pass some materials while concentrating others. The following table summarizes the capabilities of the various crossflow technologies and their possible uses.

Crossflow Separation Process	Membrane Pore Size	Fluid Component Retained	Transmembrane Pressure	Sample Applications
Reverse Osmosis	5 to 15 Å	Ions and most organics over 100 molecular weight	1,400 to 6,900 kPa [200 to 1,000 psi(g)]	Metal ion recovery BOD & COD reduction in waste streams
Nanofiltration	10 to 90 Å	Organics with 300 to 1,000 molecular weight	1,400 to 5,500 kPa [200 to 800 psi(g)]	Water softening Desalting of organics
Ultrafiltration	45 to 1,100 Å	Most organics over 1,000 molecular weight, including pyrogens, viruses, bacteria, and colloids	70 to 1,400 kPa [10 to 200 psi(g)]	Concentration and recovery of industrial organics and dilute suspended oils
Microfiltration	400 to 20,000 Å	Very small suspended particles, some emulsions, and some bacteria	7 to 170 kPa [1 to 25 psi(g)]	High volume removal of bacteria and small suspended solids

The following entries in Section 5.4 are examples of filtration systems: **Environment Canada's Reverse Osmosis Units** (No. 137), **Environment Canada's Microfiltration/Ultrafiltration Unit** (No. 139), **Koch Tubular Ultrafiltration (UF) System** (No. 145), **Osmonics Crossflow Membrane Filtration Systems** (No. 148), and **TriWaste Micro-flo Mobile Water Treatment System** (No. 152).

Flocculation/Precipitation

Flocculation (or coagulation) and precipitation are processes that prepare solids for subsequent removal. Flocculation is used to destabilize small suspended solids (colloids) so that they will agglomerate and settle. Precipitation converts dissolved solids into suspended solids that can be removed by subsequent processes. Both flocculation and precipitation rely on subsequent solid/liquid separation to actually remove solids. Flocculation and precipitation are often integrated into a single water treatment system, in which case, chemical storage and feed systems vary with the type of chemicals used. The mix tanks should provide rapid and uniform dispersion of the chemicals in the process fluid, while flocculation will provide the energy to bring small solids together, creating larger solids. The solids formed by coagulation or precipitation are removed by solid/liquid separation. For dilute solids streams, the separator may be a settling tank or air flotation unit. For more concentrated slurries, a centrifuge may be used.

Flocculation - Flocculation and coagulation are both terms that describe the mechanism of agglomerating suspended particles into larger particles. No differentiation is made between the two terms. Flocculation is more commonly used in North America since most manufacturers market coagulants as flocculating agents.

Due to the nature of water and colloid surface chemistry, most colloidal solids in dilute aqueous solutions have a net negative charge. As a result, these solids tend to repel each other and settle slowly because of their size. Flocculation is the addition of chemicals that reduce the surface charge and allow the solids to agglomerate. As the solids agglomerate and increase in size, they settle faster and are easier to remove.

Commonly used inorganic flocculents include alum, ferric chloride, and lime. To use alum (aluminum sulphate) as an example, fluffy, gelatinous "floc" of aluminum hydroxide is formed upon mixing with wastewater. Due to its large surface area, this floc enmeshes smaller particles and creates larger ones. Alum (aluminum sulphate) is effective in clarifying both organic and inorganic suspensions. When using alum, the pH should be controlled in the range of 6.5 to 7.5. Alum dosages of 100 to 1,000 mg/L (.013 to 0.134 oz/gal) should be effective for dilute suspensions. Suspensions low in alkalinity may require the addition of lime or caustic to produce the final pH range of 6.5 to 7.5. As it is difficult to determine the precise chemical dose necessary for agglomeration, many operators simply increase the chemical dose. The additional chemical often results in the precipitation of inorganic polymers that form large particles and entrap colloidal solids.

Ferric chloride is effective in clarifying both organic and inorganic suspensions. Treating dilute suspensions with ferric chloride requires dosages of approximately 50 to 500 mg/L (.007 to .067 oz/gal), although larger dosages may be required for concentrated or highly alkaline suspensions. The pH should be above 6 for best results and can be controlled by adding lime or caustic to produce the required pH. Excessive doses of ferric chloride result in a brown-coloured effluent and should be avoided.

Organic polyelectrolytes consisting of long-chain, water-soluble polymers, such as polyacrylamides, are more recently developed flocculents. They are available in anionic, cationic, or nonionic form and may be effective alone when flocculating suspensions of inorganic materials, such as clay, soils, colloidals, and metal salts. These polyelectrolytes are usually not effective alone when flocculating organic suspensions, but can be used with alum or ferric chloride for treating organic suspensions. Dosages vary with both the type of charge on the polymer and the type of suspension to be treated. Cationic polyelectrolytes are generally added in higher dosages, 1 to 10 mg/L in dilute situations (less than 100 mg/L suspended solids), and anionic or nonionic polymers are added at approximately 0.5 to 5 mg/L. When the solution is concentrated and the concentration of suspended solids is greater than 1,000 mg/L, 1 to 300 mg/L of a cationic polyelectrolyte or 1 to 100 mg/L of an anionic or nonionic compound are added.

Selecting the best flocculent to use and determining the correct dosage must be done on a case-by-case basis. Small tests conducted in a mason jar often assist in this process.

Precipitation - Dissolved solids, which may be present as either cations or anions, must be precipitated before they can be removed. Like flocculation, precipitation involves the addition of chemicals, which vary depending on the ionic species present in the wastewater. Depending on concentration, most dissolved solids precipitate easily. The subsequent precipitates may settle or require additional treatment.

Ion Exchange

Ion exchange is a process that removes unwanted ions from wastewater by transferring them to a solid resin material "in exchange" for an equivalent number of innocuous ions such as H^+, OH^-, Na^+, which are held by electrostatic forces to functional groups on the surface of a solid ion exchanger material. Ion exchange material is usually a synthetic organic resin that can be a weakly or strongly acidic cationic exchanger or a basic ionic exchanger. Natural substances, such as zeolite, clay, and protein, have also been used. The ion exchanger has a limited capacity for storage of ions and eventually becomes saturated. Resin regeneration consists of "elution" of the adsorbed materials with an organic solvent or inorganic salt solution containing innocuous ions, which replace the accumulated undesirable ions, returning the exchange material to a usable condition.

Ion exchange can be used to remove or concentrate the following groups of contaminants that may be found in spill situations. The upper practical limit of exchangeable ions is about 2,500 to 4,000 mg/L (0.33 to 0.53 oz/gal).

- Inorganics - all metallic elements when present as soluble species, either cationic or anionic, and anions such as halides, sulphate, nitrate, and cyanides.
- Organics (water-soluble and ionic) - acids, such as carboxylics, sulphonics, and some phenols, at a pH sufficiently alkaline, and amines when the solution is acidic enough to form the corresponding acid salt.

Ion exchange is not suitable for treating nonionic compounds, highly concentrated [>4,000 mg/L (>0.53 oz/gal)] waste streams, or streams high in suspended solids or oxidants. Pretreatment for removal of suspended solids may be necessary for longer resin service. Frequent resin regeneration may also be required for very concentrated waste. Certain organics, particularly aromatics, that may be present in a waste stream can be irreversibly sorbed by resins, thereby decreasing their capacity. While the residue produced must be further treated before disposal, the effluent seldom exceeds 10% of the original waste volume. As the units take little time or effort to be put into or taken out of operation, ion exchange systems are convenient for use in mobile treatment systems.

Neutralization

Neutralization is simply the pH adjustment of a solution to bring it back to an acceptable range. A pH range of 6 to 9 is representative of natural waters. Strong bases are the most economical reagents for neutralizing strong acids and vice versa. However, excess application of strong neutralizers can result in extreme pH levels (i.e., outside the range of 6 to 9) that may be as hazardous as the original spilled chemical. Consequently, sodium dihydrogen phosphate (NaH_2PO_4) has been identified as the most promising agent for neutralizing basic or alkaline spills, and sodium bicarbonate ($NaHCO_3$) as the most promising neutralizing agent for acid spills.

Misapplication of 1.0 M sodium dihydrogen phosphate could result in a pH as low as 4.5. This is safer than a pH of 0.3, which is possible with an over-application of 1.0 M sulphuric acid, or a pH of 2.4 for 1.0 M acetic acid. Similarly, an over-application of 1.0 M sodium bicarbonate solution would yield a pH of 8.3 compared to 12.4 for a 0.0024 M solution of lime. While lime is less expensive and more available than sodium bicarbonate, the characteristics of lime could lead to excessive amounts being applied to a spill when sodium bicarbonate could more safely neutralize the spill.

Neutralization of basic spills in streams or ponds using sodium dihydrogen phosphate may result in an increase in algal growth as phosphate is a growth nutrient for aquatic plants. Natural dispersion of streams or ponds dilutes most chemical spills, but the consequences of not neutralizing acidic or alkaline spills or of overdosing such spills with strong acids or bases can be burns to plants, possible increases in the Chemical Oxygen Demand (COD), and resolubilization of heavy metals.

Photochemical Oxidation/Reduction

Photochemical Oxidation/Reduction, which is also referred to as **Advanced Oxidation**, is a destructive process that takes place at ambient temperature. It uses hydroxyl radicals to oxidize organic compounds into carbon dioxide, water, and in the presence of chlorinated organic compounds, salt. An ultraviolet (UV) light source is commonly used to generate these highly reactive hydroxyl radicals from hydrogen peroxide, ozone, or titanium dioxide. In general, the rates of destruction increase with increasing residence time, oxidant dosage, and the intensity of UV light. Destruction rates are highly dependent on chemical structure and the presence of inorganic solids. Typical contaminant classes amenable to treatment include PCBs, pesticides, chlorinated solvents, nitrotoluene, benzenes, toluenes, xylenes, chlorophenols, cyanides, organic acids, ketones, ammonia, ethers, phenols, PAHs, nitrosamines, and dioxins and furans. These systems are best applied when the contaminant loading is less than 1% (10,000 ppm).

These systems are commercially available for on-site treatment of wastewater and contaminated groundwater. The following entries in Section 5.4 are examples of this form of treatment: **Advanced Oxidation Water Treatment System** (No. 126), **Brinecell Oxidizers** (No.132), and **Cyanide Destruction System** (No.136).

The table on page 250 summarizes information for some of the treatment technologies presented in this section (U.S. EPA, 1988).

REFERENCES

(1) Goldman, J.C. and P.T. Bowen, "Exploring Wastewater Treatment - A Treasure Chest of Technologies", in: *Pollution Engineering, 24(15)*, Reed Publishing, Newton, MA (September,1992).

(2) Martin, E.J. and J.H. Johnson, *Hazardous Waste Management Engineering*, Van Nostrand Reinhold Company, New York, NY (1987).

(3) Tedder, D.W., "Separations in Hazardous Waste Management", in: *Hazardous Waste Management, 21(1)*, 23-74, Marcel Dekker Inc. (1992).

(4) U.S. EPA (United States Environmental Protection Agency), "EPA's Handbook Responding to Sinking Hazardous Substances", Pudvan Publishing Company, Northbrook, IL (January, 1988).

Summary of Wastewater Treatment Technologies

Technique	Applications	Limitations	Secondary Impacts	Relative Cost
Activated Carbon Adsorption	Removal of a broad range of dissolved organics from aqueous streams with TOC concentrations of 10,000 ppm or less. Best suited for compounds with low solubility and polarity. Also used to remove some inorganic solute.	Not cost-effective for waste streams with TOCs in excess of 1%. Not suitable for treating waste streams high in suspended solids (>50 ppm) or oil and grease (>10 ppm). Not effective for compounds with low molecular weight or highly soluble compounds.	Process generates an "exhausted" carbon that must be regenerated. Regeneration is usually conducted off site where adequate controls are taken to avoid secondary impacts. Backwash streams must be treated to remove high concentration solids.	Medium to high
Biodegradation	Degradation of oxidizable organics present at non-inhibitory levels.	Not suitable for highly concentrated waste streams or streams containing inhibitory concentrations of metals. Not suitable for treatment of aliphatics; chlorinated polyaromatics are degraded slowly. Subject to failure from shockloads. Initiation time can be long.	Limited localized air emissions may result. Process generates a biomass sludge that contains high concentrations of toxic compounds and sludge must be dewatered and treated before disposal.	Low to medium
Ion Exchange	Removal of ion, both organic and inorganic, at concentrations from 2,500 to 5,000 mg/L.	Not suitable for highly concentrated waste streams or streams high in suspended solids or oxidants.	No significant impacts.	High
Neutralization	Adjustment of pH for acidic or alkaline waste stream.	No significant limitations when properly applied.	Potential for air pollution problems.	Low to high
Precipitation	Removal of dissolved metals from waste streams. No concentration limit.	Difficult to obtain minimum solubility of a metal due to such factors as formation of organometallic complexes and the tendency of each metal to have its minimum solubility at a different pH level.	Process generates a large volume of sludge that must be treated before disposal.	Low to high
Flocculation	Agglomeration of particles into larger particles that are subsequently settled by sedimentation.	No significant limitations when applied properly.	Process generates a large volume of sludge that must be treated before disposal.	Low
Chemical Reduction/Oxidation	Degradation of organics that are difficult to oxidize, organometallic compounds, and reduced inorganics.	Not suitable for high strength waste streams or waste containing suspended solids.	No significant impacts.	High
Air/Steam Stripping	Removal of VOCs, gases, and odours from groundwater, industrial wastewater, accidental spills, process effluents, and runoff from landfills.	With air stripping, vented air containing VOCs may require further treatment. Carbon adsorption is often used for this purpose, which increases cost of system. Not as many compounds can be removed with air stripping as with steam stripping.	Potential for air pollution problems with air stripping.	Medium to high

Section 5.4

Treatment/Disposal - Aqueous Treatment Systems

3L FILTERS
OIL/WATER SEPARATORS

No. 125

DESCRIPTION

Separators with granular activated carbon (GAC) modules for removing hydrocarbon contaminants from groundwater, wash water, and other liquid wastes.

OPERATING PRINCIPLE

In the oil separation unit, suspended solids and hydrocarbon contaminants are removed from the process stream. The GAC unit works both as a polishing phase to remove final levels of emulsified hydrocarbon contamination and to physically adsorb other contaminants to the carbon media.

STATUS OF DEVELOPMENT AND USAGE

Commercially available product in general use.

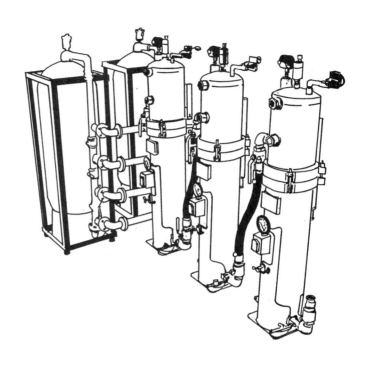

TECHNICAL SPECIFICATIONS

Physical

The number of separation stages (housings) is determined by each specific application. A typical system consists of a pre-filter which removes suspended solid (particulate) contamination in the first stage. In the second-stage, an oleophilic pad removes the large oil droplets and hydrophilic fibreglass coalescing elements separate out the fine micronic oil particles. Oil droplets collect on the outside of the coalescer and gravitate to the top of the vessel to be purged from the unit either manually or with automatic controls. The cartridge remains free of hydrocarbon buildup, reducing the need for cartridge replacement. In the final stage, the GAC adsorbers remove chemical contaminants and emulsified oil.

Operational

The special filtration elements and media remove contaminants that cannot be efficiently removed by sand or carbon filters alone. The entire configuration provides longer life to the granular activated carbon units. The systems can be designed for both fixed and mobile installations and have been successfully used in a variety of applications.

Power/Fuel Requirements

The process stream must be supplied to the basic systems at the appropriate pressure. Units with accessories, such as pumps, controls, pH meters, and turbidity meters, require an electric power supply.

Transportability

Units are available as skid packages, mobile units, or completely housed in prefabricated buildings. Mobile units are operational under Certificate of Approval number A650037 (MOE) for temporary "pump and treat" remedial projects.

CONTACT INFORMATION

Manufacturer

3L Filters Limited
427 Elgin Street North
Cambridge, Ontario
N1R 8G4

Telephone (519) 621-9949
Facsimile (519) 621-3371

REFERENCES

(1) Manufacturer's Literature.

ADVANCED OXIDATION WATER TREATMENT SYSTEMS

No. 126

DESCRIPTION

Advanced Oxidation Processes (AOPs) are radical initiated means of oxidizing small amounts of toxic organic compounds present in groundwater and wastewaters. The following such processes are outlined here: **Nulite TiO_2 Photocatalytic Water Treatment System, Rayox Advanced Oxidation Water Treatment System**, and **Ultrox Advanced Oxidation Water Treatment System**.

OPERATING PRINCIPLE

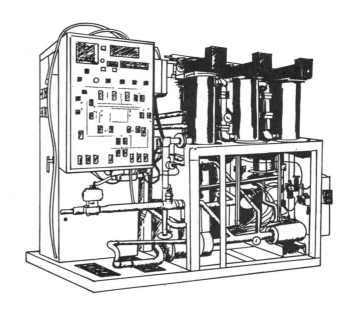

These processes usually involve the use of an oxidant, such as hydrogen peroxide or ozone, with or without a catalyst, e.g., titanium dioxide. In most cases, the resulting solution is irradiated with ultraviolet (UV) light. In the **Nulite TiO_2 Photocatalytic System**, organics in water are converted into carbon dioxide and water, as well as halide ions, if the organic pollutants contain halogen atoms. Contaminated water is piped into a water reactor where it flows over mesh coated with titanium dioxide. This mesh is exposed to a selected band of light which excites the TiO_2 catalyst causing the formation of the hydroxyl radical (•OH) which attacks and breaks down the hazardous organic compounds.

The **Rayox System** uses a proprietary chemical reactor design in which hydrogen peroxide or ozone is used as the primary oxidant. Proprietary ultraviolet lamps and enhanced oxidation additives are used to improve the oxidation process, which ultimately results in destruction of contaminants to carbon dioxide, water, and, in the case of chlorinated contaminants, chloride ion.

The **Ultrox System** uses a combination of UV radiation, ozone, and hydrogen peroxide to oxidize organic compounds in water. Hydrogen peroxide is combined with the contaminated water. Ozone is generated and injected into the treatment tank, into which contaminated water is pumped and irradiated with UV light. The light reacts with the ozone gas and hydrogen peroxide, producing hydroxyl radicals which destroy organic contaminants. Residual ozone in the offgas is converted to oxygen by a catalytic decomposer, eliminating any release of ozone. Treated water flows to discharge.

STATUS OF DEVELOPMENT AND USAGE

Commercially available systems that can be used to destroy or reduce total organic carbon in pure water, drinking water, groundwater, and plant process water.

TECHNICAL SPECIFICATIONS

Physical

Nulite TiO_2 Photocatalytic System - The stainless steel reactor cells and electrical controls are housed in a weather-tight cabinet. Systems are custom-built to meet the needs of specific sites. This entails adding "wafer" panels containing a series of reactor cells, to allow larger flows to be processed.

	6 Cell System	12 Cell System	12 Wafer System
Power	0.6 kW (0.8 hp)	1.2 kW (1.6 hp)	7.7 kW (10 hp)
Weight	45 kg (100 lb)	68 kg (150 lb)	545 kg (1,200 lb)
Dimensions	3.2 × 3.2 × 2 m (1 × 1 × 6.5 ft)	3.2 × 0.5 × 2 m (1 × 1.7 × 6.5 ft)	1.2 × 1 × 2 m (4 × 3.2 × 6.5 ft)
Connections	13 mm (0.5 in) Swagelock	13 mm (0.5 in) Swagelock	51 mm (2 in) NPT

5.4 Treatment/Disposal - Aqueous Treatment Systems

Rayox System - This system is modular which allows design flexibility. The system is available as a permanent installation or as a mobile unit for smaller scale remediation projects.

Ultrox System - System specifications are developed on the basis of extensive laboratory and commercial application database and bench-scale treatability studies conducted at the manufacturer's facilities. Skid-mounted pilot plant units are also available for use at the customer's site to acquire additional design data. The major components of this system are the UV/oxidation reactor module, which contains the UV bulbs and uniformly diffuses ozone gas into the feed water, an air compressor/ozone generator module, a chemical feed system which introduces hydrogen peroxide into the influent using an in-line static mixer, and a catalytic ozone decomposition unit which uses a proprietary nickel-based catalyst to convert ozone to oxygen. Each major component is skid-mounted.

Operational

Nulite TiO$_2$ Photocatalytic System - These systems operate at ambient temperature and are most efficient at concentrations below 1,000 ppm. This technology treats organic-contaminated waters to less than 1 ppb total organic carbon (TOC) content. The reactor cells are pressure tested to 690 kPa (100 psi).

	6 Cell System	12 Cell System	12 Wafer System
Maximum flow (depends on application)	3 L/min (11 gpm)	6 L/min (23 gpm)	40 L/min (150 gpm)
Operating pressure drop	210 kPa (45 psi)	210 kPa (45 psi)	210 kPa (45 psi)

Rayox System - Typical contaminants that can be treated include PCBs, pesticides, chlorinated solvents, nitrotoluenes, benzenes, toluenes, xylenes, chlorophenols, cyanides, organic acids, ketones, ammonia, ethers, phenols, PAHs, nitrosamines, dioxins, and furans. The system is best used when the contaminant loading is less than 1% (10,000 ppm). According to the manufacturer, the process can provide greater than 99.999% destruction efficiency for waterborne organic contaminants. Use of UV and enhanced oxidation (ENOX) additives improves reaction rates for reactive compounds with otherwise slow oxidation rates, such as tetrachloride, chloroform, and trichloroethane. The following are typical operating parameters.

Flow rate	7.6 to 3,800 L/min (2 to 1,000 gpm)
Energy use	4 to 1,000 kW (15 to 1,340 hp)
Residence time	6 to 60 seconds
pH	3 to 10
Hydrogen peroxide dosage	10 to 100 mg/L
Ozone dosage	0.5 to 2 kg/1,000 L (4.4 to 18 lb/1,000 gal)
ENOX dosage	10 to 500 mg/L

Ultrox System - A low-pressure system that operates full-time or intermittently in a continuous or batch treatment mode with a microprocessor that controls and automates the treatment process. The following is the range of available sizes for the reactor, ozone generator, and ozone decomposition units. The reactors are available in sizes from 280 to 14,760 L (75 to 4,000 gal) and with flow rates from 3.8 to 3,780 L/min (1 to 1,000 gpm). The capacity of the ozone generator ranges from 9.5 to 110 kg/day (21 to 250 lb/day). The ozone decomposition unit has flow rates up to 4,700 m^3/s (1,000 cfm). The following are typical operating parameters for the system.

Influent pH	5.2 to 7.2
Retention time	20 to 60 minutes
Ozone dose	38 to 110 mg/L
Hydrogen peroxide dose	13 to 38 mg/L

Power/Fuel Requirements

All systems require electricity. The 12 Wafer **Nulite TiO$_2$ Photocatalytic System** requires 240 VAC and draws 32 amps. The **Rayox System** requires 4 to 1,000 kW (5.4 to 1,340 hp) of electric power and hydrogen peroxide, ozone, and ENOX additives are consumables required for operation. The **Ultrox System** also requires a supply of hydrogen peroxide for operation. Contaminated water must be supplied to all systems at an appropriate flow rate.

Transportability

The systems are compact and relatively lightweight. Most components can be transported with a standard pickup truck.

PERFORMANCE

Environment Canada's Emergencies Engineering Division (EED) is involved in research and development of advanced oxidation systems for application to grroundwater cleanup. The division has a laboratory-scale **Nulite TiO$_2$ Photocatalytic System** as well as a lab-scale and mobile Solarchem **Rayox unit**. The lab unit is a batch process system, while the mobile unit can be used as a flowthrough or batch system. Various chemicals and mixtures of chemicals in water have been tested using the lab-scale units. The mobile unit has been extensively tested on a wide variety of targeted compounds and some organometallic complexes. All tests have demonstrated the successful destruction of targeted contaminants, particularly aromatic and alkane compounds, chlorinated and non-chlorinated. Results of typical field trials conducted by EED are shown in the following table.

		Concentration (mg/L)	
Contaminant(s)	**Flow Rate**	**Influent**	**Effluent**
1,4-dioxane	19 to 114 L/min (5 to 30 gpm)	100 ppm	<10 ppb
MeCL	19 L/min (5 gpm)	130 to 730 ppb	3.1 ppb
Trichloroethylene		9.7 to 20 ppm	0.4 ppb
Trans-dichloroethylene		6 to 13 ppm	<0.1 ppb
Vinyl chloride		10 to 1,010 ppb	0.5 ppb
Chloroethane		10 ppb	<0.3 ppb
Nitrate esters	15 L/min (4 gpm)	~1,000 to 5,000 ppm	<1 ppm
Explosives			
Trichloroethane	30 L/min (7.9 gpm)	1.6 ppm	not detected
Benzene		0.23 ppm	not detected
Chloroform		0.08 ppm	0.04 ppm
Chlorobenzene		0.05 ppm	not detected
1,2 dichloroethane		0.01 ppm	not detected
Benzene	batch, 30 minutes	36 ppm	0.63 ppm
Toluene		54 ppm	0.03 ppm
Xylene		28 ppm	not detected
Cyanide	N/A	6 ppm	2 ppm
Trans-dichloroethylene	batch, 5 minutes	0.5 ppm	not detected
Dichloroethane		5 ppm	not detected
Benzene		3 ppm	0.009 ppm
Trichloroethylene	27 L/min (7 gpm)	30,000 ppm	0.4 ppb
Trans-dichloroethylene		20,000 ppm	<0.1 ppb
Vinyl chloride		500 ppm	0.5 ppb
Chloroethane		10 ppm	<0.3 ppb
VOCs	batch	20 ppm	0.15 ppm
VOCs	N/A	85 ppm	4.5 ppm

5.4 Treatment/Disposal - Aqueous Treatment Systems

The **Rayox** process has been used to treat contaminated groundwater in Canada, the United States, Europe, and the Far East. The following are performance data from these applications compiled by the manufacturer.

Concentration (mg/L)

Contaminant(s)	Influent	Effluent
Trichloroethylene	110	<0.005
1,4-Dioxane	120	<0.002
PCBs	0.3	<0.000020
Trichloroethylene	1	<0.000020
Pentachlorophenol	8	<0.001
Total petroleum	15	<0.001
Total BTEX	5	<0.001
Trichloroethane	2	<0.005
Dichloroethane	1	<0.005
Chloroform	0.4	<0.010
Atrazine	5	<0.020
Phenol	1	<0.001
PAHs	2	<0.001
Freon-112	0.08	0.002
MTBE	50	<0.010
Acetone	62	<0.005
Nitrosamines	0.02	<0.000003

A U.S. EPA Superfund Innovative Technology Evaluation (SITE) demonstration was conducted at a former drum-recycling facility in San Jose, California for two weeks in 1989. Approximately 50,000 L (13,000 gal) of groundwater contaminated with volatile organic compounds (VOCs) were treated in the **Ultrox System** during 13 test runs. During the first 11 runs, the five operating parameters were adjusted to evaluate the system. The last two runs were conducted under the same conditions as Run No. 9 to verify the reproducibility of the system's performance. The following are the key findings of the demonstration.

- The groundwater treated by the Ultrox system met the applicable standards of the U.S. National Pollutant Discharge Elimination System (NPDES) at the 95% confidence level. Success was obtained by using a hydraulic retention time of 40 minutes, an ozone dose of 110 mg/L, a hydrogen peroxide dose of 13 mg/L, operating all 24 UV lamps, and with influent pH at 7.2 (unadjusted).

- No volatile organics were detected in the exhaust from the ozone decomposition unit.

- The ozone decomposition unit destroyed ozone in the reactor off-gas to levels less than 0.1 ppm. The ozone destruction efficiencies were observed to be greater than 99.99%.

- The Ultrox system achieved removal efficiencies as high as 90% for the total VOCs present in the groundwater at the site. The removal efficiencies for trichloroethylene (TCE) were greater than 99%. However, the maximum removal efficiency for 1,1-dichloroethane (1,1-DCE) was about 65% and for 1,1,1-trichloroethane (1,1-TCA) was about 85%. Removal efficiencies are summarized in the following table.

	Mean Influent (μg/L)	Mean Effluent (μg/L)	Removal (%)
Run No. 9			
TCE	65	1.2	98
1,1-DCA	11	5.3	54
1,1,1-TCA	4.3	0.75	83
Total VOCs	170	16	91
Run No. 12			
TCE	52	0.55	99
1,1-DCA	11	3.8	65
1,1,1-TCA	3.3	0.43	87
Total VOCs	150	12	92

Run No. 13

TCE	49	0.63	99
1,1-DCA	10	4.2	60
1,1,1-TCA	3.2	0.49	85
Total VOCs	120	20	83

- Within the treatment system, the removal of 1,1-DCA and 1,1,1-TCA appears to be due to both chemical oxidation and stripping. Specifically, stripping accounted for 12 to 75% of the total removals for 1,1-TCA, vinyl chloride, and other VOCs.

- The organics analyzed by Gas Chromatography (GC) methods represent less than 2% of the total organic carbon (TOC) present in the water. Very low TOC removal occurred, which implies that partial oxidation of organics (and not complete conversion to carbon dioxide and water) took place in the system.

CONTACT INFORMATION

Manufacturer

Nulite TiO$_2$ Photocatalytic System

Matrix Photocatalytic Incorporated
Main Level, 511 McCormick Boulevard
London, Ontario
N5W 4C8

Telephone (519) 457-2963
Facsimile (519) 457-1676

Rayox System

Solarchem Environmental Systems
130 Royal Crest Court
Markham, Ontario
L3R 0A1

Telephone (905) 477-9242
Facsimile (905) 477-4511

Ultrox System

Ultrox, A division of Resource Conservation Company
2435 South Anne Street
Santa Anna, CA 92704-5308
U.S.A.

Telephone (714) 545-5557
Facsimile (714) 557-5396

OTHER DATA

A system similar to the **Nulite TiO$_2$ Photocatalytic System** and using the same technology is also available from Matrix Photocatalytic Incorporated for treating air streams contaminated with organics. The manufacturer of the **Rayox System** also offers a PCB-destruction process (**Solvolox**) for remediation of soils contaminated with PCBs and other nonvolatile chlorinated hydrocarbons. Environment Canada's Emergencies Engineering Division owns a bench-scale Solvolox unit for testing and evaluation.

REFERENCES

(1) Manufacturer's Literature.
(2) U.S. EPA (United States Environmental Protection Agency), "Demonstration Bulletin: Ultraviolet Radiation and Oxidation", EPA Report No. 540/M5-89/012, Cincinnati, OH (November, 1989).

AIR PLASTICS - AIR STRIPPING TOWERS   No. 127

DESCRIPTION

Custom-designed, corrosion-resistant, vertical, counterflow packed tower air strippers for removing volatile organic compounds (VOCs) from groundwater or other contaminated water.

OPERATING PRINCIPLE

Contaminated water is cascaded down through the packed column countercurrent to a flow of air from the blower at the bottom. As the air comes into contact with the contaminated water, the VOCs are transferred from the water phase to the air and exit the column with the air stream.

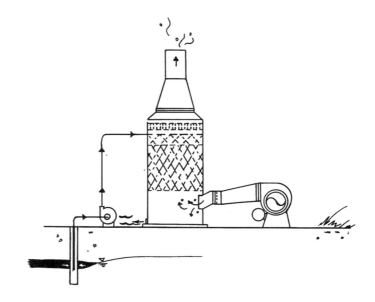

STATUS OF DEVELOPMENT AND USAGE

Commercially available products marketed for removing VOCs, gases, and odours from groundwater, industrial wastewater, accidental spills, process effluents, and runoff from landfills.

TECHNICAL SPECIFICATIONS

Physical

Standard stripping units consist of a vertical packed tower, a circulating pump, a fresh air blower, and a strategically selected packed bed in capacities of 19 to 11,355 L/min (5 to 3,000 gpm). A typical stripper system sized to handle 378 L/min (100 gpm) requires a 1.5 to 1.8 m (5 or 6 ft) diameter tower about 4.6 to 6 m (15 to 20 ft) high. Standard-design systems can be customized to fit most applications. All stripping towers have unitized assembly, corrosion-resistant construction, and monitoring instrumentation.

Operational

Air Plastics stripping towers accept contamination levels of 5 ppm and treat it down to 1 ppb, with up to +99% removal efficiency. Cleansed water is discharged at the bottom of the column and the air containing the contaminants is discharged to the atmosphere or to an optional carbon adsorber. No regeneration or replacement of the packed bed is required and maintenance is minimal. In many cases, the vented air may need additional treatment.

Power/Fuel Requirements

Contaminated water must be supplied to the system at an appropriate flow rate and pressure. A power source is required to power the blowers, pumps, and controls. The power requirements are design-specific.

Transportability

All Air Plastics strippers are modular in construction and are therefore easy to transport.

CONTACT INFORMATION

Manufacturer

Air Plastics Incorporated
1224 Castle Drive
Mason, OH 45040   Telephone (513) 398-8081
U.S.A.   Facsimile (513) 398-8082

REFERENCES

(1) Manufacturer's Literature.

ALGASORB PROCESS   No. 128

DESCRIPTION

A proprietary, algal-based material used commercially to remove and recover heavy metal ions from contaminated groundwaters or other wastewater streams.

OPERATING PRINCIPLE

This process is based on the natural, very strong affinity of biological materials, such as the cell walls of plants and microorganisms, for heavy metal ions. Biological materials, primarily algae, have been immobilized in a polymer to produce a "biological" ion exchange resin called AlgaSorb. AlgaSorb can be packed into columns through which waters containing heavy metal ions are passed. The heavy metal ions are adsorbed to AlgaSorb and water that is almost metal-free exits the column for reuse or discharge.

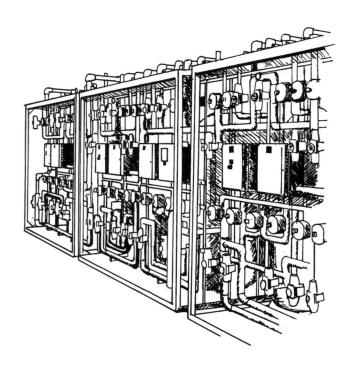

STATUS OF DEVELOPMENT AND USAGE

AlgaSorb is a proprietary material used commercially to treat industrial wastewater and groundwater and to recover precious metals from aqueous solutions. The process upon which AlgaSorb is based is a commercially available, patented technology.

TECHNICAL SPECIFICATIONS

Physical

The AlgaSorb ion exchange resins consist of a nonliving biomass immobilized in a silica gel polymer. The process systems that use the AlgaSorb resins are available in system modules that are integrated to produce a custom system for each application.

Operational

AlgaSorb is a hard material that can be packed into columns which, when pressurized, exhibits flow characteristics that enhance the adsorption of the metal ions to the AlgaSorb. This technology works well for removing heavy metal ions from groundwaters containing high levels of either dissolved solids or organic contaminants, or both. Metal concentrations can be reduced to low ppb levels. The reaction of heavy metal ions with the AlgaSorb forms complexes composed of the algal cell and the metal ions. Once the AlgaSorb is saturated with metal ions, the metals can be stripped from the AlgaSorb, which is then ready for reuse. Metal ions have been sorbed and stripped over many cycles with no noticeable loss in efficiency in the AlgaSorb. An advantage of the algal matrix over standard ion exchange resin is that the components of hard water (Ca^{2+} and Mg^{2+}) on monovalent cations (Na^+ and K^+) do not significantly interfere with the binding of the toxic heavy metal ions. AlgaSorb is also effective for removing heavy metals from water containing organic residues. Organics often foul synthetic ion exchange resins, which limits their use for many wastewater treatment applications, including groundwater treatment. Conditions can be adjusted so that only one or two types of metal ions are adsorbed from a solution containing several metal ions, or so that a variety of metal ions can be sorbed from solution and selectively stripped from the algal cell, one metal at a time, enhancing recycle/reuse possibilities.

Power/Fuel Requirements

An electric power source is required. There may be other power/fuel requirements depending on the particular system configuration.

5.4 Treatment/Disposal - Aqueous Treatment Systems

Transportability

The systems are designed to be transported in modules and constructed on site.

PERFORMANCE

Studies done as part of the U.S. EPA's Superfund Innovative Technology Evaluation (SITE) demonstrated that the process is effective for removing mercury from groundwaters contaminated with mercury at levels near 1 ppm and with a total dissolved solid content of over 11,000 ppm. Mercury was removed to levels below the discharge limit of 10 µg/L.

As part of the U.S. Department of Energy's (DOE's) Applied Research and Development Program, the process was also shown to be effective for removing mercury and uranium from contaminated groundwaters at DOE sites at Oakridge and Savannah River. Initial mercury concentrations at these sites were 30 µg/L and 10 µg/L, respectively. After treatment, effluents contained mercury levels that were within the U.S. allowable drinking water standard of 2.0 µg/L.

CONTACT INFORMATION

Manufacturer

Resource Management & Recovery
4980 Baylor Canyon Road
Las Cruces, NM 88011
U.S.A.

Telephone (505) 582-9228
Facsimile (505) 582-9228 (Voice or Fax)

REFERENCES

(1) Darnall, D.W. and H.D. Feiler, "Recovery of Heavy Metals from Contaminated Groundwaters", from a paper presented at the Hazardous Materials Control '91 (SUPERFUND) Exhibition, Washington, D.C. (Dec., 1991).
(2) Manufacturer's Literature.
(3) U.S. EPA (United States Environmental Protection Agency), "Emerging Technologies: Removal and Recovery of Metal Ions from Groundwater", EPA Report No. 540/5-90/005a, Cincinnati, OH (Aug., 1990).

ANDCO MOBILE TREATMENT SYSTEM No. 129

DESCRIPTION

A mobile treatment system that uses electrochemical technology to remove heavy metals, dyes, pigments, and some dissolved and suspended solids from contaminated wastewater and groundwater streams.

OPERATING PRINCIPLE

A DC current across consumable carbon steel electrodes generates an insoluble iron matrix which adsorbs and coprecipitates heavy metals, dyes, pigments, and other contaminants from the water. The insoluble constituents are then separated from the aqueous stream by clarification and dewatering.

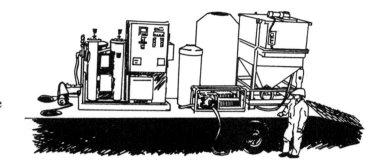

STATUS OF DEVELOPMENT AND USAGE

Several pilot plants, which are used for on-site treatability studies, are available for short-term remediation programs on a rental or lease basis.

TECHNICAL SPECIFICATIONS

Physical

Full-scale plants, sized from 3.8 L/min (1 gpm) to over 7,570 L/min (2,000 gpm) are custom-designed to meet the requirements of specific sites. The pilot plants are small-scale plants used for on-site treatment feasibility studies, low volume cleanup of process or batching operations, and remediation of contaminated spill sites. The pilot plants are available in three sizes. The following are specifications for the three models.

	3 GPM Unit	10 GPM Unit	50 GPM Unit
Dry weight	55 kg (125 lb)	450 kg (1,000 lb)	900 kg (2,000 lb)
Shipping weight	90 kg (200 lb)	450 kg (1,000 lb)	900 kg (2,000 lb)
Dimensions			
Width	0.5 m (1.8 ft)	1.5 m (5 ft)	2.2 m (7.3 ft)
Depth	0.4 m (1.2 ft)	1.5 m (5 ft)	1.4 m (4.7 ft)
Height	1.2 m (4 ft)	1.3 m (4.3 ft)	2.2 m (7.3 ft)
Nominal contamination capacity/flow*	75 ppm @ 0.8 L/min (0.2 gpm) 5 ppm @ 11 L/min (3 gpm)	75 ppm @ 2.3 L/min (0.6 gpm) 4.5 ppm @ 38 L/min (10 gpm)	150 ppm @ 11 L/min (3 gpm) 9 ppm @ 190 L/min (50 gpm)

* higher levels are possible with multiple passes

Operational

The Andco electrochemical system removes heavy metals from contaminated groundwater, surface water, or leachate. The system removes a single heavy metal or a mixture of heavy metals including hexavalent chrome. The process also removes colour from water caused by dyes or pigments and significantly reduces COD, BOD, TDS, and TSS. The system requires no chemical additions. Automated systems can be designed that can be monitored by modem for remote operation and data logging.

5.4 Treatment/Disposal - Aqueous Treatment Systems

Power/Fuel Requirements

	3 GPM Unit	**10 GPM Unit**	**50 GPM Unit**
Treatment Skid			
Electricity	110 VAC/ 50 to 60 Hz/15 amps	110 VAC/ 50 to 60 Hz/40 amps	208, 240, or 480 VAC/ 50 to 60 Hz/30, 60 amps
Water	not available	150 L/min (40 gpm) maximum	not available
Sludge Pump			
Pneumatic	not available	210 kPa/1.6 m^3/min (30 psi/55 cfm)	not available

Transportability

The pilot-plant systems are skid-mounted and can easily be placed on a flatbed trailer.

PERFORMANCE

The following table shows the demonstrated performance of the Andco systems.

	Pb	**Sn**	**Cd**	**Ag**	**Pd**	**Mo**	**Zn**	**Cu**	**Ni**	**Fe**	**Cr**	**V**	**Al**	**Sb**	**As**
Influent (mg/L)*	13	<5.0	20	10	4	1.5	230	32	33	69	62	43	20	55	7
Effluent (mg/L)**	0.1	<0.2	0.002	<0.02	0.02	<0.05	0.03	0.01	0.05	0.05	0.05	0.02	0.2	<0.05	0.01
Simultaneously removed metals	Cu Cr Ni Pd	Ag Cu Ni Pd	Cu Zn	Cu Ni Pb Sn	Ag Cu Ni Pb	-	Cd Cu Fe Sn	Cr Ni Zn	Cu Cr Zn	Cd Cu	Mo V	-	Cu Cr	Cd Pb	Zn Pb

* Influent pH range, 6 to 9
** Not necessarily lower limit

CONTACT INFORMATION

Manufacturer

Andco Environmental Processes Incorporated
595 Commerce Drive
Amherst, NY 14228-2380
U.S.A.

Telephone (716) 691-2100
Facsimile (716) 691-2880

REFERENCES

(1) Manufacturer's Literature.

BIOACCELERATION TREATMENT SYSTEM   No. 130

DESCRIPTION

A skid-mounted biological treatment system that uses an amended microbial population to achieve biological degradation of water containing organic contaminants.

OPERATING PRINCIPLE

Contaminated water enters a mixing tank where the pH is adjusted if necessary and inorganic nutrients are added. The water then flows to the reactor chambers where organic contaminants are biodegraded.

STATUS OF DEVELOPMENT AND USAGE

This is a commercially available, patented system suitable for treating a wide range of wastewaters, including groundwater, holding ponds, and process effluents. The system was tested by the U.S. Environmental Protection Agency (U.S. EPA) as part of the Superfund Innovative Technology Evaluation (SITE) program. More than 50 full-scale systems are now in use.

TECHNICAL SPECIFICATIONS

Physical

The system features multi-stage designs which are skid-mounted, pre-piped, and pre-wired. The following are specifications for the standard models.

Model No.	Width	Length	Height	Dry Weight	Wet Weight	Flow Rate
1/2K-2	1.5 m (5 ft)	2.9 m (9.4 ft)	1.8 m (6 ft)	950 kg (2,100 lb)	3,220 kg (7,100 lb)	19 L/min (5 gpm)
1K-2	1.5 m (5 ft)	4.1 m (14 ft)	1.8 m (6 ft)	1,540 kg (3,400 lb)	6,120 kg (13,500 lb)	42 L/min (11 gpm)
2K-2	2.1 m (7 ft)	4.1 m (14 ft)	2.4 m (8 ft)	2,400 kg (5,300 lb)	12,340 kg (27,200 lb)	80 L/min (21 gpm)
4K-3	2.1 m (7 ft)	5.3 m (18 ft)	3 m (10 ft)	4,400 kg (9,700 lb)	23,860 kg (52,600 lb)	180 L/min (48 gpm)
6K-4	2.7 m (9 ft)	6.6 m (22 ft)	2.4 m (8 ft)	6,030 kg (13,300 lb)	31,250 kg (68,900 lb)	240 L/min (64 gpm)
8K-4	2.7 m (9 ft)	8 m (27 ft)	3 m (10 ft)	7,080 kg (15,600 lb)	44,315 kg (97,700 lb)	360 L/min (96 gpm)
10K-4	2.7 m (9 ft)	8 m (27 ft)	3 m (10 ft)	7,940 kg (17,500 lb)	50,350 kg (111,000 lb)	405 L/min (107 gpm)
12K-4	2.7 m (9 ft)	8.5 m (28 ft)	3 m (10 ft)	9,525 kg (21,000 lb)	60,420 (133,200 lb)	485 L/min (128 gpm)
14K-5	2.7 m (9 ft)	11 m (36 ft)	3 m (10 ft)	11,110 kg (24,500 lb)	70,535 kg (155,500 lb)	605 L/min (160 gpm)
16K-5	3 m (10 ft)	11 m (36 ft)	3 m (10 ft)	12,250 kg (27,000 lb)	78,470 kg (173,000 lb)	680 L/min (180 gpm)
18K-5	3.4 m (11 ft)	11 m (36 ft)	3 m (10 ft)	13,380 kg (29,500 lb)	86,865 kg (191,500 lb)	760 L/min (200 gpm)

*   Flow rates are based on typical wastestreams below 250 ppm COD. A different reactor model may be required for higher influent concentrations, certain contaminants, and extremely low effluent concentration objectives.

Operational

Systems are available to treat contaminated water at a flow rate of 3.8 to 760 L/min (1 to 200 gpm). Contaminants found to be amenable to treatment include pentachlorophenol (PCP), creosote constituents, gasoline and fuel oil, chlorinated hydrocarbons, phenolics and solvents. The system uses an amended microbial population to achieve biological degradation. The system is considered to be amended when a specific microorganism is added to the indigenous microbial population in the wastewater to optimize degradation of a particular pollutant. The microorganisms that perform the degradation are immobilized on a highly porous packing in a multi-cell, submerged fixed-film bioreactor. The biological growth is first developed during a one- to two-week acclimation period. Air is supplied by fine bubble membrane diffusers mounted at the base of each cell. The system can also be operated under anaerobic conditions. As the water flows through the bioreactor, contaminants are degraded to carbon dioxide, water, and, in the case of chlorinated organics, chloride ions. The reactor is enclosed which controls vapour emissions.

Power/Fuel Requirements

An electric power source is required and the wastewater stream must be provided to the system at an appropriate flow rate.

Transportability

The systems are fully integrated and mounted on skids or trailers for transportability.

PERFORMANCE

This process was demonstrated by the U.S. Environmental Protection Agency as part of the Superfund Innovative Technology Evaluation (SITE) program. Demonstrations took place on groundwater contaminated with pentachlorophenol (PCP) at a wood preserving facility in New Brighton, Minnesota for 6 weeks in 1989. A 19 L/min (5 gpm) unit was operated for 2 weeks at each of the three throughput rates of 3.8, 11, and 19 L/min (1, 3, and 5 gpm), after an initial 2-week acclimation period. It was concluded that the system:

(1) successfully reduced PCP concentrations from 45 to 1 ppm or less in one pass;
(2) achieved 96 to 99% removal of PCP;
(3) produced minimal sludge and no PCP air emissions;
(3) successfully mineralized chlorinated phenolics;
(4) eliminated biotoxicity present in the groundwater;
(5) appeared to be unaffected by low concentrations of oil/grease (approximately 50 ppm) and heavy metals found in the water; and
(6) required minimal operator attention.

CONTACT INFORMATION

Manufacturer

Bio Trol Incorporated
10300-T Valley View Road
Suite 107
Eden Prairie, MN 55344-3546
U.S.A.

Telephone (612) 942-8032
Facsimile (612) 942-8526

REFERENCES

(1) Manufacturer's Literature.
(2) United States Environmental Protection Agency (U.S. EPA), "Demonstration Bulletin; Aqueous Biological Treatment Systems", EPA Report No. 540/M5-91/001, Cincinnati, OH (May, 1991).

BREEZE AIR STRIPPING SYSTEM

No. 131

DESCRIPTION

A compact aeration tank coupled with a blower designed to quickly and easily remove VOCs and dissolved gases from contaminated groundwater.

OPERATING PRINCIPLE

Water is pumped into the tank and the blower provides air to diffusers inside the tank, which release specifically sized air bubbles. Rising bubbles and turbulence provide an air-to-water interface which promotes phase transfer and results in the VOCs being transferred from the water phase to the air. The VOCs then exit the tank in the air stream.

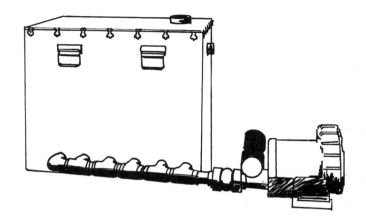

STATUS OF DEVELOPMENT AND USAGE

Commercially available product marketed to remove VOCs and dissolved gases from contaminated groundwater, leaking storage tanks, hazardous waste sites, potable water, and chemical plants.

TECHNICAL SPECIFICATIONS

Physical

The tank is available in polypropylene or stainless steel and is equipped with an inlet, an outlet, a vent connection, and a removable cover. The remote air blower is available with an explosion-proof motor.

Operational

Without extensive maintenance or downtime, the system strips BTEX, PCE, TCE, Radon, benzene, H_2S, and other VOCs and dissolved gases even at low concentrations. Removal rates vary and can reach as high as 99.9%. Individual units can treat flows of up to 568 L/min (150 gpm) and multiple units can treat larger flows. As it features variable air-to-water ratios, it is adaptable to any application. It can operate with either gravity flow, pumped discharge, or pumped inlet and can easily be interfaced with other treatment systems for oil and grease removal. The equipment is low profile and inconspicuous. In many applications, the vented air containing the VOCs may require additional treatment.

Power/Fuel Requirements

An electric power supply is required for the blower and contaminated water at an appropriate flow rate and pressure.

Transportability

The system is compact and can easily be carried by two people using the handles on the tank.

CONTACT INFORMATION

Manufacturer

Aeromix Systems Incorporated
2611 North Second Street
Minneapolis, MN 55411-1634

Telephone (612) 521-8519
Facsimile (612) 521-1455

OTHER DATA

In many applications, the vented air containing the VOCs may have to be treated with an additional treatment system.

REFERENCES

(1) Manufacturer's Literature.

BRINECELL OXIDIZERS   No. 132

DESCRIPTION

Electrocatalytic oxidizers that can be used to oxidize, deodorize, and sterilize any oxidizable organic contaminants.

OPERATING PRINCIPLE

Salt (NaCl) is mixed with the contaminated liquid to produce a solution of 1 to 10% NaCl. Submersible pumps then circulate the liquid effluent through an electrocatalytic cell, where a DC current is imparted to the solution. Electrolysis separates the water and salt into their basic elements, liberating ozone, nascent chlorine, and their respective hydroxyl radicals. These very powerful oxidants will oxidize, in minutes, all oxidizable organics and reduce BOD and COD in the liquid.

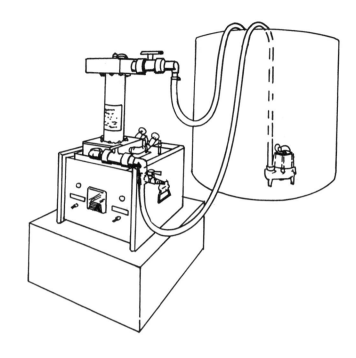

STATUS OF DEVELOPMENT AND USAGE

Commercially available product in use by many industries to eliminate oxidizable organics and to reduce BOD and COD in effluents. BrineCells are used by the textile industry to bleach fabrics and the pulp and paper industry to bleach pulp. Effluents containing ammonia from meat-processing plants and cyanide from electroplating may also be treated using BrineCells. Organic contaminants from soil and groundwater may be oxidized with the systems.

TECHNICAL SPECIFICATIONS

Physical

The BrineCell systems consist of a submersible pump(s), an oxidizer cell, and a DC power supply. The electrodes are made using a patented process in which solid metals are fused to produce electrodes with an expected lifespan of several years and the ability to carry voltages as high as 250 VDC and power up to 100,000 watts. Systems are available with recirculation rates from 40 to 8,000 L/min (10 to 2,000 gpm). The units come mounted inside indoor/outdoor enclosures, which are skid-mounted. The largest unit measures $1 \times 1.5 \times 2$ m ($3.3 \times 5 \times 6.7$ ft) and weighs 816 kg (1,800 lb).

Operational

BrineCells operate by recirculating the effluents for 2 to 60 minutes, depending on the contaminants. The units can also operate on a continuous flow basis by flowing from tank-to-tank or by installing them in a series formation, from cell-to-cell. Operation can be low voltage and high salt, or high voltage and low salt. If salt is needed, rock salt may be used or concentrated saline may be injected through the saline injection valves on the cells.

Power/Fuel Requirements

The indoor/outdoor, skid-mounted units require a 3-phase electric power supply of 240 VAC and a supply of salt (NaCl).

Transportability

The tanks are skid-mounted, portable, and can be moved with a forklift.

PERFORMANCE

The following is a partial list of preliminary laboratory results supplied by the manufacturer. Complete results of the lab tests are available from the manufacturer.

Contaminant	Influent (mg/L)	Treatment Time (minutes)	Effluent (mg/L)
phenols	26,000	60	20
toluene	5	20	0.2
chloroform	12	15	2
carbon tetrachloride	13	15	0
pentachlorophenol	20	3	0
lindane	500	30	0
TKN	23	60	8
ammonia	201	60	0.35
cyanide	5,400	60	1,340
dursban	420	60	0

CONTACT INFORMATION

Manufacturer

BrineCell Incorporated
P.O. Box 27488
Salt Lake City, UT 84127
U.S.A.

Telephone (801) 973-6400
Facsimile (801) 973-6463

OTHER DATA

The manufacturer will process samples to determine the system's effectiveness for specific applications.

REFERENCES

(1) Manufacturer's Literature.

CALGON ACTIVATED CARBON ADSORPTION TREATMENT SYSTEMS

No. 133

DESCRIPTION

Calgon Dual Adsorption Systems and **Calgon Disposorb Activated Carbon Adsorption Canisters** are treatment systems for removing dissolved organic wastes from contaminated water. The activated carbon adsorption canisters are designed for treating low flows of liquids. They are modular, compact, constructed of polyethylene, and designed to be disposed of when treatment is complete.

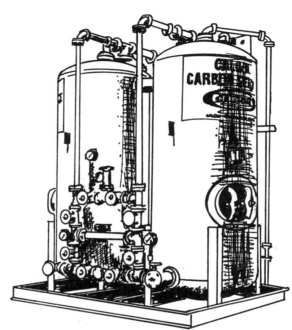

OPERATING PRINCIPLE

Dual Adsorption Systems - Wastewater is pumped through two lined carbon steel adsorber vessels containing beds of activated carbon which adsorb the dissolved organics. Effluent lines carry the treated water to the desired point of discharge. Spent carbon can be replaced with fresh carbon for continued operation, with the spent carbon being returned to a carbon regeneration facility or delivered to an appropriate waste management facility.

Disposorb Canisters - Liquids contaminated with organics are percolated through activated carbon inside the canister, where the contaminants are physically adsorbed to the carbon media.

STATUS OF DEVELOPMENT AND USAGE

Commercially available products in general use.

TECHNICAL SPECIFICATIONS

Physical

Dual Adsorption Systems - All systems consist of pre-piped dual adsorber vessels constructed of carbon steel with corrosion-resistant epoxy lining. The following models are available.

Model	Vessel Size (diameter)	Carbon Quantity Per Vessel
Model 4	1.2 m (4 ft)	907 kg (2,000 lb) pre-loaded
Model 6	1.8 m (6 ft)	2,720 kg (6,000 lb) pre-loaded
Model 8	2.4 m (8 ft)	4,540 kg (10,000 lb)
Model 10	3 m (10 ft)	9,070 kg (20,000 lb)

Disposorb Canisters - Two sizes are available: 208 L (55 gal) with a carbon capacity of 75 kg (165 lb) and 1,325 L (350 gal) with a carbon capacity of 450 kg (1,000 lb).

Operational

Dual Adsorption Systems - Model 4 handles flows up to 490 L/min (130 gpm), Model 6 handles flows up to 1,700 L/min (450 gpm), and Model 8 handles flows up to 1,325 L/min (350 gpm), all at pressures up to 620 kPa (90 psi). Model 10 handles flows up to 2,650 L/min (700 gpm) with 7.5 minutes of contact time per vessel. All systems can be designed with backwashing capabilities and can operate in a series or in parallel configuration. One hundred and fifty grades of activated carbon are available for the dual adsorption systems, many of which are designed for specific liquid purification objectives.

Disposorb Canisters - Disposorb units are designed to handle flows from 4 to 114 L (1 to 30 gpm) at pressures up to 230 kPa (33 psi). The manufacturer can fill the units with Filtrasorb 300- or 400-grade granular activated carbon.

Transportability

All **Dual Adsorption Systems** are designed for transport on a flatbed truck/trailer. The **Disposorb Canisters** weigh 75 kg (165 lb) or 450 kg (1,000 lb) and must be handled by a drum hoist, hand truck, forklift and pallet, or other drum-handling equipment.

PERFORMANCE

Treatment technologies using granular activated carbon have a proven ability to remove organic contaminants to non-detectable levels. The degree of removal varies, however, in relation to factors such as flow rates, operating pressures, carbon dosage and type, contact time, influent pH, temperature, and viscosity, and if applicable, liquid pre-treatment to remove oils, greases, or suspended solids from the influent which may impair the system's ability to operate adequately. In the 1980s, a study was conducted of 31 operating plants that used granulated activated carbon to remove toxic organic compounds from groundwater. The plants were treating contaminated flows ranging from 19 to 8,517 L/min (5 to 2,250 gpm). Contamination was caused by industrial accidents at 22 of the sites and at the remaining sites by leachate from lagoons and dumpsites, and spills from railroad and truck accidents. Data from these treatment systems showed that granular activated carbon adsorption systems removed a wide range of organic compounds to below their detection levels. This study is detailed in O'Brien and Fisher (1983), referenced below.

OTHER DATA

Calgon provides an off-site carbon reactivation service that allows carbon to be transferred to and from the system's two adsorber vessels without exposing workers. The spent carbon is then returned to a Calgon Carbon reactivation centre for reactivation and reuse. Carbon regeneration facilities and other suppliers of activated carbon are listed in trade journals such as *Fraser's Canadian Trade Directory* (777 Bay Street, Toronto, Ontario M5W 1A7, Tel. 416-596-5086, Fax 416-593-3201) and the *Thomas Register of American Manufacturers* (Thomas Publishing Company, Five Penn Plaza, New York, NY 10001, Tel. 212-695-0500, Fax 212-290-7206).

CONTACT INFORMATION

Manufacturer

Calgon Carbon Corporation
P.O. Box 717
Pittsburgh, PA 15230-0717
U.S.A.

Telephone (412) 787-6700
Toll-free (800) 422-7266
Facsimile (412) 787-4523

Canadian Supplier

STANCHEM, Inc.
43 Jutland Road
Etobicoke, Ontario
M8Z 2G6

Telephone (416) 259-8231
Facsimile (416) 259-6175

REFERENCES

(1) Hager, D.G., "Industrial Wastewater Treatment by Granular Activated Carbon", *Industrial Water Engineering*, (January/February, 1974).
(2) Manufacturer's Literature.
(3) O'Brien, R.P. and J.L. Fisher, "There is an Answer to Groundwater Contamination", reprinted from *Water/Engineering & Management*, File No. 27-86, Calgon Carbon Corporation, Pittsburgh, PA (May, 1983).
(4) Rizzo, J.L. and A.R. Shepard, "Treating Industrial Wastewater with Activated Carbon", *Chemical Engineering*, McGraw-Hill, New York, N.Y. (January, 1977).
(5) Ying, W., E.A. Dietz, and G.C. Woehr, "Adsorptive Capacities of Activated Carbon for Organic Constituents of Wastewaters", *Environmental Progress 9(1)* (February, 1990).

CARBTROL L-1, L-4, AND L-5 ACTIVATED CARBON ADSORPTION CANISTERS

No. 134

DESCRIPTION

Water purification systems that use activated carbon to remove dissolved organics from groundwater, wastewater, or liquid process streams.

OPERATING PRINCIPLE

Liquids contaminated with organics are percolated through activated carbon inside the canister, where the contaminants are physically adsorbed to the carbon media.

STATUS OF DEVELOPMENT AND USAGE

Commercially available products in general use.

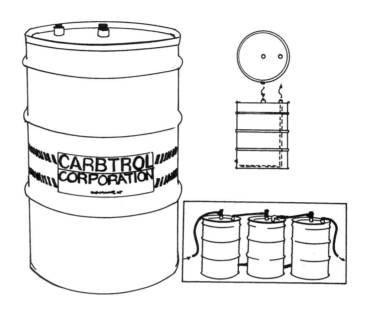

TECHNICAL SPECIFICATIONS

Physical

The L-1 canister is constructed of mild steel with epoxy phenolic internal coating with polyethylene liner and PVC internal piping. The following kits are available as options: an Interconnecting Piping Kit which includes flexible 32-mm (1.3-in) PVC tubing with hose clamps, inlet pressure gauge, and an intermediate sample valve, and a Suspended Solids Removal Kit consisting of a basket filter mounted on a support frame.

Model	Vessel Size	Carbon Capacity	Shipping Weight
L-1	0.9 m (3 ft) high 0.6 m (2 ft) diameter	90 kg (200 lb)	115 kg (250 lb)
L-4	1.7 m (5.5 ft) high 1.2 m (3.8 ft) in diameter	455 kg (1,000 lb)	680 kg (1,500 lb) dry carbon 1,040 kg (2,300 lb) drained carbon
L-5	2.3 m (7.5 ft) high 1.2 m (3.8 ft) diameter	815 kg (1,800 lb)	1,090 kg (2,400 lb) dry carbon 1,900 kg (4,200 lb) drained carbon

Operational

Carbtrol L-1 - The L-1 Canister handles flows up to 38 L/min (10 gpm). Activated carbon supplied in the L-1 is a Carbtrol high activity virgin carbon. Large 32-mm (1.3-in) internal piping results in low pressure drops, allowing the operation of three canisters in series. Three L-1 Canisters arranged in a series results in a total contact time of 10 minutes. Canisters can also be manifolded in parallel for higher flows. Manifolds for up to 228 L/min (60 gpm) are kept in stock.

Carbtrol L-4 and L-5 - Both canisters handle flows up to 190 L/min (50 gpm) with maximum pressures of 62 kPa (9 psi).

Power/Fuel Requirements

Contaminated water must be pumped through the adsorption system at appropriate flow rate and pressures.

Transportability

A drum hoist, hand truck, forklift and pallet, or other drum-handling equipment are required to handle the L-1 canister.
The L-4 and L-5 canisters must be moved by forklift. All canisters are approved by Transport Canada for shipping spent hazardous carbon.

PERFORMANCE

Treatment technologies using granular activated carbon have a proven ability to remove organic contaminants to non-detectable levels. The degree of removal varies, however, in relation to such factors as flow rates, operating pressures, carbon dosage and type, contact time, influent pH, temperature, and viscosity, and if applicable, liquid pre-treatment to remove oils, greases, or suspended solids from the influent which may impair the system's ability to operate adequately.

CONTACT INFORMATION

Manufacturer

Carbtrol Corporation
51 Riverside Avenue
Westport, CT 06880
U.S.A.

Telephone (203) 226-5642
Facsimile (203) 226-5322

OTHER DATA

Carbtrol offers full carbon regeneration and "Take-Back" services.

Carbon regeneration facilities and other suppliers of activated carbon are listed in trade journals such as *Fraser's Canadian Trade Directory* (777 Bay Street, Toronto, Ontario M5W 1A7, Tel 416-596-5086, Fax 416- 593-3201) and the *Thomas Register of American Manufacturers* (Thomas Publishing Company, Five Penn Plaza, New York, NY 10001, 212-695-0500, Fax 212-290-7206).

REFERENCES

(1) Hager, D.G., "Industrial Wastewater Treatment by Granular Activated Carbon", *Industrial Water Engineering*, (January/February, 1974).
(2) Manufacturer's Literature.
(3) O'Brien, R.P. and J.L. Fisher, "There is an Answer to Groundwater Contamination", reprinted from *Water/Engineering & Management*, File No. 27-86, Calgon Carbon Corporation, Pittsburgh, PA (May, 1983).
(4) Rizzo, J.L. and A.R. Shepard, "Treating Industrial Wastewater with Activated Carbon", *Chemical Engineering*, McGraw-Hill, New York, N.Y. (January, 1977).
(5) Ying, W., E.A. Dietz, and G.C. Woehr, "Adsorptive Capacities of Activated Carbon for Organic Constituents of Wastewaters", *Environmental Progress, 9(1)* (February, 1990).

CARBTROL DIFFUSED AIR STRIPPERS

No. 135

DESCRIPTION

Low profile air strippers designed to remove volatile organic compounds (VOCs) from water. Strippers are pre-piped and wired for ease of installation. The **Multi-stage Diffused Air Stripper** consists of multiple turbulent diffused aeration cells, blowers, motors, starter, and controls. The **Counterflow Plate Stripper** consists of multiple stacked aeration plates, with air provided by a centrifugal fan.

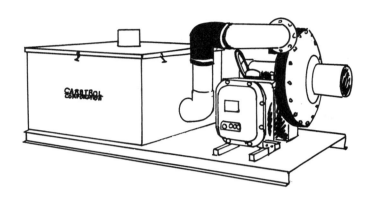

OPERATING PRINCIPLE

Air is bubbled through the contaminated water, promoting phase transfer of the contaminants, which results in the VOCs being transferred from the water phase to the air. The VOCs then exit the column with the air stream. If required, the VOCs are then captured by carbon adsorption canisters. The influent water is also polished by an activated carbon adsorption canister.

STATUS OF DEVELOPMENT AND USAGE

Commercially available products in general use.

TECHNICAL SPECIFICATIONS

Physical

Multi-stage Diffused Air Stripper - The following are specifications for the four models available.

	MD-4	MD-6	MD-8	MD-12
Dimensions	1.2 × 1.2 m (4 × 4 ft)	1.2 × 2.3 m (4 × 7.5 ft)	1.2 × 2.8 m (4 × 9.3 ft)	1.2 × 3.8 m (4 × 13 ft)
Water flow	38 L/min (10 gpm)	38 L/min (10 gpm)	38 L/min (10 gpm)	38 L/min (10 gpm)
Air flow	5.7 m^3/min (200 cfm)	8.5 m^3/min (300 cfm)	11 m^3/min (400 cfm)	17 m^3/min (600 cfm)
Blower power	1.5 kW (2 hp)	2.2 kW (3 hp)	3.7 kW (5 hp)	5.6 kW (7.5 hp)
No. of stages	4	6	8	12

Counterflow Plate Stripper -This 4-stage unit is less than 2.4 m (8 ft) high and constructed of corrosion-resistant, high density polyethylene. The aeration plates are easy to access for cleaning without disassembly. Full motor and control options are available and the system is skid-mounted for ease of installation.

Operational

Low profile design makes these systems wind-resistant, and easy to set up and clean. The air-to-water ratio is 66 to 1, when operating at 38 L/min (10 gpm). The **Counterflow Plate Stripper** provides counter current stripping for highest efficiency in a low profile package. The number of plates required is based on inlet concentrations, discharge concentration limits, flow rate, and groundwater temperature.

Power/Fuel Requirements

Contaminated water must be supplied to the system at an appropriate flow rate and pressure. An electric power supply is required.

Transportability

The systems are all skid-mounted and can be trailer-mounted as an option.

PERFORMANCE

The **Multi-stage Diffused Air Strippers** are capable of removing from 97 to 98% of VOCs. The diffuser, in conjunction with the activated carbon modules for effluent polishing and off-gas treatment, can achieve final removal of BTEX to 1 ppm or lower. The **Counterflow Plate Stripper** is designed to provide the highest efficiency stripping in a low profile package, with up to 99.99% VOC removal.

CONTACT INFORMATION

Manufacturer

Carbtrol Corporation
51 Riverside Avenue
Westport, CT 06880
U.S.A.

Telephone (203) 226-5642
Facsimile (203) 226-5322

REFERENCES

(1) Manufacturer's Literature
(2) Punt, M. and H. Whittaker, "A Comparison of Steam Stripping and Air Stripping for the Removal of Volatile Organic Compounds from Water", in: *Proceedings of the Eighth Technical Seminar of Chemical Spills*, Environment Canada, Ottawa, Ont. (June, 1991).
(3) Whittaker, H., "A Review of Countermeasures for the Removal of Volatile Organic Compounds from Groundwater", in: *Proceedings of the Fifth Technical Seminar of Chemical Spills,* Environment Canada, Ottawa, Ont. (February, 1988).

CYANIDE DESTRUCTION SYSTEM   No. 136

DESCRIPTION

Cleans up groundwater contaminated with cyanide.

OPERATING PRINCIPLE

Cyanide hydrolysis, a reaction that destroys cyanide, occurs naturally even at ambient temperatures, but takes too long to be economically feasible. This system applies "thermal hydrolysis", during which the temperature is increased and the pressure rigidly controlled in a specially designed reaction vessel, to quickly destroy all cyanide complexes to well below compliance levels. The reaction is as follows: $CN^- 2H_2O \rightarrow NH_3 + HCOO^-$.

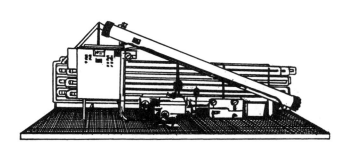

STATUS OF DEVELOPMENT AND USAGE

A commercially available system designed to clean up groundwater. Canadian and U.S. patents are pending.

TECHNICAL SPECIFICATIONS

Physical - Cyanide hydrolysis installations are skid-mounted, take up little space, and can be used for small and large operations. Smaller operations producing less than 780 mL/min (300 gpd) can rely on the batch process and a continuous operating system is available for larger requirements.

Operational - The Cyanide Destruction System is the only known process capable of destroying complex cyanides to levels consistently below compliance levels. The process requires no dangerous treatment chemicals and has specifically designed control loops for safety. Any process variable falling outside its operating range results in a complete system shutdown and alarm.

Power/Fuel Requirements - About 700 kW·h of electricity is required per m^3 treated.

Transportability - The systems are modular and easily transportable by truck.

PERFORMANCE

With financial support from the American Electroplaters Society, the Ontario Research Foundation conducted a multi-year laboratory and field test program on the feasibility of cyanide destruction through "thermal hydrolysis". The results showed that all cyanides, including iron, nickel, and copper complexes, could be reduced to below acceptance limits. Cyanide Destruct Systems has four processes in operation treating groundwater: one in Ontario at a former electroplating plant and three at aluminum smelters in Quebec, Texas, and Oregon. The company also has about 25 batch and continuous flow plants in operation throughout Canada, the United States, and South Africa.

CONTACT INFORMATION

Manufacturer

Cyanide Destruct Systems Incorporated
383 Elmira Road
Guelph, Ontario   Telephone (519) 837-1899
N1K 1H3   Facsimile (519) 837-1622

REFERENCES

(1) Manufacturer's Literature.

ENVIRONMENT CANADA'S REVERSE OSMOSIS UNITS No. 137

DESCRIPTION

Environment Canada's Emergencies Engineering Division (EED) owns three reverse osmosis (RO) units: a lab-scale unit, a bench-scale unit, and a mobile field-scale unit. The mobile unit is fully self-contained with its own diesel power source and electrical generator.

OPERATING PRINCIPLE

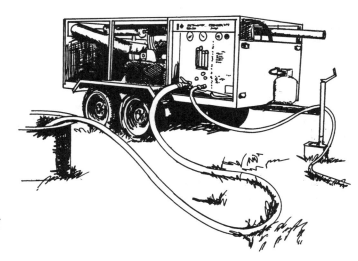

Reverse osmosis is a crossflow membrane technology. The process is termed crossflow since the influent, designated as feed, and the concentrate streams flow parallel to the membrane instead of perpendicular to it. Membrane filtration is the application of pressure and flow across a semipermeable membrane to separate the components of a fluid. The pressure required to drive the process is determined by the specific nature of the feed solution and the pore size of the membrane. If the solution contains significant levels of ionic solutes, the operating pressure must first overcome the osmotic pressure, which is the natural energy potential between the more concentrated and the less concentrated streams.

In the crossflow process, the feed stream is separated into two effluent streams: the permeate, which passes through the membrane, and the concentrate, which retains the dissolved and suspended solids rejected by the membrane.

STATUS OF DEVELOPMENT AND USAGE

Environment Canada's Emergencies Engineering Division develops and maintains a wide range of prototype cleanup equipment. In cooperation with the private sector and public organizations, these pieces of equipment are used to demonstrate and adapt innovative methods for on-site mitigation of water or soil contamination caused by pollution emergencies or insecure hazardous waste sites. Since the early 1980s, the Division has designed, developed, and evaluated bench-scale, pilot-scale, and field-scale membrane units. The latest have been used in a variety of spill cleanups across the country.

TECHNICAL SPECIFICATIONS

Physical

The mobile unit is fully self-contained with its own diesel power source and electric generator. It is mounted on a highway trailer and the complete system measures 4.3 × 1.8 × 1.4 m (14 × 6 × 4.5 ft) with 0.5 m (1.5 ft) of ground clearance. The system weighs approximately 1,270 kg (2,800 lb).

Operational

Crossflow filtration is considered to be a self-cleaning process since the solutes and solids are swept away with the concentration stream, eliminating the need to frequently change the filter media. Crossflow membranes produce precise separations at the ionic, molecular, and macromolecular levels. Systems can be designed to concentrate all feed stream materials, purify solvents, or selectively pass some materials while concentrating others.

Environment Canada's mobile RO unit can treat up to 2,000 L/h (530 gal/h) of contaminated water. Since 1983, the RO units have treated approximately 8 million L of contaminated water.

5.4 Treatment/Disposal - Aqueous Treatment Systems

The separation spectrum of various crossflow filtration processes is presented in the following table.

Crossflow Separation Process	Membrane Pore Size	Fluid Component Retained	Transmembrane Pressure	Sample Applications
Reverse Osmosis	5 to 15 Å	Ions and most organics over 100 molecular weight	1,400 to 6,900 kPa [200 to 1,000 psi(g)]	Metal ion recovery BOD and COD reduction in waste streams
Nanofiltration	10 to 90 Å	Organics with 300 to 1,000 molecular weight	1,400 to 5,500 kPa [(200 to 800 psi(g)]	Water softening Desalting of organics
Ultrafiltration	45 to 1,100 Å	Most organics over 1,000 molecular weight, including pyrogens, viruses, bacteria, and colloids	70 to 1,400 kPa [(10 to 200 psi(g)]	Concentration and recovery of industrial organics and dilute suspended oils
Microfiltration	400 to 20,000 Å	Very small suspended particles, some emulsions, and some bacteria	7 to 170 kPa [(1 to 25 psi(g)]	High volume removal of bacteria and small suspended solids

Unlike microfiltration and ultrafiltration, pore size is not the only criterion for rejection of compounds by reverse osmosis. One theory is that the liquid in contact with the membrane tends to dissolve into it and cause the membrane to swell. Swelling tends to alter the membrane properties and generally leads to higher permeability and lower selectivity for the permeating species. Substances that have been successfully removed by reverse osmosis include: polychlorinated biphenyls, glycol, solvents, chlorinated hydrocarbons, fish waste, potato wash water, and pesticides. Reverse osmosis technology removes a large range of organics from water, but in some cases, only at low concentrations. It cannot treat water with solvent concentrations greater than a few hundred ppm.

Power/Fuel Requirements

Diesel fuel is required to power the onboard generator which supplies all power requirements for the unit.

Transportability

The mobile system is trailer-mounted and can be attached to a truck for transport.

PERFORMANCE

The mobile RO unit has been successfully used to clean up a wood preservative spill in British Columbia and an underground PCB spill in Saskatchewan. The lab-scale unit was used at a pesticide plant in Nova Scotia to treat aqueous effluent resulting from firefighting and cleanup of the building, and to concentrate heavy metals from groundwater in Chalk River.

The mobile RO unit has been used to treat groundwater at four different locations, including testing at a chemical landfill site in Gloucester, Ontario in 1984 and 1986. The site contained many types of diluted organic chemicals and the testing was useful for evaluating the capability of various commercial RO membranes to remove organic chemicals on the Environment Canada priority list.

In a joint trial with the Ontario Department of Highways in 1985, the RO unit treated 450,000 L (119,000 gal) of groundwater contaminated with road salt. In 1986, groundwater at a 50-year-old petrochemical distribution terminal was treated. In 1988, in a joint trial with the U.S. EPA, the RO unit was used in conjunction with an ultraviolet enhanced oxidation unit to treat groundwater at a leaking landfill site in Oswego, New York.

In terms of industrial waste, the RO unit has treated pesticide container waste, potato wash water, fish waste, glycol in water, and general wastewaters containing various organic chemicals including solvents and chlorinated hydrocarbons.

Typical performance data values for membrane field trials are presented in the following table.

Component	Concentration (ppm) Feed	Concentration (ppm) Permeate	Process	Removal (%)	Comments
Trial Reference 1					
chlorophenol	510.00	0.01	RO	>99.9	- spiral Toray RO membrane
	1,655.00	0.02	RO	>99.9	- pre-filter (5 to 25 microns)
					- pressure of 350 to 7,000 kPa
Trial Reference 2					
boron	1.81	<0.01	RO	>99.9	- 6 membranes in series
calcium	35.70	0.49	RO	98.9	- Filtech FT-30 membrane
sodium	8,900.00	154	RO	98.3	- pH = 11 to 12
hydroxide	362.10	85.8	RO	76.3	- pressure of 5,600 kPa
carbonate	204.30	58.3	RO	71.5	
TOC	1,800.00	31	RO	98.3	
chloride	18,000.00	61	RO	99.6	
phenolics	270.00	100	RO	63.0	
Trial Reference 3					
acetone	26.81	15.99	RO	40.3	- none
t-1,2-dichloroethane	22.08	14.436	RO	34.8	
bromoform	31.03	0.529	RO	98.4	
o-xylene	13.35	1.703	RO	87.3	
benzoic acid	10.92	0.399	RO	96.3	
Trial Reference 4					
TOC	90.00	2.75	RO	97.0	- spiral RO membrane
	74.20	16		78.4	- Filmtec FT-30 SW 30-440
chloride	95.30	7.14	RO	92.5	- pressure of 5,600 kPa
sulphate	18.80	0.056	RO	99.7	
magnesium	10.73	0.89	RO	91.7	
sodium	45.67	6.46	RO	85.9	
potassium	4.02	0.74	RO	81.6	
barium	0.11	0.04	RO	63.6	
boron	0.09	0.06	RO	33.3	
iron	9.24	0.12	RO	98.7	
manganese	1.10	0.04	RO	96.4	
silicon	4.21	0.26	RO	93.8	
strontium	0.36	0.03	RO	91.7	
zinc	0.24	0.01	RO	95.8	
benzene	1.31	0.47	RO	64.1	
xylene	2.74	0.75	RO	72.6	
toluene	0.09	0.03	RO	66.7	
ethylbenzene	0.23	0.13	RO	43.5	
Trial Reference 5					
benzene	6.90	0.80	MF/RO	88.4	- precipitation was performed with lime
toluene	120.00	14.00	MF/RO	88.3	- pH = 8 to 10
ethylbenzene	22.00	2.80	MF/RO	87.3	- pressure = 2,760 kPa
phenol	300.00	66.00	MF	78.0	- spiral membrane
	300.00	91.00	RO	69.7	- MW cutoff = 200,000
acenaphthene	32.00	24.00	RO	25.0	
phenanthrene	14.00	7.80	RO	44.3	
COD	446.00	62.40	MF/RO	86.0	
BOD	215.00	15.00	MF/RO	93.0	

TOC	210.00	23.50	MF/RO	88.8
iron	143.57	0.04	MF/RO	>99.9
calcium	70.00	3.30	MF/RO	95.3
magnesium	10.98	9.60	MF/RO	12.6
barium	0.26	0.01	MF/RO	98.9
strontium	0.44	0.01	MF/RO	97.5

CONTACT INFORMATION

Environment Canada

Emergencies Engineering Division
Environmental Technology Centre
3439 River Road
Ottawa, Ontario
K1A 0H3

Telephone (613) 998-8541
Facsimile (613) 991-1673

REFERENCES

(1) Straforelli, J and H. Whittaker, "Reverse Osmosis for Clean-up of Spilled Wood Protection Solution", *Proceedings of the Second Annual Technical Seminar on Chemical Spills*, Toronto, Ont. (February, 1985).

(2) Tremblay, J. "Applications de la Technologie des Membranes dans l'oeust Canadien", *Proceedings of the Third Annual Technical Seminar on Chemical Spills*, Montreal, Que. (February, 1986).

(3) Whittaker, H., T. Kady, R. Evangelista, and C. Goulet, "Reverse Osmosis and Ultraviolet Photolysis/Ozonation Testing at the Pas Site - Oswego, N.Y.", *Proceedings of the Sixth Technical Seminar on Chemical Spills*, Calgary, Alta. (June, 1989).

(4) Zenon Environmental Inc., "Development and Demonstration of a Mobile Reverse Osmosis Adsorption Treatment System for Environmental Emergency Clean-ups", Contract No. 52SS.KEI 45-5-0715 (October, 1987).

(5) Zenon Environmental Inc., "Field Demonstration of Membrane Technology for Treatment of Landfill Leachate", RAC Project No. 439C (December, 1991).

ENVIRONMENT CANADA'S STEAM STRIPPING UNITS No. 138

DESCRIPTION

Environment Canada's Emergencies Engineering Division (EED) has one lab-scale and one mobile steam stripping unit suitable for removing volatile and some semi-volatile organics from water.

OPERATING PRINCIPLE

Steam stripping is a separation process that operates based on the differences in thermodynamic properties of liquids. In the process developed by EED, water and steam contaminated with organics are fed counter-currently through a packed column, causing the contaminant(s) to transfer from the water phase to the vapour phase. The driving force for the separation is the concentration differential of the organic component(s) between the liquid and vapour phases. Two streams are generated in this process: the treated effluent (bottoms) and the concentrated contaminant(s) (tops). The bottoms are sent to a holding tank before discharge to ensure that the water discharged complies with the applicable regulations. The tops are further treated or disposed of in an appropriate manner.

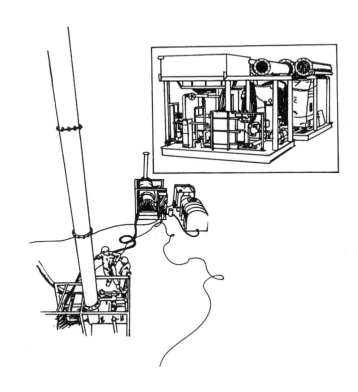

STATUS OF DEVELOPMENT AND USAGE

Environment Canada's Emergencies Engineering Division develops and maintains a wide range of prototype cleanup equipment. In cooperation with the private sector and public organizations, this equipment is used to demonstrate and adapt innovative methods for on-site mitigation of water or soil contamination caused by pollution emergencies or insecure hazardous waste sites. The Division is experienced in developing and evaluating lab- and pilot-scale steam stripping units. The mobile steam stripper unit was added to EED's inventory of treatment methods in 1989. Both the lab-scale and mobile units have successfully treated contaminated groundwater at leaking landfill sites and former process chemical plants.

TECHNICAL SPECIFICATIONS

Physical

The mobile unit was designed and constructed by CH$_2$M Hill Engineering Limited and IP Contractors, both of Calgary, Alberta. The system consists of a 9-m (30-ft) column packed with steel pall rings, a boiler, a fan condenser, and required pumps and heat exchangers. It can treat 2,000 L/h (530 gal/h).

Boiler Module - The boiler is a forced draft, low pressure steam boiler designed to run on No. 1, 2, or 3 oil. The boiler is also equipped with a blower, a fuel pump, and a Burks feedwater, close-coupled turbine pump (model 5CT7M). A panel mounted on the boiler controls the operation to the boiler as well as the feed pump. The overall dimensions of the boiler module are 2.8 m long × 1.8 m wide × 2.1 m high (9.2 × 5.9 × 6.9 ft). The following are other specifications.

Model	Saskatoon Boiler Mfg. Co. Ltd. FLL-30
Size	22 kW (30 hp)
Heating surface	14 m^2 (150 ft^2)
Shipping weight	1,360 kg (3,000 lb)
Operating weight	1,930 kg (4,250 lb)

Process Module - The process module consists of two key process elements: a packed stainless steel column and an overhead accumulator/decanter. Four small heat exchangers are used to optimize heat recovery. The tower is 31 cm

(12 in) in diameter and 6 m (20 ft) high. The overall dimensions of the process module are 2.3 m long × 1.8 m wide × 2.1 m high (7.5 × 5.9 × 6.9 ft). The process module is as high as the top of the overhead condenser when it is laying flat on the skid for transport.

Operational

Steam stripper technology is useful for removing volatile and semi-volatile organic compounds from contaminated water. Some compounds that have been successfully treated include chlorinated solvents, such as chlorobenzene, chloroform, dichloromethane, and trichloroethylene, and light aromatic compounds, such as benzene, toluene, ethylbenzene, and xylenes. Removal efficiency varies with the properties of contaminants, such as Henry's Law Constant (which is a function of the compound's solubility in the liquid phase and its volatility), and the system's operating parameters, such as temperature of the water and steam-to-water ratio. An increase in any of these parameters or properties normally produces a corresponding increase in organic removal efficiency, assuming all other parameters remain constant. Other factors influencing removal efficiencies include the size and type of the column packing and the ratio of the column diameter to packing diameter. The presence of inorganic solids will also cause column fouling which would reduce the throughput of the unit and could affect organic removal efficiencies.

Steam stripping is good for use at industrial sites as surplus supplies of steam are often available to operate the stripper unit. The Emergencies Engineering Division lab-scale steam stripper is rated at a 300 mL/min (4.8 gph) capacity, while the mobile steam stripper is rated at 33 L/min (8.6 gpm). Laboratory- and pilot-scale results show removal efficiencies of greater than 98% for many compounds, with influent concentrations as high as 20,000 ppm.

Power/Fuel Requirements

Process water must be supplied to the unit at an appropriate rate. The boiler is fired by No. 1, 2, or 3 oil, and requires a water supply to make up its steam output of 470 kg/h (1,040 lb/h). The main electric power supply is 115 VAC, single-phase, 60 Hz.

Transportability

The mobile steam stripping unit can be disassembled and packaged on two skids. These skids can be transported in a 5.5 m (18 ft) cube van to a spill site.

PERFORMANCE

The following are typical field-scale results using steam stripping.

Specific Compound	Flow rate	Concentration of Compound Influent	Effluent	Removal (%)	Reference
Chlorinated Solvents					
chlorobenzene	30 L/min (8 gpm)	26 ppb	<0.80 ppb	>97.0	Jacquemot et al. (1991)
		116 ppb	<1.6 ppb	>98.6	Jacquemot et al. (1991)
chloroform	30 L/min (8 gpm)	15 ppb	<0.23 ppb	>98.4	Jacquemot et al. (1991)
		158 ppb	<2.8 ppb	>98.2	Jacquemot et al. (1991)
dichloromethane	15 L/min (4 gpm)	2,987 ppm	<0.1 ppm	>99.9	Ladanowski et al. (1992)
	38 L/min (10 gpm)	1.0%	9 ppm	>99.9	APV Crepaco Inc. (1990)
	46 L/min (12 gpm)	1.5%	1 ppm	>99.9	APV Crepaco Inc. (1990)
1,2-dichloroethane	30 L/min (8 gpm)	26 ppb	3 ppb	88.4	Jacquemot et al. (1991)
1,1,2-trichloroethane	23 L/min (6 gpm)	1,000 ppm	1 ppm	>99.9	APV Crepaco Inc. (1990)
trichloroethylene	30 L/min (8 gpm)	24 ppb	<2.8 ppb	>88.5	Jacquemot et al. (1991)
	30 L/min (8 gpm)	2,237 ppb	11 ppb	99.5	Jacquemot et al. (1991)
Light Aromatic Compounds					
benzene	30 L/min (8 gpm)	700 ppb	6.2 ppb	99.1	Jacquemot et al. (1991)
	23 L/min (6 gpm)	1.6 ppm	<0.01 ppm	>99.4	Ladanowski (1993)
toluene	30 L/min (8 gpm)	5.7 ppb	0.92 ppb	84.0	Jacquemot et al. (1991)
	23 L/min (6 gpm)	0.26 ppm	<0.01 ppm	>96.1	Ladanowski (1993)
ethylbenzene	23 L/min (6 gpm)	0.35 ppm	<0.01 ppm	>97.1	Ladanowski (1993)
xylenes	23 L/min (6 gpm)	11 ppm	<0.02 ppm	>99.8	Ladanowski (1993)
	38 L/min (10 gpm)	1,000 ppm	100 ppm	90.0	APV Crepaco Inc. (1990)

Specific Compound	Flow rate	Concentration of Compound Influent	Effluent	Removal (%)	Reference
Alcohols					
methyl	38 L/min (10 gpm)	11,000 ppm	100 ppm	99.1	APV Crepaco Inc. (1990)
ethyl	380 L/min (100 gpm)	12,000 ppm	100 ppm	99.2	APV Crepaco Inc. (1990)
	95 to 285 L/min (25 to 75 gpm)	300,000 ppm	100 ppm	>99.9	APV Crepaco Inc. (1990)
	23 L/min (6 gpm)	160,000 ppm	50 ppm	>99.9	APV Crepaco Inc. (1990)
isopropyl	38 L/min (10 gpm)	20,000 ppm	100 ppm	99.5	APV Crepaco Inc. (1990)
	23 L/min (6 gpm)	200,000 ppm	50 ppm	>99.9	APV Crepaco Inc. (1990)
butanol	27 L/min (7 gpm)	60,000 ppm	200 ppm	90.7	APV Crepaco Inc. (1990)
Ketones					
acetone	38 L/min (10 gpm)	12,000 ppm	1 ppm	>99.9	APV Crepaco Inc. (1990)
	46 L/min (12 gpm)	15,000 ppm	1 ppm	>99.9	APV Crepaco Inc. (1990)
Miscellaneous Solvents (including dioxane, tetrahydrofuran, oxygen and nitrogen compounds)					
ammonia	15 L/min (4 gpm)	300,000 ppm	1 ppm	>99.9	APV Crepaco Inc. (1990)
ethyl ether	30 L/min (8 gpm)	231 ppb	1.9 ppb	99.2	Jacquemot et al. (1991)
tetrahydrofuran	30 L/min (8 gpm)	29 ppb	0.55 ppb	98.1	Jacquemot et al. (1991)
		890 ppb	5.8 ppb	99.4	Jacquemot et al. (1991)
		1,370 ppb	20 ppb	98.5	Jacquemot et al. (1991)

CONTACT INFORMATION

Environment Canada

Emergencies Engineering Division
Environmental Technology Centre
3439 River Road
Ottawa, Ontario
K1A 0H3

Telephone (613) 998-8541
Facsimile (613) 991-1673

REFERENCES

(1) APV Crepaco Inc. "Stream Stripping VOCs from Waste Water", Tonawanda, N.Y. (1990).
(2) Geisbrecht, G., F. Laperrière, and D. Peterson, "Mobile Steam Stripper as an Environmental Emergencies Countermeasure", in: *Proceedings of the Fifth Technical Seminar on Chemical Spills*, Montreal, Que. (February, 1988).
(3) Jacquemot, S., L. Keller, and M. Punt, "Comparison of Mobile Treatment Technologies at the Gloucester Landfill", Emergencies Engineering Division, Environment Canada, In-house report (1991).
(4) Ladanowski, C., "Field-scale Demonstration of Technologies for Treating Groundwater at the Gulf Strachan Gas Plant", Canadian Association of Petroleum Producers' Publication (April, 1993).
(5) Ladanowski, C., M. Punt, P. Kerr, and C. Adams, "Efficacy of Steam Stripping in the Removal of Dichloromethane from Groundwater", in: *Proceedings of Industrial Waste Management Conference*, Vienna, Austria (April, 1992).

ENVIRONMENT CANADA'S MICROFILTRATION/ULTRAFILTRATION UNIT

No. 139

DESCRIPTION

In 1986, Environment Canada's Emergencies Engineering Division (EED) purchased its first mobile microfiltration/ultrafiltration (MF/UF) unit. This field unit is self-contained and is located inside an aluminum frame. The membrane system uses either MF or UF modules.

OPERATING PRINCIPLE

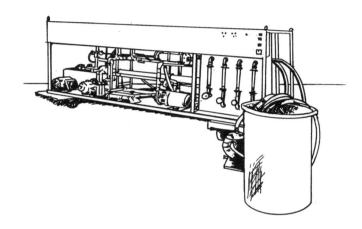

Microfiltration and ultrafiltration are crossflow membrane technologies. The process is termed crossflow since the influent, designated as feed, and the concentrate streams flow parallel to the membrane instead of perpendicular to it. Membrane filtration is the application of pressure and flow across a semipermeable membrane to separate the components of a fluid. The pressure required to drive the process is determined by the specific nature of the feed solution and the pore size of the membrane. If the solution contains significant levels of ionic solutes, the operating pressure must first overcome the osmotic pressure, which is the natural energy potential between the more concentrated and less concentrated streams. In the crossflow process, the feed stream is separated into two effluent streams: the permeate, which passes through the membrane, and the concentrate, which retains the dissolved and suspended solids rejected by the membrane.

STATUS OF DEVELOPMENT AND USAGE

Environment Canada's Emergencies Engineering Division develops and maintains a wide range of prototype cleanup equipment. In cooperation with the private sector and public organizations, these pieces of equipment are used to demonstrate and adapt innovative methods for on-site mitigation of water or soil contamination caused by pollution emergencies or insecure hazardous waste sites.

TECHNICAL SPECIFICATIONS

Physical

The complete system consists of two 3-m (9.8-ft) membranes and an automatic backwash system which is mounted on a 3.0 × 1.2 × 1.2 m (9.8 × 3.9 × 3.9 ft) skid.

Operational

Crossflow filtration is considered to be a self-cleaning process since the solutes and solids are swept away with the concentration stream, eliminating the need to frequently change the filter media. Crossflow membranes produce precise separations at the ionic, molecular, and macromolecular levels. Systems can be designed to concentrate all feed stream materials, purify solvents, or selectively pass some materials while concentrating others.

The MF membrane pore size is 400 to 20,000 Å, while the UF pore size ranges from 45 to 1,100 Å. The MF unit can process up to 40 L/min (11 gpm) of permeate from a feed solution containing up to 1,500 ppm of oil and grease. It can also be operated as an independent treatment system in which the MF/UF unit is used upstream of other processes to remove the larger inorganic compounds as well as oil and grease, which are potential foulants for other systems.

The separation spectrums of the microfiltration (MF) and ultrafiltration (UF) processes, as compared to other crossflow filtration processes, are presented in the following table.

Crossflow Separation Process	Membrane Pore Size	Fluid Component Retained	Transmembrane Pressure	Sample Applications
Reverse Osmosis	5 to 15 Å	Ions and most organics over 100 molecular weight	1,400 to 6,900 kPa [200 to 1,000 psi(g)]	Metal ion recovery BOD and COD reduction in waste streams
Nanofiltration	10 to 90 Å	Organics with 300 to 1,000 molecular weight	1,400 to 5,500 kPa [200 to 800 psi(g)]	Water softening and desalting of organics
Ultrafiltration	45 to 1,100 Å	Most organics over 1,000 molecular weight, including pyrogens, viruses, bacteria, and colloids	70 to 1,400 kPa [10 to 200 psi(g)]	Concentration and recovery of industrial organics and dilute suspended oils
Microfiltration	400 to 20,000 Å	Very small suspended particles, some emulsions, and some bacteria	7 to 170 kPa [1 to 25 psi(g)]	High volume removal of bacteria and small suspended solids

Power/Fuel Requirements

An electric power supply is required to power the system's pumps and controls.

Transportability

The system is mounted on a skid and can be transported by truck.

PERFORMANCE

Since acquiring the MF/UF unit, the Emergencies Engineering Division has used it ahead of the RO unit as a pre-treatment system. The following are typical performance data values for combined MF/RO membrane field trials.

Component	Concentration (ppm) Feed	Permeate	Process	Removal (%)	Comments
benzene	6.90	0.8	MF/RO	88.4	- precipitation was performed with lime
toluene	120.00	14	MF/RO	88.3	- pH = 8 to 10
ethylbenzene	22.00	2.8	MF/RO	87.3	- pressure = 2,760 kPa
phenol	300.00	66	MF	78.0	- spiral membrane
	300.00	91	RO	69.7	- MW cutoff = 200,000
COD	446.00	62.40	MF/RO	86.0	
BOD	215.00	15.00	MF/RO	93.0	
TOC	210.00	23.50	MF/RO	88.8	
iron	143.57	0.04	MF/RO	>99.9	
calcium	70.00	3.30	MF/RO	95.3	
magnesium	10.98	9.60	MF/RO	12.6	
barium	0.26	0.01	MF/RO	98.9	
strontium	0.44	0.01	MF/RO	97.5	

5.4 Treatment/Disposal - Aqueous Treatment Systems

The MF membrane unit was used in western Canada, combined with granular and powdered activated carbon, to remove pesticides from wash water. In many cases, removals of 96.5 to 99.9% were achieved (Tremblay, 1986). UF membranes were used at a trial to study the removal of petrochemicals from groundwater in Toronto. The UF was used as a pretreatment to RO to remove "slugs" of free, aged hydrocarbon product which appeared periodically while pumping the groundwater (Keller, 1988). UF was also used alongside the lab RO unit to remove PCB from contaminated wash water at a spill in the Ottawa area. The concentration of PCBs was reduced from a range of 10 to 641 ppm to 3 ppb using both units (Whittaker and Clark, 1985).

CONTACT INFORMATION

Environment Canada

Emergencies Engineering Division
Environmental Technology Centre
3439 River Road
Ottawa, Ontario
K1A 0H3

Telephone (613) 998-8541
Facsimile (613) 991-1673

REFERENCES

(1) Keller, L., "Recent Application of Membrane Technology in the Automotive Industry", Emergencies Engineering Division, Environmental Technology Centre, Environment Canada, Ottawa, Ont., internal report (1988).
(2) Tremblay, J. "Membrane Technology Applications in Western Canada", *Proceedings of the Third Annual Technical Seminar on Chemical Spills,* Montreal, Que. (1986).
(3) Whittaker, H. and R. Clark, "Cleanup of PCB Contaminated Groundwater by Reverse Osmosis", *Proceedings of the Second Annual Technical Seminar on Chemical Spills*, Toronto, Ont. (1985).

ESI CASCADE LP500 SERIES LOW PROFILE AIR STRIPPER

No. 140

DESCRIPTION

A modular, low profile air stripper designed to remove volatile organic compounds (VOCs) from water.

OPERATING PRINCIPLE

Air is bubbled through the contaminated water promoting phase transfer of the contaminants which transfers the VOCs from the water phase to the air. The system consists of a sump, a series of stacked aeration trays, a blower, and all necessary controls and gauges. Contaminated water enters the top tray, flows back and forth across a series of baffles along the length of the tray, and drops to the tray below. The water is aerated through bubblers between the baffles. Each bubbler is supplied with clean air from the header system.

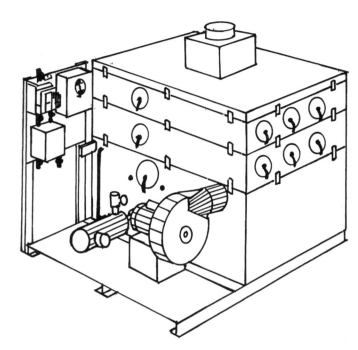

STATUS OF DEVELOPMENT AND USAGE

Commercially available product in general use to treat contaminated groundwater and process waters.

TECHNICAL SPECIFICATIONS

Physical

Each tray of this low-profile system is 1.4 m high × 1.7 m wide × 1.8 m long (4.8 × 5.5 × 6 ft), is made of 16-gauge epoxy-coated carbon steel, and weighs approximately 560 kg (1,240 lb). Flexible aeration tubes suspended above the bottom of the tray inject air downward into the tray. This action provides high turbulence and excellent foul resistance in the presence of high amounts of inorganic compounds. Clean-out ports with pop-off aeration tubes provide quick and easy maintenance. An inlet on the side of the tray allows easy inspection without disturbing liquid piping.

Operational

As trays can be added or removed to ensure efficient treatment, this system is designed to accommodate field changes, such as seasonal groundwater fluctuation and hydrocarbon concentration. The air stripper's low profile simplifies the hydraulics and may eliminate secondary pumping. This system is also easy to install with no cranes required, and easy to start up. The units are shipped completely assembled.

Power/Fuel Requirements

Contaminated water must be supplied to the system at an appropriate flow rate and pressure. An electric power supply is required for the blower.

Transportability

The system is small and modular in design and can easily be handled by two people.

PERFORMANCE

The system's efficiency depends on the number of aeration plates, the influent flow rate, and the air flow. Some examples of removal rates for benzene are provided in the following table.

5.4 Treatment/Disposal - Aqueous Treatment Systems

Removal Data for Benzene at 15.5 °C (60 °F)

Flow rate	1 tray	2 tray	3 tray	4 tray
19 L/min (5 gpm)	.99995			
38 L/min (10 gpm)	.9994	.99995		
76 L/min (20 gpm)	.975	.9994	.99998	
114 L/min (30 gpm)	.915	.9928	.9994	.99995
152 L/min (40 gpm)	.83	.9764	.9972	.9995
190 L/min (50 gpm)	.83	.9734	.9960	.9995

CONTACT INFORMATION

Manufacturer

Ejector Systems Incorporated (ESI)
910 National Avenue
Addison, IL 60101
U.S.A..

Telephone (708) 543-2214
Facsimile (708) 543-2014

OTHER DATA

Options available for the Cascade LP500 Series Low Profile Strippers include NEMA 4-blower motor starter, explosion-proof motor starter, blower and transfer pump, TEFC transfer pump, hi/low air switch, and high effluent switch.

REFERENCES

(1) Manufacturer's Literature.

GLE SLANT RIB COALESCING (SRC) OIL/WATER SEPARATOR

No. 141

DESCRIPTION

Coalescing-type separators for removing non-emulsified oils from water in such applications as groundwater remediation or ultrafilter pre-treatment.

OPERATING PRINCIPLE

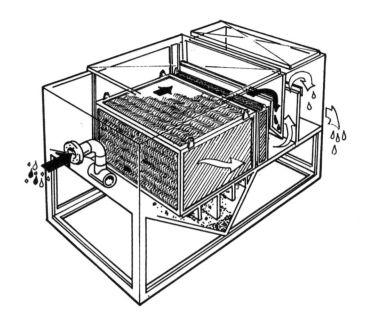

A series of plastic-ribbed plates are arranged within a rectangular tank which also contains special baffles and weirs designed to direct flow, skim oil, and control liquid levels within the separator. The oil-contaminated water is forced through the plastic plates where oil droplets attracted to the oleophilic plates coalesce or merge until they are large enough to break free and rise to the surface. The separated oil accumulates at the surface where it displaces the water. As the oil layer increases, oil spills over a weir into an oil reservoir where it can flow by gravity or be pumped automatically to remote storage tanks. Oil and water are discharged through individual outlets.

STATUS OF DEVELOPMENT AND USAGE

Commercially available product in use in applications such as groundwater remediation, ultrafilter pre-treatment, floor drain filtration, capture of steam condensate tramp oil, and vehicle wash water treatment.

TECHNICAL SPECIFICATIONS

Physical - The standard **Model SRC** is available in either steel or fibreglass construction, in 21 standard sizes from 55 to 19,000 L/min (15 to 5,000 gpm). The **Model SRC-M** is constructed as a single piece of molded fibreglass and is available in six sizes from 7.6 to 190 L/min (2 to 50 gpm).

Operational - The standard **Model SRC** coalesces and removes oils and hydrocarbons from wastewater. Effluent quality is not affected by variable flow rates or sporadic slugs of 100% oil. The separator has a sloped sludge chamber directly under the oil-coalescing media which collects and isolates settled solids from the flow. An internal oil reservoir provides temporary storage for separated oil. Six types of oil-coalescing media are available. The **Model SRC-M** is designed to remove oil from low flow wastewater when minimal solids are present. Both models produce effluent with 10 mg/L or less of non-emulsified oil and both are ineffective against oil emulsions.

Power/Fuel Requirements and Transportability - The process stream must be supplied to separators at an appropriate flow rate. The **Model SRC-M** weighs only from 36 to 118 kg (80 to 260 lb) and is person-portable. The standard **Model SRC** units weigh from 570 to 1,815 kg (1,260 to 4,000 lb) and require lifting equipment to move. Lifting lugs are provided on both the media packs and on the separators of the standard SRC models.

CONTACT INFORMATION

Manufacturer

Great Lakes Environmental Incorporated
315 South Stewart
Addison, IL 60101, U.S.A.

Telephone (708) 543-9444
Facsimile (708) 543-1169

REFERENCES

(1) Manufacturer's Literature.

HAZELTON MAXI-STRIP SYSTEMS

No. 142

DESCRIPTION

A venturi-type stripper designed to remove VOCs from contaminated water, this unit is excellent for spill sites with high concentrations of suspended and dissolved solids.

OPERATING PRINCIPLE

The Maxi-Strip uses highly turbulent jets of water to shear and accelerate fluid films in an open bore. The microturbulence achieved in the bore creates a large surface area. The same films and jets also aspirate air, promoting phase transfer which results in the VOCs being transferred from the water phase to the air. The system discharges into a sump, where the off gas separates and is collected. The VOCs then exit the column with the air stream.

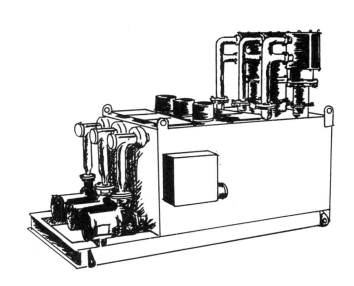

STATUS OF DEVELOPMENT AND USAGE

Commercially available products marketed for the removal of VOCs, gases, and odours from groundwater, industrial wastewater, accidental spills, process effluents, and runoff from landfills.

TECHNICAL SPECIFICATIONS

Physical

This is a modular system allowing flexibility in configuration. The number of modules is determined by the flow through the system and the required quality of the effluent. Each module is equipped with a stripper, tank, and small pump.

Model	Width	Length	Height
100	0.4 m (1.2 ft)	2.5 m (8.3 ft)	1.7 m (5.7 ft)
350	0.5 m (1.8 ft)	4.1 m (13.5 ft)	2.8 m (9.2 ft)
500	0.6 m (2.2 ft)	4.3 m (14 ft)	2.8 m (9.2 ft)

Operational

This technology was originally developed to oxidize and precipitate high concentrations of metals from mine drainage and metallurgical industries. The Maxi-Strip, which was designed for iron and biological concentrations associated with groundwater remediation and highly loaded waste streams, has no packing to foul when water contains iron, sediment, or biological contaminants. The system is low profile and inconspicuous in residential areas. As no blower is used, the system is quiet and suitable for use in noise-sensitive areas. Engineered with a "gravity in/gravity out" flow concept, the sump tank's two-baffle design allows infinite turndown. This means the system strips a defined performance curve from the hydraulic limit of the tank down to 3.8 L/min (1 gpm) without flooding or channelling.

Model	Flow Rate	Air Flow
100	3.8 to 378 L/min (1 to 100 gpm)	0.106 m^3/s (225 cfm)
350	3.8 to 1,300 L/min (1 to 350 gpm)	0.50 m^3/s (1,060 cfm)
500	3.8 to 1,900 L/min (1 to 500 gpm)	0.76 m^3/s (1,620 cfm)

Larger systems are available.

Power/Fuel Requirements

Typical systems require 460/230 or 575 volt, three-phase (230 volt, single-phase).

Transportability

All systems are skid-mounted. Trailer-mounted rental systems are available.

PERFORMANCE

Performance is quickly predicted via a computer mass transfer model based on empirical data.

CONTACT INFORMATION

Manufacturer

Hazelton Environmental
125 Butler Drive
Hazelton, PA 18201
U.S.A.

Telephone (717) 454-7515
Facsimile (717) 454-7520

OTHER DATA

If treatment of off gas is required, systems can be configured to concentrate off gas. This configuration ensures off gas treatment size/cost is minimized.

REFERENCES

(1) Dempsey, B.A., "Hazelton Air-Stripping System Test Results and Modelling", Department of Civil Engineering, Pennsylvania State University (July, 1989).
(2) Manufacturer's Literature.

HUDSON CS SERIES COALESCING OIL/WATER SEPARATORS

No. 143

DESCRIPTION

Coalescing-type separators for removing free and moderately emulsified oils from waters.

OPERATING PRINCIPLE

Water/oil solution comes in through the influent standpipe and enters the coalescing module which consists of PVC plates thermo-bonded into a dense module. As the PVC is oleophilic (attractive to oil), it attracts the oil droplets which coalesce into larger droplets and rise to form a layer of oil on the surface. A continuous-acting rotary pipe oil skimmer removes the oil from the surface and deposits it into the oil separation tube. Water removed with the oil separates back out and continues through a series of underflow and overflow baffles before discharging from the unit into the waste stream.

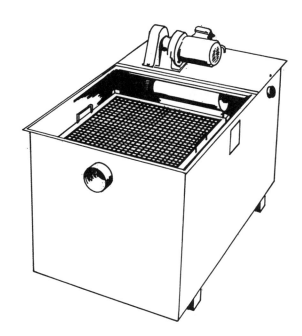

STATUS OF DEVELOPMENT AND USAGE

Commercially available product in use for groundwater remediation, ultrafilter pre-treatment, floor drain filtration, capture of steam condensate tramp oil, and vehicle wash water treatment.

TECHNICAL SPECIFICATIONS

Physical

The separators are constructed of heavy, mild steel plate of 1/8, 3/16, 1/4, or 3/8 in, with phenolic epoxy interior coating, industrial enamel exterior finish, NPT-threaded fittings, stainless steel hardware, and bronze, ball-type drain valves. Steep vertical flutes in the coalescing media reduce buildup of internal sludge and solids. Standard material is PVC rated to 60°C (140°F) and CPVC rated to 88°C (190°F), with stainless steel media for higher temperatures.

Operational

Standard systems are available for flows of 3.8 to 20,820 L/min (1 to 5,500 gpm). The separators remove oil to 10 mg/L or less, can accommodate 100% oil slugs on a limited basis, and are fully automatic.

Power/Fuel Requirements

The process stream must be supplied to the separators at an appropriate flow rate. The equipment can be supplied to operate on a broad range of voltages.

Transportability

The smaller systems are person-portable, while the larger systems require lifting equipment to move. Lifting lugs are provided on both the media packs and the separators of the larger models.

Manufacturer

Hudson Industries
Box 2212
Hudson, OH 44236

Telephone (216) 487-0668
Toll-free (800) 487-9668
Facsimile (216) 487-0811

REFERENCES

(1) Manufacturer's Literature.

HYDRO-STRIPPER AIR STRIPPER   No. 144

DESCRIPTION

Pre-engineered, skid-mounted, packed column air strippers for removing VOCs from groundwater.

OPERATING PRINCIPLE

Contaminated water is cascaded down through the packed column, air and water are thoroughly mixed, and the VOCs are transferred from the water to the air stream. The VOCs exit the column with the air stream.

STATUS OF DEVELOPMENT AND USAGE

Commercially available systems.

TECHNICAL SPECIFICATIONS

Physical - Systems have a stainless steel shell, contain polypropylene media, and are equipped with a water distributor, effluent booster pump, controls, and a low-pressure blower. Systems are 7.5 m (25 ft) tall, with a bed depth of 5.2 m (17 ft), and are mounted on a skid measuring 2.4 × 1.5 m (8 × 5 ft). The following are other specifications for the complete line.

	PHS-8.17	PHS-12.17	PHS-15.17	PHS-23.17
Capacity	49 L/min (13 gpm)	98 L/min (26 gpm)	170 L/min (45 gpm)	340 L/min (90 gpm)
Diameter	20 cm (8 in)	30 cm (12 in)	38 cm (15 in)	58 cm (23 in)
Blower power	.22 kW (0.3 hp)	0.6 kW (0.8 hp)	0.6 kW (0.8 hp)	1.1 kW (1.5 hp)
Pump power	1.5 kW (2 hp)	1.5 kW (2 hp)	2.2 kW (3 hp)	3.7 kW (5 hp)
Weight	1,360 kg (3,000 lb)	1,475 kg (3,250 lb)	1,520 kg (3,350 lb)	1,680 kg (3,700 lb)

Operational - The system's efficiency depends on packing type and height, tower diameter, and a correct air-to-water ratio. Each packed tower module can remove more than 99% of select compounds. By combining modules in series, treatment efficiencies can exceed 99.99%. The air stripping systems are designed for flexibility in application. Two or more units can be operated in series, packed bed depth can be increased in a single tower, and variable flow rates can be treated through a single unit. Because of the thin film mass transfer technology used, low pressure blowers can be used, lowering operating costs and system pressure drops. Options include low pressure water distributor, mist eliminator, off-gas treatment, explosion-proof system, liquid feed pump, packing cleaning system, and a winterization package.

Power/Fuel Requirements and Transportability - Contaminated water must be supplied to the system at an appropriate flow rate and pressure. An electric power source is required for the blowers, pumps, and controls. All Hydro-Strippers are skid-mounted and designed to be transported to remote sites.

Manufacturer

Hydro Group, Inc.
97 Chimney Rock Road
Bridgewater, NJ 08807, U.S.A.

Telephone (908) 563-1400
Facsimile (908) 563-1396

REFERENCES

(1) Manufacturer's Literature.
(2) Punt, M. and H. Whittaker, "A Comparison of Steam Stripping and Air Stripping for the Removal of Volatile Organic Compounds from Water", in: *Proceedings of the Eighth Technical Seminar of Chemical Spills*, Environment Canada, Ottawa, Ont. (June, 1991).

KOCH TUBULAR ULTRAFILTRATION (UF) SYSTEM

No. 145

DESCRIPTION

Systems for purifying contaminated water by ultrafiltration, consisting of tubular cartridges, a cleaning tank, and circulation pump and designed for waste treatment in metal working, plating, chemical process, and general industrial applications.

OPERATING PRINCIPLE

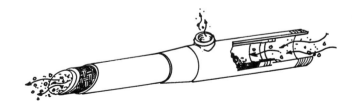

The process solution is pumped under pressure through the inside or lumen of the tube. The membrane is cast on the inside tubular surface of a reinforced fibreglass backing structure. Separation occurs at the liquid/membrane interface, and rejected solutes and particles remain inside the tube. The flow along the membrane surface can be controlled to produce high shear forces to break away rejected species and force them downstream. The ultrafiltrate that permeates through the membrane is collected in a plastic sleeve and then ducted into a permeate storage tank. The cartridges are cleaned by forward flushing and spongeballing.

STATUS OF DEVELOPMENT AND USAGE

Commercially available systems.

TECHNICAL SPECIFICATIONS

Physical

Koch Membrane Systems' UF-4VA - The UF-4VA system is designed for both pilot and demonstration applications and for use as a fully operational system for treating a variety of feed streams. Standard equipment includes: four 1.5 m (5 ft) Abcor® tubular membranes with 0.4 m^2 (4.4 ft^2) of membrane area; a 190-L (50-gal) process/cleaning tank; one centrifugal circulation pump, mounted on a baseplate and coupled to a standard NEMA frame and TEFC motor; permeate line direct-reading flowmeter; two pressure gauges; high-temperature switch and low-temperature switch, both interlocked to circulation pump; all valves, piping, and internal wiring; and a control panel. Options include all stainless steel construction, explosion-proof construction, and an expandable membrane area to increase system capacity. Membranes are available in a wide range of molecular cutoffs. Larger tubular ultrafiltration systems are also available from Koch Membrane Systems. The following are dimensions for the standard UF-4VA system.

Length   1.7 m (5.5 ft)
Width   0.6 m (2 ft)
Height   2 m (7 ft), expandable to 3.7 m (12 ft)
Weight   180 kg (400 lb)

Operational

Ultrafiltration (UF) is a low pressure [1.4 to 22 kPa (10 to 150 psi)] membrane process for separating suspended solids and high molecular weight dissolved materials from liquids. Fluid flows across the membrane surface at high velocity. This crossflow characteristic differs from the perpendicular flow of ordinary filtration, in which a "cake" builds up on the surface of the filter, requiring frequent filter replacement or cleaning. Crossflow prevents buildup of filter-cake and results in high filtration rates that can be maintained almost continuously. The UF-4VA system can treat from 475 to 2,440 L/day (125 to 645 gpd), depending on the stream. By expanding available membrane area, the capacity can be increased to treat from 950 to 4,900 L/day (250 to 1,300 gpd), depending on the characteristics of the feed stream.

UF-4VA Daily Capacity

Membrane Area	Typical Coolant, Oily Waste, and High ppm Waste	Chemical and Low ppm Waste	Fine Particle Separation (metal hydroxides, pigments)
0.4 m^2 (4.4 ft^2)	470 L/day (125 gpd)	1,130 L/day (300 gpd)	2,440 L/day (645 gpd)
0.8 m^2 (8.8 ft^2)	950 L/day (250 gpd)	2,270 L/day (600 gpd)	4,880 L/day (1,300 gpd)

The following systems with higher processing capabilities are also available.

System	Capacity
UF-18V	1.9 to 9.8 m^3/day (500 to 2,600 gpd)
UF-18/36	1.9 to 20 m^3/day (500 to 5,200 gpd)
UF-70	7.6 to 40 m^3/day (2,000 to 10,400 gpd)
KONSOLIDATOR 78	17 to 90 m^3/day (4,500 to 23,400 gpd)
KONSOLIDATOR 150	30 to 150 m^3/day (7,900 to 39,000 gpd)
KONSOLIDATOR 252	60 to 300 m^3/day (15,000 to 80,600 gpd)
UF-1200 RM	130 to 670 m^3/day (34,000 to 176,800 gpd)

Power/Fuel Requirements

Contaminated water must be supplied to the system at an appropriate flow rate. An electric power supply is required for the system's pumps.

Transportability

All systems are skid-mounted and can be transported by truck.

CONTACT INFORMATION

Manufacturer

Koch Membrane Systems Inc.
455 East Eisenhauer Parkway
Suite 150
Ann Arbor, MI 48108
U.S.A.

Telephone (313) 761-3836
Facsimile (313) 761-3844

OTHER DATA

Koch Membrane Systems also manufacture **Romicon Hollow Fiber** and **Spira-Pak Spiral Wound Modules and Systems** using the same UF crossflow method of separation.

REFERENCES

(1) Manufacturer's Literature.

LIQUID-MISER
ACTIVATED CARBON ADSORPTION CANISTERS

No. 146

DESCRIPTION

Water purification systems that use activated carbon to remove dissolved organics from groundwater, wastewater, and liquid process streams.

OPERATING PRINCIPLE

Liquids contaminated with organics are percolated through activated carbon inside the canister, where the contaminants are physically adsorbed to the carbon media.

STATUS OF DEVELOPMENT AND USAGE

Commercially available product in general use.

TECHNICAL SPECIFICATIONS

Physical

The canisters contain high-grade, liquid-phase, activated carbon and are constructed of coated carbon steel for maximum corrosion resistance.

	LM-7.5	LM-15	LM-25	LM-50
Diameter	0.6 m (1.9 ft)	0.6 m (2.2 ft)	0.9 m (3 ft)	1.2 m (4 ft)
Height	0.9 m (3 ft)	0.97 m (3.2 ft)	1.3 m (4.3 ft)	1.4 m (4.7 ft)
Bed depth	0.7 m (2.3 ft)	0.7 m (2.3 ft)	1.1 m (3.8 ft)	1.1 m (3.8 ft)
Adsorbent volume	0.17 m^3 (6.2 ft^3)	0.23 m^3 (8.3 ft^3)	0.70 m^3 (25 ft^3)	0.42 m^3 (50 ft^3)
Carbon weight	75 kg (165 lb)	100 kg (225 lb)	305 kg (675 lb)	615 kg (1,355 lb)
Shipping weight	135 kg (300 lb)	170 kg (375 lb)	425 kg (935 lb)	815 kg (1,800 lb)

Operational

Liquid-Misers are available to handle flow of up to 190 L/min (50 gpm).

	LM-7.5	LM-15	LM-25	LM-50
Maximum flow rate	28 L/min (7.5 gpm)	57 L/min (15 gpm)	95 L/min (25 gpm)	190 L/min (50 gpm)
Maximum pressure drop	1 kPa (4 in H$_2$O)	2.7 kPa (11 in H$_2$O)	2.2 kPa (9 in H$_2$O)	2.5 kPa (10 in H$_2$O)
Minimum contact time	6 minutes	4 minutes	7.5 minutes	7.5 minutes

Power/Fuel Requirements

Contaminated water must be supplied to the canister at an appropriate flow rate and pressure.

Transportability

A drum hoist, hand truck, forklift and pallet, or other drum-handling equipment are required to handle the canisters.

PERFORMANCE

Treatment technologies using granular activated carbon have a proven ability to remove organic contaminants to non-detectable levels. The degree of removal varies, however, in relation to such factors as flow rates, operating pressures, carbon dosage and type, contact time, influent pH, temperature, and viscosity, and if applicable, liquid pretreatment to remove oils, greases, or suspended solids from the influent which may impair the system's ability to operate adequately.

CONTACT INFORMATION

Manufacturer

Westport Environmental Systems
Forge Road, P.O. Box 217
Westport, MA 02790-0217
U.S.A.

Telephone (508) 636-8811
Facsimile (508) 636-2088

OTHER DATA

Westport Environmental Systems also manufactures **Odor-Miser Carbon Adsorption Canisters**, a vapour phase purification system for removing dissolved organics from air streams (listed in Section 5.5, No. 159).

REFERENCES

(1) Hager, D.G., "Industrial Wastewater Treatment by Granular Activated Carbon", *Industrial Water Engineering* (January/February, 1974).
(2) Manufacturer's Literature.
(3) O'Brien, R.P. and J.L. Fisher, "There is an Answer to Groundwater Contamination", reprinted from *Water/Engineering & Management*, File No. 27-86, Calgon Carbon Corporation, Pittsburgh, PA (May, 1983).
(4) Rizzo, J.L. and A.R. Shepard, "Treating Industrial Wastewater with Activated Carbon", *Chemical Engineering*, McGraw-Hill, Inc., New York, N.Y. (January, 1977).
(5) Ying, W., E.A. Dietz, and G.C. Woehr, "Adsorptive Capacities of Activated Carbon for Organic Constituents of Wastewaters", *Environmental Progress: 9(1)* (February, 1990).

ORS LO-PRO AIR STRIPPERS

No. 147

DESCRIPTION

A series of modular air strippers designed to remove volatile organic compounds (VOCs) from water. Unique, vacuum-induced, multi-stage counterflow aeration removes up to 99.99% of organic contaminants in a compact unit. The patented design makes it easy to use, and with low operating and maintenance costs.

OPERATING PRINCIPLE

Air is bubbled through the contaminated water promoting phase transfer of the contaminants, which results in the VOCs being transferred from the water phase to the air. The VOCs then exit the column with the air stream.

STATUS OF DEVELOPMENT AND USAGE

Commercially available product in general use.

TECHNICAL SPECIFICATIONS

Physical

	Model I	Model II	Model III
Application	Explosion-proof unit for use in classified areas	For use in non-classified areas	High-flow unit for use in non-classified areas
Dimensions	1.7 m long × 0.7 m wide × 1.3 to 2.7 m high (5.5 × 2.3 × 4.4 to 8.8 ft)	same as Model I	1.8 m long × 1.5 m wide × 1.4 to 2.1 m high (6 × 5 × 4.8 to 6.9 ft)
Tray size	17 cm and 25 cm high (7 in and 10 in)	same as Model I	33 cm high (13 in)
Water flow	4 to 76 L/min (1 to 20 gpm)	same as Model I	4 to 230 L/min (1 to 60 gpm)
Air flow	Up to 5.7 m^3/min (200 cfm)	Up to 5.1 m^3/min (180 cfm)	Up to 20 m^3/min (700 cfm)
Blower power	2.2 or 3.7 kW (3 or 5 hp)	same as Model I	7.4 or 11 kW (10 or 15 hp)

Operational

These systems can be fine-tuned for different influent concentrations and removal efficiencies by adding or removing filter plate modules. The systems use either an explosion-proof or non-explosion-proof vacuum blower. Optional equipment that can be incorporated on the systems include a sump probe with high-high, high, and low actuation points to control effluent transfer pump and interlock influent pumps, high pressure switch to interlock stripper blower if the trays become fouled, low pressure switch to interlock influent pump if the blower shuts down, signal junction box to allow connection of probes and switches to control panel, SITEPRO 2000 air stripper/transfer pump control panel, transfer pump assembly, and water flow meter. Constructed of high-density polyethylene and stainless steel, the LO-PRO can withstand corrosion from acids, bases, and most solvents. The system is wind-resistant and freestanding.

Power/Fuel Requirements

Contaminated water must be supplied to the system at an appropriate flow rate and pressure. The blower is 230 V, 3-phase with an optional conversion to single phase available.

Transportability

The system can be moved by two people.

PERFORMANCE

Efficiency of the system depends on the number of filter plates, the influent flow rate, and the air flow. The following are contaminant removal efficiencies at two different flow rates for common VOCs.

19 L/min @ 4.2 m^3/min (5 gpm @ 150 cfm)

	Influent (ppb)	Effluent (ppb)	Removal (%)
Benzene	700	0.2	99.97
Toluene	7,000	1.6	99.98
Ethyl benzene	1,900	0.5	99.97
Xylenes (total)	9,400	1.9	99.98
BTEX (total)	19,000	4.22	99.98

38 L/min @ 4.2 m^3/min (10 gpm @ 150 cfm)

	Influent (ppb)	Effluent (ppb)	Removal (%)
Benzene	700	1.0	99.86
Toluene	7,000	7.0	99.80
Ethyl benzene	1,900	2.2	99.88
Xylenes (total)	9,400	8.2	99.91
BTEX (total)	19,000	18.4	99.90

CONTACT INFORMATION

Manufacturer

ORS Environmental Systems
A Division of Sippican, Inc.
32 Mill Street
Greenville, NH 03048
U.S.A.

Telephone (603) 878-2500
North America toll-free (800) 228-2310
Facsimile (603) 878-3866

REFERENCES

(1) Manufacturer's Literature
(2) Punt, M. and H. Whittaker, "A Comparison of Steam Stripping and Air Stripping for the Removal of Volatile Organic Compounds from Water", in: *Proceedings of the Eighth Technical Seminar of Chemical Spills*, Environment Canada, Ottawa, Ont. (June, 1991).
(3) Whittaker, H., "A Review of Countermeasures for the Removal of Volatile Organic Compounds from Groundwater", in: *Proceedings of the Fifth Technical Seminar of Chemical Spills*, Environment Canada, Ottawa, Ont. (February, 1988).

OSMONICS CROSSFLOW MEMBRANE FILTRATION SYSTEMS — No. 148

DESCRIPTION

Filter systems, including equipment and filtration membranes, designed to purify water through reverse osmosis (RO), also called hyperfiltration, nanofiltration (NF), ultrafiltration (UF), or microfiltration (MF).

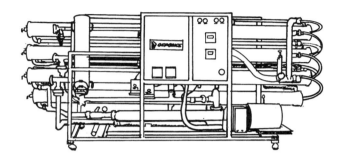

OPERATING PRINCIPLE

Crossflow membrane filtration is the separation of the components of a fluid by semipermeable membranes through applying pressure and flow. The influent or "feed" stream is separated into two effluent streams, the "permeate", which passes through the membrane, and the "concentrate", which retains the dissolved and suspended solids rejected by the membrane. The process is termed "crossflow" because the feed and the concentrate flow parallel to the membrane instead of perpendicular to it. The pressure required to drive the process is determined by the specific nature of the feed solution and the pore size of the membrane. If the solution contains significant levels of ionic solutes, the operating pressure must first overcome the osmotic pressure - the natural energy potential between the more concentrated and the less concentrated streams.

STATUS OF DEVELOPMENT AND USAGE

Commercially available systems.

TECHNICAL SPECIFICATIONS

Physical

Osmonics filtration systems are housed within rugged steel frames and mounted on skids. They are equipped with patented stainless steel separator housing TONKAFLO® multi-stage centrifugal pumps, and a patented flow configuration. A range of instrumentation and accessories is available. The following are general specifications for the different systems.

Type	Permeate Rate	Pressure	Features
B Series*	>3,105 m³/d (>820,000 gpd)	170 to 6,900 kPa (25 to 1,000 psi)	Custom-engineered to application. Advanced electronics, diagnostic alarms
CHF/CVF Series*	>820 m³/d (>216,000 gpd)	170 to 4,140 kPa (25 to 600 psi)	Horizontal or vertical configuration, comprehensive alarms.
C Series*	>34 m³/d (>9,000 gpd)	170 to 2,070 kPa (25 to 300 psi)	Wall-mount or freestanding vertical design
Process Evaluation Systems*	variable	170 to 5,520 kPa (25 to 800 psi)	Wide flexibility for evaluating applications
Orion	0.6 to 4 m³/d (150 to 1,000 gpd)	>1,720 kPa (>250 psi)	Cabinet design with patented fluid control system and modular electrical controls
Econopure	0.4 to 2 m³/d (100 to 500 gpd)	>1,720 kPa (>250 psi)	Portable, economical pure water machine

* machines are available with RO, NF, UF, or MF membrane

Operational

Crossflow filtration is a self-cleaning process. The solutes and solids are swept away with the concentration stream, which eliminates frequent filter media changes or resin regeneration. Crossflow membranes produce precise separations at the ionic, molecular, and macromolecule levels. Systems can be designed to concentrate all feed stream materials, purify solvents, or selectively pass some materials while concentrating others.

The following is a summary of the capabilities and applications of the various crossflow technologies.

Crossflow Separation Process	Membrane Pore Size	Fluid Component Retained	Transmembrane Pressure	Sample Applications
Reverse Osmosis	5 to 15 Å	Ions and most organics over 100 molecular weight	1,400 to 6,900 kPa [(200 to 1,000 psi(g)]	Metal ion recovery BOD and COD reduction in waste streams
Nanofiltration	10 to 90 Å	Organics with 300 to 1,000 molecular weight	1,400 to 5,500 kPa [(200 to 800 psi(g)]	Water softening and desalting of organics
Ultrafiltration	45 to 1,100 Å	Most organics over 1,000 molecular weight, including pyrogens, viruses, bacteria, and colloids	70 to 1,400 kPa [(10 to 200 psi(g)]	Concentration and recovery of industrial organics and dilute suspended oils
Microfiltration	400 to 20,000 Å	Very small suspended particles, some emulsions, and some bacteria	7 to 170 kPa [(1 to 25 psi(g)]	High volume removal of bacteria and small suspended solids

Power/Fuel Requirements

Contaminated water must be supplied to the system at an appropriate flow rate. An electric power supply is required for the pumps.

Transportability

All systems are skid-mounted and can be transported by truck.

CONTACT INFORMATION

Manufacturer

Osmonics Incorporated
5951 Clearwater Drive
Minnetonka, MN 55343-8995
U.S.A.

Telephone (612) 933-2277
Facsimile (612) 933-0141

REFERENCES

(1) Manufacturer's Literature.

PACE OIL/WATER SEPARATORS   No. 149

DESCRIPTION

Self-contained oil/water separators with a patented mechanical process that breaks emulsions and handles any combination of fresh water, salt water, free oil, mechanical emulsions, and chemical emulsions. PACE stands for Positive Accelerated Coalescence of Emulsions.

OPERATING PRINCIPLE

Wastewater is drawn into the primary separator by the suction pump. Free oil is separated and rises to the top of the primary separator. Chemical and mechanical emulsions pass through the bag filter into the second stage of treatment and are pumped across the surface of the membrane at high velocity and moderate pressure. Water, salts, and detergents are forced through the membrane and oil is rejected. The concentrated emulsion then passes through the reservoir where excess oil is stripped and returned to the primary separator. The remainder passes back to the pump suction, so that the pumping rate is much greater than the system output. High velocity flows across the membrane surface shear off excess oil and

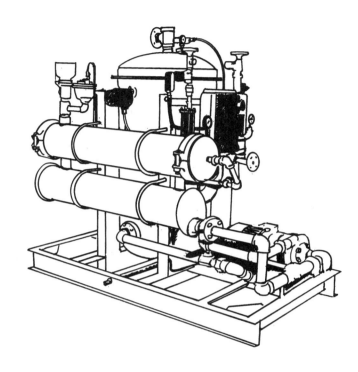

minimize fouling. Discharge of clean water from the system produces the vacuum required to draw wastewater into the system. The primary separator is full of liquid at all times. As oil is accumulated, emulsion is displaced and the oil/water interface moves lower. When this interface reaches the interface sensor, it triggers the oil-ejection cycle. The pump stops, the clean water discharge solenoid valve closes, and the water inlet solenoid valve opens. Water under regulated line pressure ejects the accumulated oil through the oil discharge check valve fitted on the primary separator. When a preset quantity of oil has been ejected, the pump starts, the water inlet closes, and the clean water discharge opens.

STATUS OF DEVELOPMENT AND USAGE

Commercially available systems used as a pre-treatment process in many industries.

TECHNICAL SPECIFICATIONS

Physical

The systems are skid-mounted, assembled, piped, wired, and tested. All valves, machinery, and controls necessary for automatic operation, in-place washing, filling, and draining are included. Several standard models are available.

	S-1	S-2	S-3	S-4	S-5	S-6	S-7
Rated capacity	3 m³/d (800 gpd)	1.5 m³/d (4,000 gpd)	30 m³/d (8,000 gpd)	72 m³/d (19,000 gpd)	160 m³/d (42,000 gpd)	212 m³/d (56,000 gpd)	315 m³/d (83,000 gpd)
Length	1.7 m (5.6 ft)	1.8 m (6.1 ft)	2.1 m (7.0 ft)	2.9 m (9.5 ft)	4.2 m (13.8 ft)	4.3 m (14.0 ft)	4.5 m (15.0 ft)
Width	.70 m (2.3 ft)	29 m (4.1 ft)	1.5 m (5.0 ft)	1.8 m (5.8 ft)	2.6 m (8.7 ft)	3.2 m (11 ft)	3.5 m (12 ft)
Height	1.5 m (4.8 ft)	1.7 m (5.5 ft)	1.8 m (5.9 ft)	2.1 m (6.8 ft)	2.3 m (7.7 ft)	2.5 m (8.3 ft)	2.7 m (8.9 ft)
Weight	408 kg (900 lb)	770 kg (1,700 lb)	1,360 kg (3,000 lb)	2,086 kg (4,600 lb)	2,650 kg (8,700 lb)	5,170 kg (11,400 lb)	6,350 kg (14,000 lb)

Operational

Any combination of fresh water, salt water, free oil, mechanical emulsions, and chemical emulsions can be treated and any combination of petroleum oils up to a specific gravity of 0.94 can be processed. Any viscosity that can be pumped through the piping is treatable. The PACE system is a variable-flow unit. The throughput depends on the total dynamic head that the system must overcome, the temperature and contaminant level of the wastewater, and the characteristics of the membrane. Maximum suction lift for all units is 3 m (10 ft) and the maximum oil discharge head is 9.1 m (30 ft). The clean water discharge normally has a hydrocarbon content below 15 ppm and the oil blocked from the emulsion is typically concentrated to about 60 to 70%.

Power/Fuel Requirements

The systems require electric power in voltages of 208, 230/460/575 VAC, 3-phase, 60 Hz.

Transportability

The systems are skid-mounted and can be moved with standard lifting equipment and trucks.

PERFORMANCE

The following are results of testing conducted at a U.S. Coast Guard Designated Testing Facility and witnessed by representatives of the U.S. Coast Guard and the Canadian Water Pollution Control Directorate. Copies of the Test Report are available from the manufacturer. Limits set by the U.S. Coast Guard and the International Maritime Organization (IMO) for the discharge of these contaminants from ships are set at 15 ppm.

Contaminant	Maximum Oil Content in Effluent	Average Content in Effluent
Heavy fuel oil	not detected	not detected
Light fuel oil	not detected	not detected
Chemical emulsion	6.5 ppm	1.0 ppm

CONTACT INFORMATION

Manufacturer

Scienco/Fast Divisions of Smith & Loveless, Incorporated
3240 North Broadway
St. Louis, MO 63147-3515
U.S.A.

Telephone (314) 621-2536
Facsimile (314) 621-1952

REFERENCES

(1) Manufacturer's Literature.

PACT® BATCH WASTEWATER TREATMENT SYSTEM

No. 150

DESCRIPTION

A batch activated sludge/powdered carbon process for treating nonbiodegradable and/or biodegradable organics in wastewater, groundwater, and leachate. PACT® is a registered trademark of Zimpro Environmental, Inc.

OPERATING PRINCIPLE

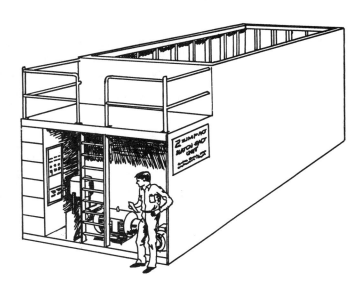

Wastewater is pumped into the aeration tank where it comes in contact with a mixture of biological solids and powdered activated carbon. During aeration, the biodegradable portion of the waste is treated biologically, while the nonbiodegradable contaminants are adsorbed on the carbon particles. When the aeration cycle is complete, tank contents are allowed to settle. Solids go to the bottom of the tank, while the clarified effluent is drawn off and discharged to the receiving stream for disposal. The solids are retained in the tank for use with the next batch of waste, except for the portion that is periodically withdrawn for dewatering, disposal, or regeneration if necessary. Additional powdered carbon is added directly to the tank as needed.

STATUS OF DEVELOPMENT AND USAGE

A commercially available system for treating industrial process wastes, landfill leachates, surface runoff, and contaminated groundwater.

TECHNICAL SPECIFICATIONS

Physical

Batch PACT plants are prefabricated and shipped in five sizes which are ready to hook up and operate. The following are specifications for the five models.

	Model B13	Model B45	Model B70	Model B90	Model B140
Overall Plant					
Height	3.6 m (12 ft)	3.6 m (12 ft)	3.6 m (12 ft)	3.6 m (12 ft)	3.6 m (12 ft)
Width	3.6 m (12 ft)	3.6 m (12 ft)	3.6 m (12 ft)	3.6 m (12 ft)	3.6 m (12 ft)
Length	5.5 m (18 ft)	6.4 m (21 ft)	8.8 m (29 ft)	12 m (38 ft)	17 m (55 ft)
Aeration Tank					
Height	3.6 m (12 ft)	3.6 m (12 ft)	3.6 m (12 ft)	3.6 m (12 ft)	3.6 m (12 ft)
Width	2.4 m (8 ft)	3.6 m (12 ft)	3.6 m (12 ft)	3.6 m (12 ft)	3.6 m (12 ft)
Length		4.9 m (16 ft)	7.3 m (24 ft)	9.8 m (32 ft)	15 m (48 ft)
Volume	14 m^3 (500 ft^3)	57 m^3 (2,000 ft^3)	85 m^3 (3,000 ft^3)	113 m^3 (4,000 ft^3)	170 m^3 (6,000 ft^3)

Operational

	Model B13	Model B45	Model B70	Model B90	Model B140
Maximum design flow	50,000 L/d (13,000 gpd)	170,000 L/d (45,000 gpd)	265,000 L/d (70,000 gpd)	340,000 L/d (90,000 gpd)	530,000 L/d (140,000 gpd)
Maximum COD loading	45 kg/day (100 lb/day)	170 kg/day (380 lb/day)	260 kg/day (580 lb/day)	340 kg/day (750 lb/day)	520 kg/day (1,150 lb/day)
Maximum batches/day	5	5	5	5	5
Minimum batch time	4.8 hrs	4.8 hrs	4.8 hrs	4.8 hrs	4.8 hrs

Power/Fuel Requirements

An unspecified electric power source is required.

Transportability

The systems are designed to be transported by highway to the treatment site.

PERFORMANCE

The performance data in the following table were obtained using a Batch PACT Wastewater Treatment Plant of an unspecified capacity. This table is from a paper, "Biophysical Treatment of Leachate Containing Organic Contaminants", presented at the Forty-first Purdue Industrial Waste Conference in 1986.

	Inorganic Chemicals	Organic Chemicals	Contaminated Surface Runoff	Leachate
Influent (mg/L)				
BOD	187	745	94	1,530
COD	2,210	9,910	199	3,200
DOC	300	1,690	60	0
Total Toxic Organics	32.0	0	0	7.5
Effluent (mg/L)				
BOD	30	16	14	31
COD	84	124	27	123
DOC	28	28	11	0
Total Toxic Organics	0.02	0	0	<0.4
Removal Efficiencies %				
BOD	84	98	85	98
COD	96	99	86	96
DOC	91	98	82	0
Total Toxic Organics	>99.9	N/A	N/A	>95

Organic contaminants included MEK, tetrahydrofuran, dichloroethane, and methylene chloride.

CONTACT INFORMATION

Manufacturer

Zimpro Environmental, Inc.
301 West Military Road
Rothschild, WI 54474
U.S.A.

Telephone (715) 359-7211
Facsimile (715) 355-3219

REFERENCES

(1) Manufacturer's Literature.

SHALLOWTRAY® LOW PROFILE AIR STRIPPERS   No. 151

DESCRIPTION

Two lines of low profile, modular air strippers, one constructed of polyethylene and one of stainless steel, for removing volatile organic compounds (VOCs) from water. The systems consist of a sump, a series of stacked aeration trays (1 to 5 trays), a blower, and all necessary controls and gauges.

OPERATING PRINCIPLE

Contaminated water is sprayed into the inlet chamber through a coarse mist spray nozzle. The water flows over a flow distribution weir and along the baffled aeration tray. Clean air is blown up through 0.5 cm (0.2 in) diameter holes in the aeration tray. The air forms a froth of bubbles approximately 15 cm (6 in) deep on the aeration tray, generating a large mass transfer surface area where the contaminants are volatilized. The VOCs then exit the column with the air stream.

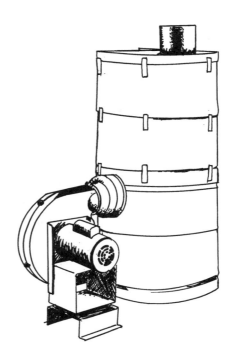

STATUS OF DEVELOPMENT AND USAGE

Commercially available product in general use to treat contaminated groundwater and process water.

TECHNICAL SPECIFICATIONS

Physical

Physical dimensions vary from model to model, ranging from 1 to 2.3 m (3.3 to 7.5 ft) wide, 0.8 to 3.8 m (2.5 to 13 ft) long, and 1 to 2.6 m (3.3 to 8.5 ft) high. The systems are compact and low in profile. More than 40 air stripper configurations are available offering effective treatment rates from 0.1 to 250 m^3/h (0.5 to 1,100 gpm).

Operational

ShallowTray® systems are resistant to fouling problems as treatment trays have large diameter aeration holes and the turbulent action of the froth scours the tray surfaces, reducing the buildup of oxidized iron. If cleaning or inspection is required, trays can be easily removed. The following are flow rates for the eight different series of polyethylene (P) and stainless steel air strippers. Each series consists of four or five models.

Series	Flow Rate
1300	0.1 to 6.8 m^3/h (0.5 to 30 gpm)
1300-P	0.1 to 3.4 m^3/h (0.5 to 15 gpm)
2300	0.2 to 11 m^3/h (1 to 50 gpm)
2300-P	0.2 to 11 m^3/h (1 to 50 gpm)
2600	0.5 to 26 m^3/h (2 to 115 gpm)
3600	0.7 to 36 m^3/h (3 to 160 gpm)
31200	1.4 to 97 m^3/h (6 to 425 gpm)
41200	1.8 to 125 m^3/h (8 to 550 gpm)

Power/Fuel Requirements

Contaminated water must be supplied to the system at an appropriate flow rate and pressure, and an electric power supply is required for the blower.

Transportability

The smaller systems are easily handled by two people, while the larger systems require lifting equipment and a truck for transportation. ShallowTray® air strippers are also ideal as trailer-mounted, portable strippers for pump tests, pilot studies, short-term cleanup, or emergency response.

PERFORMANCE

The system's efficiency varies with the number of trays, the influent flow rate, and the air flow. In fact, performance increases as the flow rate is reduced. Also, the system's modular design allows trays to be added, which increases the percent removal of contaminants.

Removal efficiencies of 99.99% can be achieved for BTEX, MTBE, and TCE, with both polyethylene and stainless steel models. The following are some contaminants successfully stripped by the ShallowTray® low profile air stripper.

1,1,1-Trichloroethane	c-1,2-Dichlorethylene	o-Xylene
1,1,2,2-Tetrachloroethane	Carbon tetrachloride	p-Xylene
1,1-Dichloroethane	Chloroform	Radon
1,1-Dichloroethylene	Ethylbenzene	Styrene
1,2,3-Trimethylbenzene	Ethylene dibromide	t-1,2-Dichloroethylene
1,2,4-Trimethylbenzene	—Xylene	Tetrachloroethylene
1,2-Dichloroethane	MEK	Tetrahydrofuran
1,3,5-Trimethylbenzene	Methylene chloride	Toluene
Acetone	MIBK	TPH
Benzene	MTBE	Trichloroethylene
BTEX	Naphthalene	Vinyl chloride

CONTACT INFORMATION

Manufacturer

North East Environmental Products Incorporated
17 Technology Drive
West Lebanon, NH 03784
U.S.A.

Telephone (603) 298-7061
Facsimile (603) 298-7063

OTHER DATA

A demonstration video of the ShallowTray® process is available from the manufacturer.

Other equipment available from North East Environmental Equipment includes the **VaporMate™,** a low energy technology for treating VOCs at groundwater and soil remediation sites, the **EconoPump™**, a multi-well jet system, and the **PurgePanel™**, a UL-listed panel for communicating to hazardous areas. Additional information as well as detailed pricing and drawings on specific projects are available from the manufacturer.

REFERENCES

(1) Manufacturer's Literature.

TRIWASTE MICRO-FLO MOBILE WATER TREATMENT SYSTEM — No. 152

DESCRIPTION

A mobile, self-contained system that uses physio-chemical treatment to remove both organic and inorganic contaminants from wastewaters. This process can be used to treat and reduce the volume of liquid hazardous and nonhazardous wastes. The effluent, a concentrated material in the form of sludge, is normally handled as concentrated waste.

OPERATING PRINCIPLE

This treatment involves a combination of processes including addition of oxidants or reductants, pH adjustment, particle settling, microfiltration/ultrafiltration, and adsorption. A self-cleaning microfiltration/ultrafiltration system removes sub-micronic particles, in combination with the other pre-treatment steps.

STATUS OF DEVELOPMENT AND USAGE

Commercially available systems for treating groundwater and wastewater.

TECHNICAL SPECIFICATIONS

Physical

All components of the system, including a quality control laboratory, are mounted in an enclosed 14 m (45 ft) trailer. The laboratory is equipped with a high purity liquid chromatograph and other laboratory equipment to ensure that effluent meets the appropriate quality guidelines.

Operational

The system is fully automated and requires minimal staff. It can be set up on site within a one-day period. It is capable of contaminant removal efficiencies of up to 99.999% and can be operated in batch or continuous mode. Current units have a throughput rate of 18 to 38 L/min (5 to 10 gpm), while new units have been designed for 150 to 265 L/min (40 to 70 gpm). The concentrated waste, in the form of sludge, and the solid residues must be further treated or disposed of.

Power/Fuel Requirements

Contaminated water must be supplied at an appropriate flow rate and diesel fuel is required for the generator.

Transportability

The system, which is mounted within an enclosed highway trailer, is designed to be highly mobile.

PERFORMANCE

At the following projects in British Columbia that have used this treatment system, effluents were treated to the point that contaminants were not detectable. Approximately 160,000 L (42,000 gal) of wastewater contaminated with chlorophenols, dioxins, and furans were treated at a sawmill operation on Vancouver Island, over 152,000 L (40,000 gal) of wastewater contaminated with chlorophenols, dioxins, furans, oil, and grease were successfully treated at a sawmill operation in Squamish, and 38,000 L (10,000 gal) of Ecobrite anti-sapstain chemical were treated at a terminal in Vancouver, British Columbia. The major contaminants being treated were chlorophenols, dioxins, furans, boron, arsenic, and high pH levels. Detailed test results are available from the manufacturer.

CONTACT INFORMATION

Manufacturer

TriWaste Reduction Services Incorporated
1700, 800 - 5th Avenue S.W.
Calgary, Alberta
T2P 3T6

Telephone (403) 234-3240
Facsimile (403) 261-6737

REFERENCES

(1) Manufacturer's Literature.

WATER SCRUB UNITS - ACTIVATED CARBON ADSORPTION CANISTERS

No. 153

DESCRIPTION

A water purification system that uses activated carbon to remove dissolved organic contaminants from groundwater, wastewater, or liquid process streams.

OPERATING PRINCIPLE

Liquids contaminated with organics are percolated through activated carbon inside the canister, where the contaminants are physically adsorbed to the carbon.

STATUS OF DEVELOPMENT AND USAGE

Commercially available product in general use.

TECHNICAL SPECIFICATIONS

Physical

Models 30, 55, 85, and 110

These units are made of heavy-duty mild steel, lined with baked-on epoxy phenolic or double-layered epoxy coatings.

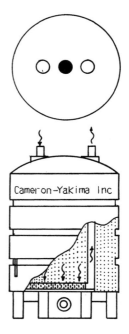

	WSU30	WSU55	WSU85	WSU110
Vessel size	0.8 m (2.5 ft) high	0.9 m (3 ft) high	1 m (3.3 ft) high	1.2 m (3.8 ft) high
	0.5 m (1.6 ft) dia.	0.6 m (2 ft) dia.	0.7 m (2.2 ft) dia.	0.8 m (2.7 ft) dia.
Carbon capacity - Weight	50 kg (110 lb)	90 kg (200 lb)	130 kg (280 lb)	205 kg (450 lb)
- Volume	110 L (3.9 ft^3)	200 L (7.1 ft^3)	280 L (10 ft^3)	450 L (16 ft^3)
Shipping weight	68 kg (150 lb)	120 kg (270 lb)	170 kg (380 lb)	290 kg (640 lb)

Model 55P

This unit is made of a high-density polyethylene body and an epoxy-coated steel frame to allow treatment of highly corrosive waste streams and to prevent bulging at high pressures.

Vessel size	1 m (3.3 ft) high
	0.8 m (2.5 ft) diameter
Carbon capacity - Weight	84 kg (185 lb)
- Volume	195 L (6.9 ft^3)
Shipping weight	100 kg (225 lb)

Models 1000S, 1500S, and 2000S

These models are made of heavy-duty mild steel, lined with a high-thickness, fusion-bonded epoxy coating.

	WSU1000S	WSU1500S	WSU2000S
Vessel size	1.6 m (5.3 ft) high	2.2 m (7.3 ft) high	2.2 m (7.3 ft) high
	1.2 m (4 ft) diameter	1.2 m (4 ft) diameter	1.2 m (4 ft) diameter
Carbon capacity - Weight	450 kg (1,000 lb)	680 kg (1,500 lb)	815 kg (1,800 lb)
- Volume	1,020 L (36 ft^3)	1,530 L (54 ft^3)	1,810 L (64 ft^3)
Shipping weight	730 kg (1,610 lb)	1,040 kg (2,300 lb)	1,180 kg (2,600 lb)

5.4 Treatment/Disposal - Aqueous Treatment Systems

Model 1000 - Poly Unit

This unit is constructed of heavy-duty, crosslinked polyethylene.

Vessel size	1.8 m (6 ft) high
	1.1 m (3.7 ft) diameter
Carbon capacity	Weight 450 kg (1,000 lb)
	Volume 960 L (34 ft^3)
Shipping weight	545 kg (1,200 lb)

Models 82HP and 270HP

These units, which are designed to handle high pressure applications, are made of one-piece molded fibreglass-reinforced plastic (FRP) with an epoxy-coated steel frame.

	WSU82HP	WSU270HP
Vessel size	1.9 m (6.3 ft) high	2.2 cm (7.3 ft) high
	0.5 m (1.8 ft) diameter	0.9 m (3 ft) diameter
Carbon capacity - Weight	135 kg (300 lb)	360 kg (800 lb)
- Volume	310 L (11 ft^3)	820 L (29 ft^3)
Shipping weight	250 kg (550 lb)	590 kg (1,300 lb)

Operational

The adsorber units in all models of Water Scrub Units consist of a proprietary PVC flow collector engineered to ensure even flow distribution and complete use of the carbon bed. Activated carbon for the units is available from Cameron-Yakima in a variety of mesh sizes, performance characteristics, and raw materials.

Model	Maximum Flow Rate	Pressure
Unit 30	38 L/min (10 gpm)	190 kPa (27 psi)
Unit 55	57 L/min (15 gpm)	190 kPa (27 psi)
Unit 85	76 L/min (20 gpm)	190 kPa (27 psi)
Unit 110	95 L/min (25 gpm)	170 kPa (25 psi)
Unit 55P	57 L/min (15 gpm)	190 kPa (27 psi)
Unit 1000S	227 L/min (60 gpm)	190 kPa (27 psi)
Unit 1500S	265 L/min (70 gpm)	190 kPa (27 psi)
Unit 2000S	303 L/min (80 gpm)	190 kPa (27 psi)
Unit 1000	227 L/min (60 gpm)	160 kPa (23 psi)
Unit 82HP	45 L/min (12 gpm)	1,140 kPa (165 psi)
Unit 270HP	125 L/min (33 gpm)	1,140 kPa (165 psi)

Power/Fuel Requirements

Contaminated water must be pumped through the adsorption system at the appropriate flow rate and at varying pressures.

Transportability

A drum hoist, hand truck, forklift and pallet, or other drum-handling equipment are required to handle the canister.

PERFORMANCE

Treatment technologies using granular activated carbon have a proven ability to remove organic contaminants to non-detectable levels. The degree of removal varies, however, in relation to such factors as flow rates, operating pressures, carbon dosage and type, contact time, influent pH, temperature, and viscosity, and if applicable, liquid pre-treatment to remove oils, greases, or suspended solids from the influent which may impair the system's ability to operate adequately.

CONTACT INFORMATION

Manufacturer

Cameron-Yakima, Inc.
1414 South First Street
P.O. Box 1554
Yakima, WA 98907
U.S.A.

Telephone (509) 452-6605
Facsimile (509) 453-9912

OTHER DATA

Carbon regeneration facilities and other suppliers of activated carbon are listed in trade journals such as *Fraser's Canadian Trade Directory* (777 Bay Street, Toronto, Ontario M5W 1A7, Tel 416-596-5086, Fax 416-593-3201) and the *Thomas Register of American Manufacturers* (Thomas Publishing Company, Five Penn Plaza, New York, NY 10001, 212-695-0500, Fax 212-290-7206).

REFERENCES

(1) Hager, D.G., "Industrial Wastewater Treatment by Granular Activated Carbon", *Industrial Water Engineering*, (January/February 1974).
(2) Manufacturer's Literature.
(3) O'Brien, R.P. and J.L. Fisher, "There is an Answer to Groundwater Contamination", reprinted from *Water/Engineering & Management*, File No. 27-86, Calgon Carbon Corporation, Pittsburgh, PA (May, 1983).
(4) Rizzo, J.L. and A.R. Shepard, "Treating Industrial Wastewater with Activated Carbon", *Chemical Engineering*, McGraw-Hill, Inc., New York, NY (January, 1977).
(5) Ying, W., E.A. Dietz, and G.C. Woehr, "Adsorptive Capacities of Activated Carbon for Organic Constituents of Wastewaters", *Environmental Progress 9(1)* (February, 1990).

ZENOGEM PROCESS   No. 154

DESCRIPTION

A biological reactor integrated with an ultrafiltration (UF) membrane system.

OPERATING PRINCIPLE

Crossflow membrane filtration is the separation of the components of a fluid by semipermeable membranes through applying pressure and flow. The influent or "feed" stream is separated into two effluent streams, the "permeate", which passes through the membrane, and the "concentrate", which retains the dissolved and suspended solids rejected by the membrane. The pressure required to drive the process is determined by the specific nature of the feed solution and the pore size of the membrane. If the solution contains significant levels of ionic solutes, the operating pressure must first overcome the osmotic pressure - the natural energy potential between the more concentrated and the less concentrated streams. In the ZenoGem process, the concentrate is recycled to the bioreactor and the permeate is discharged.

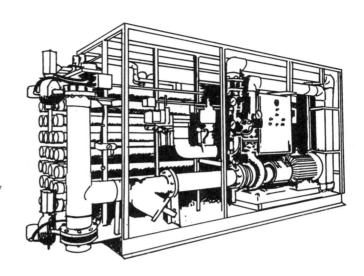

STATUS OF DEVELOPMENT AND USAGE

The ZenoGem process is in various stages of commercial operation on a wide range of wastewater streams.

TECHNICAL SPECIFICATIONS

Physical

Standard Systems - 1.9, 3.8, 38, and 150 L/min (0.5, 1, 10, and 40 gpm) turnkey, modular, skid-mounted units (batch and continuous versions).
Custom Systems - 0 to 1,890 L/min (0 to 500 gpm) on-site, modular or retrofit units.
Pilot Systems - Fleet of .38, 1.9, and 3.8 L/min (0.1, 0.5, and 1 gpm) skid- or trailer-mounted units for on-site demonstration.

Operational

The integrated biological contactor/membrane process almost totally removes oils, organics, nitrogenous compounds, metals, and all suspended solids. The following table compares the ultrafiltration process with other crossflow processes.

Crossflow Process	Membrane Pore Size	Fluid Component Retained	Transmembrane Pressure	Sample Applications
Reverse Osmosis	5 to 15 Å	Ions and most organics over 100 molecular weight	1,400 to 6,900 kPa [(200 to 1,000 psi(g)]	Metal ion recovery; BOD and COD reduction in waste streams
Nanofiltration	10 to 90 Å	Organics with 300 to 1,000 molecular weight	1,400 to 5,500 kPa [(200 to 800 psi(g)]	Water softening and desalting of organics
Ultrafiltration	45 to 1,100 Å	Most organics over 1,000 molecular weight, including pyrogens, viruses, bacteria, and colloids	70 to 1,400 kPa [(10 to 200 psi(g)]	Concentration and recovery of industrial organics and dilute suspended oils
Microfiltration	400 to 20,000 Å	Very small suspended particles, some emulsions, and some bacteria	7 to 170 kPa [(1 to 25 psi(g)]	High volume removal of bacteria and small suspended solids

Power/Fuel Requirements

Contaminated water must be supplied to the system at an appropriate flow rate. An electric power supply is required for the pumps.

Transportability

Systems are skid- or trailer- mounted and can be transported by truck.

PERFORMANCE

The following performance data were provided by the manufacturer.

Automotive Plant Effluent

Parameter	Reduction (%)	Effluent Concentration (mg/L)
BOD	98.5	17
COD	91.8	435
Total oil and grease	97.2	18
Hydrocarbon oil and grease	98.4	4.5
Total suspended solids	-	<10
TKN	93.0	4.7
NH3-N	-	0.97
Lead	83.0	0.05
Zinc	78.5	0.45

Transfer Station Oily Wastewater

Parameter	Influent Concentration (mg/L)	Effluent Concentration (mg/L)	Reduction (%)
BOD	2,964	20	99
TOC	1,188	194	84
Oil and grease	2,300	12	99

Landfill Leachates

Parameter	Influent Concentration	Effluent Concentration	Reduction (%)
BOD	315	3	99
TOC	451	113	75
Phenol (µg/L)	9,298	0.8	99.9+
BTEX (µg/L)	515	0.2	99.9+

CONTACT INFORMATION

Manufacturer

Zenon Environmental Incorporated
845 Harrington Court
Burlington, Ontario
L7N 3P3

Telephone (905) 639-6320
Facsimile (905) 639-1812

REFERENCES

(1) Manufacturer's Literature.

Section 5.5

Treatment/Disposal - Vapour Treatment Systems

CALGON VENTSORB & SWEETSTREET ACTIVATED CARBON ADSORPTION CANISTERS

No. 155

DESCRIPTION

VentSorb Canisters are charged with activated carbon and designed for treating off-gases or vapours from soil-venting or air-stripping operations. One type of VentSorb Canisters is modular, made of prefabricated steel, and designed for low flow air purification. The **VentSorb High Flow Canisters** are reusable, prefabricated steel or plastic canisters capable of purifying higher air flows. **Calgon Sweetstreet Canisters** are compact, modular canisters designed to remove sulphide and methyl mercaptan odours venting from gravity sewer manholes.

OPERATING PRINCIPLE

These units use granular activated carbon to remove organic contaminants from an air stream. Contaminated air is distributed across the activated carbon inside the canister, where the contaminants are physically adsorbed to the carbon media.

STATUS OF DEVELOPMENT AND USAGE

Commercially available products in general use.

TECHNICAL SPECIFICATIONS

Physical

VentSorb Canisters - The canister is constructed of steel with an internal phenolic coating. The 208-L (55-gal) canisters hold 90 kg (200 lbs) of carbon and contain all the elements found in a full-scale adsorption system: the vessel; activated carbon; inlet connection and distributor; and an outlet connection for the purified air stream. Air is distributed across the carbon bed with a corrosion-resistant stainless steel septum.

VentSorb High Flow Canisters - Available in plastic or steel construction, these units are prefabricated in 70, 90, and 122 cm (28, 36, and 48 in) diameters. The units can be supplied as skid-mounted, modular adsorption systems with a fan and interconnecting ductwork.

Sweetstreet Canisters - Available to fit round manholes in sizes from 56 to 86 cm (22 to 34 in) in diameter.

Operational

VentSorb Canisters - VentSorb units are designed for low flow air purification applications treating up to 47 L/s (100 cfm), while the **VentSorb High Flow Canisters** are designed for air purification up to 520 L/s (1,100 cfm). The units are filled by the manufacturer with Calgon Carbon Type BPL or IVP granular activated carbon. Subject to Calgon's acceptance criteria, most VentSorb units can be returned to the manufacturer after use and adsorbed contaminants are destroyed in the carbon reactivation process. The carbon in the canister can easily be emptied out and replaced with fresh carbon.

Sweetstreet Canisters - The units are supplied with 9 kg (20 lb) of Calgon Carbon Type IVP, which has more than twice the capacity for hydrogen sulphide than an equal quantity of standard vapour-phase carbon.

Power/Fuel Requirements

Contaminated air must be supplied to the **VentSorb Canisters** by some means such as a blower, air stripper exhaust supply, or soil vapour extraction unit. If the system is equipped with the optional blower, an appropriate supply of electricity is required to power the blower. The electrical requirement varies according to the blower specified for the application. **Sweetstreet Canisters** have no power requirements.

Transportability

The smallest **VentSorb** unit weighs approximately 90 kg (200 lb) and can be moved with conventional drum-handling methods. The larger units can be supplied skid-mounted and with forklift slots. The **Sweetstreet** units are lightweight and person-portable.

CONTACT INFORMATION

Manufacturer

Calgon Carbon Corporation
P.O. Box 717
Pittsburgh, PA 15230-0717
U.S.A.

Telephone   (412) 787-6700
Canada Toll Free (800) 422-7266
Facsimile   (412) 787-4523

Canadian Supplier

STANCHEM, INC.
43 Jutland Road
Etobicoke, Ontario
M8Z 2G6

Telephone (416) 259-8231
Facsimile   (416) 259-6175

OTHER DATA

Under a service contract, Calgon can also supply large-scale vapour phase treatment units called Vapour Pacs, which have flow rates of up to 4,700 L/s (10,000 cfm) and carbon quantities of up to 5,670 kg (12,500 lb).

REFERENCES

(1) Manufacturer's Literature.

GLOBAL VAPOUR TREATMENT MODULE - CHLORO-CAT CATALYTIC OXIDIZER

No. 156

DESCRIPTION

Destroys halogenated or mixed organic vapour contaminants that are discharged from soil vapour extraction and groundwater treatment systems during site remediation. This module is part of Global's line of soil and groundwater remediation products that are designed to work together as a comprehensive subsurface remediation system.

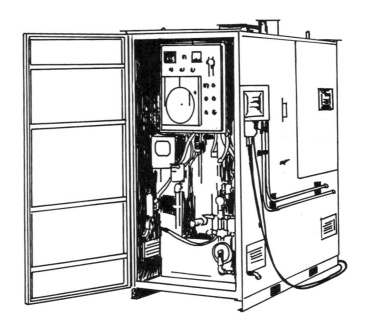

OPERATING PRINCIPLE

In oxidation, hydrocarbon and oxygen molecules combine at a given temperature to form carbon dioxide and water vapour molecules. In thermal oxidation, this occurs at very high temperatures and a relatively long residence time. Catalytic oxidation is basically the same chemical process except that the catalyst initiates and accelerates the reaction at much lower temperatures. Flameless combustion of hydrocarbons and carbon monoxide occurs at much lower temperatures [260 to 482°C (500 to 900°F)] than those typical for thermal oxidizers [760 to 1,090°C (1,400 to 2,000°F)].

STATUS OF DEVELOPMENT AND USAGE

Commercially available product designed specifically to remove VOCs from soil venting and air stripping applications.

TECHNICAL SPECIFICATIONS

Physical

Standard Chloro-Cat systems include a heat exchanger, modulating burner, fuel train, catalyst bed, spark-proof fan motor, fresh air dilution valves, system controls, LEL (Lower Explosive Level) monitor, temperature-recorded, flame-out shutdown detector, flame arrestor, and exhaust stack. The system components are housed in a weatherproof, insulated steel cabinet with three access doors that allow service to all internal parts. The following are specifications for the three standard modules.

	Model 5	Model 10	Model 20
Width	1.3 m (4.3 ft)	2.0 m (6.5 ft)	2.3 m (7.5 ft)
Height	2.5 m (8.3 ft)	2.7 m (9 ft)	3.0 m (9.5 ft)
Length	2.5 m (8.3 ft)	4.0 m (13 ft)	4.6 m (15 ft)
Weight	1,815 kg (4,000 lb)	2,500 kg (5,500 lb)	3,400 kg (7,500 lb)

Operational

The Chloro-Cat model is selected based on volume of air flow, type of contaminant, and desired destruction efficiency. During operation, VOC-laden air is drawn into the Chloro-Cat's fan and discharged into the system's heat exchanger. The air passes through the side of the heat exchanger and into the burner, where the contaminated air is raised to the catalyzing temperature. When the VOC-laden air passes through the specialty catalyst, an exothermic reaction takes place and the VOCs are converted to carbon dioxide, water vapour, and inorganic acids. The hot purified air then passes on the shell side of the heat exchanger where the energy released by the reaction is used to preheat the incoming air, thus minimizing the system's fuel costs. In fact, in many cases the Chloro-Cat is self-sustaining. The air is finally discharged into the atmosphere. A scrubber module is sometimes required to neutralize the inorganic acids.

Power/Fuel Requirements

An electric power supply is required. The burner operates on either natural gas or propane. This module is designed as part of Global's line of soil and groundwater remediation products that are designed to work together as a comprehensive subsurface remediation system, powered and controlled by Global's Power Distribution Module.

Transportability

The systems are housed in steel cabinets with forklift channels and lifting lugs for transportation. Trailer-mounted systems are also available.

CONTACT INFORMATION

Manufacturer

Global Technologies Incorporated
4927 North Lydell Avenue
Milwaukee, WI 53217
U.S.A.

Telephone (414) 332-5987
Facsimile (414) 332-4375

Canadian Distributor

Mapleleaf Environmental Equipment
36 Geneva Court
Brockville, Ontario
K6V 1N1

Telephone (613) 498-1876
Facsimile (613) 345-7633

Mapleleaf Environmental Equipment
5919 5th Street S.E.
Calgary, Alberta
T2H 1L5

Telephone (403) 255-9083
Facsimile (403) 252-8712

OTHER DATA

The **Vapour Treatment Module** is part of Global's line of soil and groundwater remediation products, which includes **Vacuum Extraction and Vapour Liquid Separator Modules** (see Section 2.4, No. 52) and **Power Distribution Modules**. Each module interfaces easily to form a comprehensive subsurface remediation system that works continuously and automatically.

REFERENCES

(1) Manufacturer's Literature.

JOHN ZINK FLARING SYSTEMS

No. 157

DESCRIPTION

Flare systems designed to completely incinerate hazardous gases to protect personnel and equipment.

OPERATING PRINCIPLE

A flare usually consists of an elevated flare burner and support structure. Flammable vapours are drawn in by the flare, ignited by a pilot light or ignitor, and burned off. The flare burner is elevated to isolate the flame from vapours at ground level. Basically, a flare is a large-capacity gas burner with a high turndown capability which operates efficiently from high to very low flows. Flares now include highly sophisticated air-blower smokeless flares, low-noise steam flares with multiple injectors, and staged flare systems that use kinetic energy from the gas to entrain air for combustion and smokeless operation. Enclosed flares are used in populated areas.

STATUS OF DEVELOPMENT AND USAGE

Commercially available product for smokeless and safe disposal of combustible waste gases or fluids. Applications include emergency tank stripping/ off-loading and venting during planned outages in refineries, steel plants, chemical/petrochemical processing facilities, tank farms, and strategic underground petroleum reservoirs and loading terminals.

TECHNICAL SPECIFICATIONS

Physical - Many types of flares are available through John Zink, including utility flares; smokeless flares (steam-assisted, air-assisted, gas-assisted, water-assisted, and high- or low-pressure); portable flares; rental flares; multi-point flares; pit flares; offshore flares; liquid flares (smokeless, utility, production testing); and landfill/biogas flares.

Operational - These portable flaring systems provide complete incineration of hazardous gases or liquids while protecting personnel and equipment. These flares have been used to flare off the contents of damaged tankers to control vapour and unload. Flares are available self-supported, supported by derricks or guy wires, or mounted on booms.

Power/Fuel Requirements - Auxiliary fuel, either propane or natural gas, is required on some flares. An electric power supply may also be required, depending on the control and burner systems incorporated into the flare.

Transportability - Trailer-mounted systems are available. A crane is required to set up some of the flares at the site.

CONTACT INFORMATION

Manufacturer

John Zink Canada West
205-9th Avenue, Suite 500
Calgary, Alberta
T2G 0R3

Telephone (403) 269-4300
Facsimile (403) 269-4303

REFERENCES

(1) Manufacturer's Literature.

NAO TRAILER-MOUNTED EMERGENCY FLARING SYSTEMS AND MOBILE VOC WAGON

No. 158

DESCRIPTION

Trailer-mounted Emergency Flaring Systems - Designed for quick response to emergency situations, as well as scheduled maintenance for existing systems, each of these mobile units is a self-contained system, designed to provide complete incineration of hazardous gases, protect personnel and equipment, and keep a plant on-line when a main flare is decommissioned for inspection or maintenance.

Mobile VOC Wagon - Designed to eliminate combustible waste gases during existing thermal oxidizer downtime, maintenance outages, or even while recovering or controlling liquid spills.

OPERATING PRINCIPLE

Trailer-mounted Emergency Flaring Systems - A flare usually consists of an elevated flare burner and support structure. Flammable vapours are supplied to or are drawn in by the flare, ignited by a pilot light or ignitor, and burned off. The flare burner is elevated to isolate the flame from vapours at ground level. In its simplest form, a flare is a large-capacity gas burner with a high turndown capability that permits it to operate efficiently from high flows down to very low flows. Flares have developed from very simple pipe flares to highly sophisticated air-blower smokeless flares, and from low-noise steam flares with multiple injectors for high, smokeless performance and turndown to the staged flare systems that use the kinetic energy of the gas and its ability to entrain its own air for combustion so that no steam or other medium is required for smokeless operation. Enclosed flares have been developed for use in populated areas.

Mobile VOC Wagon - The VOC Wagon is used in conjunction with a vapour collection system. The gas or waste vapour stream is pulled into the system with a blower. The vapours are subjected to temperatures of 760 to 980°C (1,400 to 1,800°F) and retained in the combustion chamber for at least half a second. The wastes are thus converted to innocuous vapours, such as carbon dioxide and water. When halogenated compounds are incinerated, secondary pollutants are generated. For those applications, integral scrubbers or particulate-collection subsystems are incorporated to capture evolved hydrochloric/hydrofluoric acid and other secondary pollutants.

STATUS OF DEVELOPMENT AND USAGE

Commercially available products designed for quick response to emergency situations. **Emergency Flaring Systems** include emergency tank stripping/off-loading and venting during planned outages in refineries, steel plants, chemical/petrochemical processing facilities, tank farms, strategic underground petroleum reservoirs and loading terminals. The **Mobile VOC Wagon** is designed for use in the event of a plant malfunction, equipment failure or whenever an emission control system has been rendered inoperative due to a fire or mechanical/electrical shutdown. These units have also been used for emergency spill response, maintenance outages, and during the cleaning of pipelines, storage tanks, and process vessels.

TECHNICAL SPECIFICATIONS

Physical

Emergency Flaring Systems

Available in a number of configurations including trailer-mounted emergency flaring systems, temporary flaring systems, and tripod flares.

Trailer-mounted Emergency Flaring Systems can be configured to the user's requirements. Options include air

5.5 Treatment /Disposal - Vapour Treatment Systems

injection, gas or steam assist, multi-tip, kinetic energy smokeless, shrouded tips to inject supplemental fuel, fire snuffing, all necessary piping including flexible stainless steel hook-up lines, pilot monitors, automatic ignition, up to 1,600 m (5,250 ft or almost 1 mile) of 20-cm (8-in) pipe, 800 m (0.5 mile) of 61-cm (24-in) pipe, and 31- and 46-cm (12- and 18-in) headers.

Temporary Flaring Systems can be customized and supplied with the same options as the trailer-mounted emergency flaring systems. The following are the standard sizes.

Flare Tip	Stack Height
20 cm (8 in)	24 m (80 ft)
31 cm (12 in)	31 m (100 ft)
46 cm (18 in)	46 m (150 ft)
61 or 76 cm (24 or 30 in)	67 m (220 ft)
91 cm (36 in)	91 m (300 ft)

Tripod Flares are available with 5, 7.6, or 10 cm (2, 3, or 4 in) tips and 3.7 to 6 m (12 to 20 ft) stacks.

Mobile VOC Wagon

The VOC Wagon consists of a combustion chamber, burner and ignition system, booster blower, water seal, and flame arrester, all mounted on a 5 m (16 ft) trailer.

Operational

Trailer-mounted Emergency Flaring Systems - Portable, self-contained systems that can be set up without cranes or other lifting devices. A complete flaring system with 24 m (80 ft) of guyed stack can be put into service in about 20 minutes. Outriggers from the 12 x 2.4 m (40 x 8 ft) trailer assure stability even in winds of 160 km/h (100 mph).

Temporary Flaring Systems - These temporary flares can be set up on most ground surfaces. No special foundations are required for stability, even in 160 km/h (100 mph) winds.

Tripod Flares - The tripod legs on these flares are removable for quick shipment to emergency sites, such as tankcar derailments. With NAO's waste gas eductor, these flares will pull in waste vapours for efficient incineration of the hazardous or noxious fumes. To create a vacuum and assure complete destruction of low BTU waste gases, the eductor operates on propane or natural gas. These flares can also be equipped with solar-powered control panels.

Mobile VOC Wagon - The handling capacity of this system is about 240 L/s (500 cfm), based on gasoline vapour of 65 BTU/ft^3 at 21°C (70°F). The capacity varies significantly depending on the heating value of the vapour stream. Larger, portable, trailer-mounted thermal oxidation systems are available, which can handle in excess of 708 L/s (1,500 cfm) of flammable vapour. The VOC Wagon and similar systems are designed for a minimum residence time of 0.5 seconds. Operating temperatures average 760 to 980°C (1,400 to 1,800°F) and combustion efficiency is rated at 99 to 99.9%.

Power/Fuel Requirements

Flaring Systems - Auxiliary fuel, either propane or natural gas, is required on some flares. An electric power supply may also be required, depending on the control and burner systems incorporated into the flare.

Mobile VOC Wagon - An electric power supply, compressed air, and auxiliary fuel are required.

Transportability

Flaring Systems - Trailer-mounted systems are self-contained. The temporary flaring systems are modular in design and can be easily loaded onto a flatbed trailer for transport to the site. A crane is required for setup at the site. The tripod flares are extremely portable and can be transported in the box of a standard pickup truck or by most small aircraft.

Mobile VOC Wagon - These systems are mounted on a 5 m (16 ft) trailer, which can be towed by a standard pickup truck.

PERFORMANCE

Flaring Systems - According to the manufacturer, these flares have been used to unload a butadiene tank, clean out a liquified petroleum gas (LPG) tank for re-certification, clean out an ammonia tank for decommissioning, and clean ammonia and propane railcars.

CONTACT INFORMATION

Manufacturer

NAO Incorporated
1284 East Sedgley Avenue
Philadelphia, PA 19134
U.S.A.

Telephone (215) 743-5300
Facsimile (215) 743-3018

Canadian Representatives

British Columbia, Alberta, Saskatchewan, and Manitoba

Lynum Engineering Sales
#2, 8456 - 129A Street
Surrey, British Columbia
V3W 1A2

Telephone (604) 594-0100
Facsimile   (604) 594-0105

Ontario

Robertson White Engineering Ltd.
1315 Finch Avenue West
Suite 203
North York, Ontario
M3J 2G6

Telephone (416) 636-1311
Facsimile   (416) 636-2449

Quebec and Maritimes

DRN Incorporated
8550 Delmeade
Montreal, Quebec
H4T 1L6

Telephone (514) 473-2127
Facsimile   (514) 342-2987

OTHER DATA

NAO also manufactures a skid-mounted VOC thermal oxidizer unit (**VOC Skid**), as well as **horizontal thermal oxidizers (NHTO)** and **vertical thermal oxidizers (NVTO)**. The VOC Skid was used at the site of a million-gallon (3,785 m^3) gasoline leak at a loading terminal on Long Island, NY. The leak formed a vast underground pool of gasoline and volatile hydrocarbon vapours and carcinogenic benzene vapours from this pool were threatening nearby communities. A standard pickup truck towed a trailer with a VOC skid oxidation unit to the site. The volatile vapours were sucked in by an explosion-proof, hermetically sealed booster blower, at a rate of 240 L/s (500 cfm). Operating at 980°C (1,800°F), the unit provided a combustion efficiency of 99.9%, effectively destroying the vapour-contaminated air.

REFERENCES

(1) Manufacturer's Literature.

ODOR-MISER
ACTIVATED CARBON ADSORPTION CANISTERS

No. 159

DESCRIPTION

Vapour phase purification systems that use activated carbon to remove dissolved organics from air streams.

OPERATING PRINCIPLE

Air contaminated with vapour phase organics is distributed through activated carbon inside the canister, where the contaminants are physically adsorbed to the carbon media.

STATUS OF DEVELOPMENT AND USAGE

Commercially available product in general use.

TECHNICAL SPECIFICATIONS

Physical and Operational - The canisters, which contain high grade, vapour phase, activated carbon, are constructed of carbon steel and coated for maximum corrosion resistance. Odor-Misers are available to handle flows of up to 470 L/s (1,000 cfm).

Westport Environmental Systems also manufactures the **Liquid-Miser Activated Carbon Adsorption Canister**, a water purification system for removing dissolved organics from groundwater, wastewater, and liquid process streams (see No. 146 in Section 5.4).

	OM-100	OM-250	OM-500	OM-1000
Diameter	58 cm (23 in)	66 cm (26 in)	91 cm (36 in)	122 cm (48 in)
Height	91 cm (36 in)	97 cm (38 in)	132 cm (52 in)	142 cm (56 in)
Bed depth	61 cm (24 in)	69 cm (27 in)	91 cm (36 in)	91 cm (36 in)
Adsorbent volume	0.16 m^3 (5.5 ft^3)	.23 m^3 (8.3 ft^3)	.63 m^3 (22 ft^3)	1.07 m^3 (38 ft^3)
Carbon weight	75 kg (165 lb)	100 kg (225 lb)	290 kg (640 lb)	510 kg (1,130 lb)
Shipping weight	140 kg (300 lb)	180 kg (400 lb)	410 kg (900 lb)	730 kg (1,600 lb)

	OM-100	OM-250	OM-500	OM-1000
Maximum flow rate	47 L/s (100 cfm)	120 L/s (250 cfm)	240 L/s (500 cfm)	470 L/s (1,000 cfm)
Maximum pressure drop	3.2 kPa (13 in H$_2$O)	3.2 kPa (13 in H$_2$O)	3.4 kPa (13.5 in H$_2$O)	3.5 kPa (14 in H$_2$O)
Minimum contact time	3.3 seconds	2.0 seconds	2.5 seconds	2.3 seconds

Power/Fuel Requirements - Contaminated air must be supplied to the canister at an appropriate flow rate and pressure.

Transportability - A drum hoist, hand truck, forklift and pallet, or other drum-handling equipment are required to move the canisters.

PERFORMANCE

Treatment technologies using granular activated carbon can remove organic contaminants to non-detectable levels. The degree of removal varies with such factors as flow rates, operating pressures, carbon dosage and type, and contact time.

Manufacturer

Westport Environmental Systems
Forge Road, P.O. Box 217
Westport, MA 02790-0217, U.S.A.

Telephone (508) 636-8811
Facsimile (508) 636-2088

REFERENCES

(1) Manufacturer's Literature.

TORVEX CATALYTIC OXIDATION SYSTEMS   No. 160

DESCRIPTION

A system for destroying VOCs from water-stripping effluent or soil-venting applications.

OPERATING PRINCIPLE

In oxidation, hydrocarbon and oxygen molecules combine to form carbon dioxide and water vapour molecules. Thermal oxidation occurs at higher temperatures than catalytic oxidation. Flameless combustion of hydrocarbons and carbon monoxide occurs at much lower temperatures [260 to 482°C (500 to 900°F)] than those typical for thermal oxidizers [760 to 1,090°C (1,400 to 2,000°F)].

STATUS OF DEVELOPMENT AND USAGE

Commercially available product for removing VOCs from soil-venting and air-stripping applications.

TECHNICAL SPECIFICATIONS

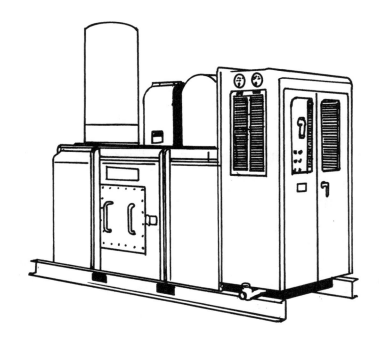

Physical - Consists of the well field air extraction system with self-actuating auto dilution subsystem, a burner, a heat exchanger, a preheat section, and the catalyst. The heat exchanger and all internals of the catalytic oxidizer except the burner are made of 304 Stainless Steel. The catalyst consists of platinum metal dispersed on a high surface area activated alumina imparted on a ceramic honeycomb structure. Systems are available skid-mounted, and with PLC and modem for remote access status updates.

Operational - The system achieves 95 to 99% VOC destruction and is available in sizes that can process up to 32,900 L/s (70,000 cfm). Soil vapour is pulled by vacuum from wells through a multi-position locking shut-off valve and flame arrestor. A feedback temperature signal from the catalytic oxidizer actuates a modulating air dilution valve which releases dilution air to the system through an inlet silencer. Any condensate in the diluted air stream is knocked out in a coalescing filter. The flow rate can be adjusted through a blower recycle line and valve. A blind flange is provided to hook up an air stripper so that the air effluent can be processed from the water stripper. The air stream now enters the catalytic oxidizer. The system operates at high VOC concentrations without excessive heat buildup due to cold-side bypass valves and lines around the heat exchanger. A raw gas or combustion air-fed burner adjusts the temperature of the incoming air so the catalytic reaction can occur. The organics are destroyed across the platinum catalyst on ceramic honeycomb. The catalysts are modular units that may occasionally need to be renewed or replaced.

Power/Fuel Requirements - 440-VAC, 3-phase, 60-cycle electric power is required for the electric components. The burner also operates on natural gas or propane. Options available for other electrical feed.

Transportability - The system is skid-mounted for portability.

CONTACT INFORMATION

Manufacturer

CSM Environmental Systems Incorporated
2333 Morris Avenue, Suite C-4
Union, NJ 07083   Telephone (908) 688-1177
U.S.A.   Facsimile (908) 688-1045

REFERENCES

(1) Manufacturer's Literature.

TUB SCRUB UNITS - ACTIVATED CARBON ADSORPTION CANISTERS

No. 161

DESCRIPTION

A carbon adsorption system for removing vapour contaminants from air streams.

OPERATING PRINCIPLE

Air contaminated with organics is distributed through activated carbon in the canister, where the contaminants are physically adsorbed to the carbon media.

STATUS OF DEVELOPMENT AND USAGE

Commercially available product in general use.

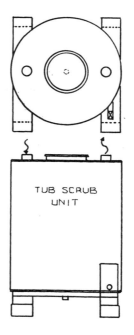

TECHNICAL SPECIFICATIONS

Physical

Models 12, 30, 55, 85, 110, and 200

All units are constructed of heavy-duty mild steel and lined with baked-on epoxy phenolic or double-layered epoxy coatings.

	TSU12	TSU30	TSU55	TSU85	TSU110	TSU200
Vessel size	0.5 m (1.8 ft) high 0.4 m (1.2 ft) dia.	0.8 m (2.5 ft) high 0.5 m (1.6 ft) dia.	0.9 m (3 ft) high 0.6 m (2 ft) dia.	1 m (3.3 ft) high 0.7 m (2.2 ft) dia.	1.2 m (3.8 ft) high 0.8 m (2.7 ft) dia.	1.3 m (4.3 ft) high 0.96 m (3.2 ft) dia.
Carbon capacity	14 kg (30 lb) 30 L (1.1 ft^3)	35 kg (80 lb) 82 L (2.9 ft^3)	70 kg (155 lb) 155 L (5.5 ft^3)	100 kg (220 lb) 224 L (7.9 ft^3)	165 kg (360 lb) 365 L (13 ft^3)	265 kg (585 lb) 590 L (21 ft^3)
Shipping weight	20 kg (47 lb)	55 kg (125 lb)	100 kg (225 lb)	140 kg (315 lb)	250 kg (550 lb)	355 kg (785 lb)

Model 55P

This model is constructed with a high-density polyethylene body and an epoxy-coated steel frame to allow treatment of highly corrosive waste streams and to prevent bulging at high pressures.

Vessel size	1 m (3.4 ft) high
	0.7 m (2.2 ft) diameter
Carbon capacity	85 kg (185 lb)
	195 L (6.9 ft^3)
Shipping weight	100 kg (225 lb)

Models 1000S, 1500S, and 2000S

These models are constructed of heavy-duty mild steel and lined with high-thickness, fusion-bonded epoxy coatings.

	TSU1000S	TSU1500S	TSU2000S
Vessel size	1.6 m (5.3 ft) high 1.2 m (4 ft) diameter	2.2 m (7.3 ft) high 1.2 m (4 ft) diameter	2.2 m (7.3 ft) high 1.2 m (4 ft) diameter
Carbon capacity	455 kg (1,000 lb) 1,000 L (35 ft^3)	680 kg (1,500 lb) 1,520 L (54 ft^3)	815 kg (1,800 lb) 1,820 L (64 ft^3)
Shipping weight	730 kg (1,600 lb)	1,045 kg (2,300 lb)	1,180 kg (2,600 lb)

Model 1000 - Poly

This model is constructed of heavy-duty, cross-linked polyethylene.

Vessel size	1.8 m (6 ft) high
	1.1 m (3.7 ft) diameter
Carbon capacity	450 kg (1,000 lb)
	960 L (34 ft^3)
Shipping weight	540 kg (1,200 lb)

Operational

The adsorber units feature a proprietary PVC flow distributor engineered to ensure even flow distribution and complete use of the carbon bed. The flow distributor can be removed and inspected without emptying the drum. Activated carbons for the adsorption unit are available from Cameron-Yakima with various performance characteristics and mesh sizes, and made from several different raw materials.

Model	Flow Rate	Pressure
Unit 12	up to 25 L/s (50 cfm)	up to 190 kPa (27 psi)
Unit 30	up to 47 L/s (100 cfm)	up to 190 kPa (27 psi)
Unit 55	up to 70 L/s (150 cfm)	up to 190 kPa (27 psi)
Unit 85	up to 85 L/s (180 cfm)	up to 190 kPa (27 psi)
Unit 110	up to 120 L/s (250 cfm)	up to 170 kPa (25 psi)
Unit 200	up to 165 L/s (350 cfm)	up to 140 kPa (21 psi)
Unit 55P	up to 47 L/s (100 cfm)	up to 140 kPa (21 psi)
Units 1000S, 1500S, and 2000S	up to 280 L/s (600 cfm)	up to 190 kPa (27 psi)
Unit 1000 - Poly	up to 165 L/s (350 cfm)	up to 160 kPa (23 psi)

Power/Fuel Requirements

Contaminated air/vapour must be pumped through the adsorption system at flow rates and pressures shown above.

Transportability

A drum hoist, hand truck, forklift and pallet, or other drum-handling equipment are required to handle the canister.

PERFORMANCE

Treatment technologies using granular activated carbon have a proven ability to remove organic contaminants to non-detectable levels. The degree of removal varies, however, in relation to factors such as flow rates, operating pressures, carbon dosage and type, and contact time.

CONTACT INFORMATION

Manufacturer

Cameron-Yakima, Inc.
1414 South First Street
P.O. Box 1554   Telephone (509) 452-6605
Yakima, WA 98907   Facsimile (509) 453-9912

OTHER DATA

The TSU55 Radial and TSU200 Radial are specially designed units with a shallow carbon bed that allows high flows to be processed at an extremely low pressure drop. Carbon regeneration facilities and other suppliers of activated carbon are listed in trade journals such as *Fraser's Canadian Trade Directory* (777 Bay Street, Toronto, Ontario M5W 1A7, Tel. 416-596-5086, Fax 416-593-3201) and the *Thomas Register of American Manufacturers* (published by the Thomas Publishing Company, Five Penn Plaza, New York, NY 10001, Tel 212-695-0500, Fax 212-290-7206).

REFERENCES

(1) Manufacturer's Literature.

Section 5.6

Treatment/Disposal - Solids Treatment Systems

ANDERSEN MOBILE SOIL DECONTAMINATION SYSTEM　　　　No. 162

DESCRIPTION

A mobile soil decontamination system with a dry scrubbing system designed to process up to 27 t/h (30 tons/h) of soil contaminated with hydrocarbons.

OPERATING PRINCIPLE

A rotary kiln agitates and heats the soil, enabling thermal desorption of the organics into the gas phase at relatively low temperatures. The kiln discharges into a dust chamber, where soil is separated from the gas stream and discharged to a drag-type conveyor which conveys the clean soil into a pug mill soil cooler where it is cooled with sprayed water. The gas stream is then fed into a dual cyclone collector which removes most of the residual dust from the gas stream. Dust removed from the gas stream is discharged to the same soil-cooling system that serves the rotary kiln. Clean gas from the cyclone collectors is then routed to an afterburner which thermally oxidizes the organics in the gas phase. Emissions from the afterburner are directed to a vertical spray cooler which introduces water or an alkaline solution to cool the gas stream before the gas is filtered in a fabric filtration baghouse.

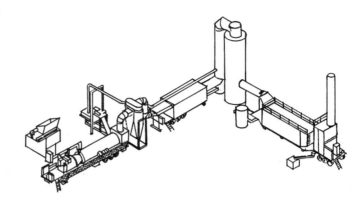

STATUS OF DEVELOPMENT AND USAGE

Commercially available product used in soil reclamation projects.

TECHNICAL SPECIFICATIONS

Physical

The soil decontamination system is an extremely large modular soil processing system designed to be transported in modules to the contaminated site where it can be assembled for on-site treatment of soils. It consists of **a rotary kiln system** measuring 19 × 3.4 m (62 × 11 ft), an **afterburner system** measuring 3.8 × 15 m (12 × 50 ft), a **spray cooler**, a **baghouse collector**, which includes fan, trailer, and auxiliary equipment and measures 21 × 3.4 m (68 × 11 ft) when assembled, and a **materials handling system** which includes a soil-feed system and clean soil handling system.

Operational

The system can treat large quantities of contaminated soils continuously, 24 hours a day, 7 days a week with a minimum number of operators. All component parts of the system are fully automated and the system is fully winterized and weather-protected to withstand routine operating temperatures from -40°C (-40°F) to as high as 54°C (130°F). The system is designed to produce a minimum soil processing temperature of 315°C (600°F) with an exit gas temperature of 370°C (700°F), assuming an incoming soil temperature of 4°C (40°F), an ambient air temperature of 20°C (70°F), a soil feed rate of 27 t/h (30 tons/h), hydrocarbon contamination at 30,000 ppm (wt) with a volatility approximating that of #2 fuel oil, and a moisture content of 15% (wt). With these feed conditions, the system is guaranteed to reduce hydrocarbon contamination to less than 100 ppm (wt) while destroying at least 99% of the hydrocarbons and achieving an outlet particulate loading not exceeding 0.05 g/m^3 (0.02 gr/sdcf). At a feed rate of 14 t/h (15 tons/h), the system can process soil with the same hydrocarbon contamination to the maximum 100 ppm (wt) level for soil containing 40% moisture.

The trailer for the rotary kiln system is equipped with a hydraulic pump which operates six hydraulic jacks that can level the trailer on site. Guy wires may be required on the spray cooler tower in earthquake-prone areas or in areas with high wind velocities if a suitable concrete pad is not provided. The exhaust stack is equipped with a sampling platform and sampling ports as well as a continuous emissions monitoring system.

Power/Fuel Requirements

The burner is fired with natural gas but can be modified to accommodate an oil-feed system upon setup. The main power supply for the system is a 460 VAC, 3-phase, 60 Hz electric power connection to the main control system. Cooling water for the spray cooler is required at approximately 150 L/min (40 gpm). Chemical additives may be needed if special treatment solutions are being used in the spray cooler.

Transportability

Rotary Kiln System - The rotary kiln system must be shipped with the kiln drum removed and transported on a separate flatbed truck. Without the drum installed, the kiln trailer weighs 27 t (30 tons). The drum weighs 21 t (23 tons) including shipping saddles needed to mount the drum on a separate trailer.

Afterburner System - The entire afterburner system is mounted on a dual-axle trailer with dual wheels and weighs approximately 23 t (25 tons).

Spray Cooler - The spray cooler is mounted on a heavy-duty highway trailer and weighs approximately 36 t (40 tons). A wide load permit is required for highway transport.

Baghouse Collector - The baghouse system is mounted on a heavy-duty highway trailer with three axles and weighs 36 t (40 tons). The stack and stack support structure must be shipped separately.

Materials Handling System - This system includes a soil-feed system and a clean soil handling system, each mounted independently on separate skids, which must be transported separately on flatbed trucks.
Other system components, such as ducting and work platforms, must be transported on separate trucks.

PERFORMANCE

Available from manufacturer on request.

CONTACT INFORMATION

Manufacturer

Andersen 2000 Inc.
306 Dividend Drive
Peachtree City, GA 30269
U.S.A.

Telephone (770) 486-2000
Facsimile   (770) 487-5066

REFERENCES

(1) Manufacturer's Literature.

ALBERTA TACIUK PROCESS (ATP) SYSTEM

No. 163

DESCRIPTION

Anaerobic thermal desorption process for remedial treatment of soils and sludges.

OPERATING PRINCIPLE

Based on pyrolysis, which is the alteration, destruction, or thermal cracking of chemical or elemental constituents in feed material when heat is applied. Contaminating organic materials are vaporized and extracted from the processor vessel as hydrocarbon vapours which are condensed to liquids outside the processor and available for reuse or destruction. Operates at temperatures up to 600°C (1,112°F).

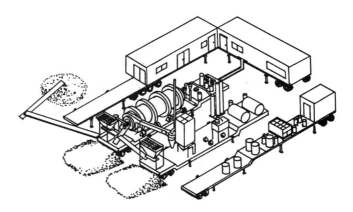

STATUS OF DEVELOPMENT AND USAGE

A commercially proven technology for treating contaminated soils.

TECHNICAL SPECIFICATIONS

Physical - The system consists of the ATP rotating process vessel, a vapour treatment and liquid recovery module, a flue gas treatment module, and a central control module. An area of about 30 × 40 m (98 × 130 ft) is required for setup. The entire system weighs 300 t (330 tons).

Operational - Operates continuously, 24 hours a day, seven days a week with a minimum number of operators. ATP plants are supplied in various sizes, depending on requirements. The manufacturer has supplied transportable plants with a capacity of 10 t/h (11 tons/h) and portable plants designed to process 5 t/h (5.5 tons/h). For waste treatment, plant feed capacity can be as high as 25 t/h (27 tons/h). All steps for extracting contaminants are confined to a single process vessel. Water and oil contaminants are recovered as separate products, and are thus available for secondary treatment, reuse, or separate disposal. Solids products are hydrocarbon-free and usually meet leachate criteria for direct disposal. A dechlorination reagent can be integrated into the process to handle chlorinated wastes.

Power/Fuel Requirements and Transportability - An electric power supply is required and an auxiliary fuel for the heat-generating burner. The system, which is designed to be mobile, is mounted on highway trailers.

PERFORMANCE

This system has been tested in batch and demonstration scales with oily soils and sludges from oil refineries; coal tars and wood preservatives, including PCBs; off-specification petrochemicals such as styrene; oil-based drilling muds and cuttings; pesticides; and halogenated organics. Four projects treating PCB-contaminated soils were done for the Superfund Innovative Evaluation Technology Evaluation (SITE) programs. Reports are available from the U.S. EPA in Ohio.

Manufacturer

UMATAC Industrial Processes
A Division of UMA Engineering Ltd.
210-2880 Glenmore Trail S.E.
Calgary, Alberta   Telephone (403) 279-8080
T2C 2E7   Facsimile (403) 236-0595

Licensee in the United States

SoilTech ATP Systems, Inc.
800 Canonie Drive
Porter, Indiana 46304
U.S.A.   Telephone (219) 926-8651
                        Facsimile (219) 926-7169

REFERENCES

(1) Manufacturer's Literature.

BENNETT MKIII TRANSPORTABLE ROTARY INCINERATOR

No. 164

DESCRIPTION

A transportable rotary kiln incinerator designed to treat materials contaminated with hydrocarbons, including sorbent pads and contaminated soils or sand.

OPERATING PRINCIPLE

The contaminated material is fed through a hopper into the primary rotary pyrolysis unit by means of a large auger/feed pipe conveying system. In the main rotary kiln unit, which operates at 700 to 870°C (1,290 to 1,600°F), hydrocarbons are vaporized and react under starved-oxygen conditions to produce combustible gases. The gases then enter the cyclone section of the afterburner where air is injected and oxidation takes place at 1,000 to 1,220°C (1,830 to 2,230°F).

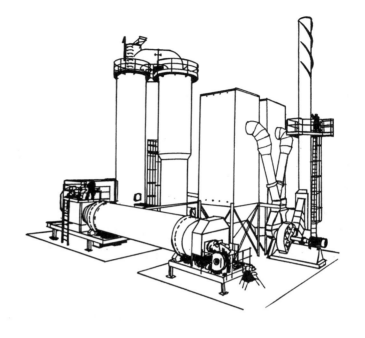

STATUS OF DEVELOPMENT AND USAGE

Originally built in 1982 under a Canadian government contract to design, build, and test a road-transportable incinerator that could be moved from site to site to dispose of oily waste. Aqua-Guard has developed a full-scale model which it now markets commercially.

TECHNICAL SPECIFICATIONS

Physical - All components are made of refractory-lined carbon steel and the stack is made of carbon steel. Bag houses are 3.1 × 3.2 m (10 × 11 ft) and 13 m (42 ft) tall and the stack is 0.9 m (3 ft) in diameter and 18 m (60 ft) tall. The following are other dimensions.

Component	Outside Diameter	Inside Diameter	Length
Primary Combustion Chamber	2.3 m (7.5 ft)	1.9 m (6.3 ft)	12 m (40 ft)
Secondary Combustion Chamber	2.6 m (8.5 ft)	2.3 m (7.5 ft)	15 m (50 ft)
Quench Tower	3.5 m (12 ft)	3.2 m (11 ft)	15 m (50 ft)

Operational - Designed to handle up to 20 t/h (22 tons/h) of contaminated oily waste materials, containing up to 20% hydrocarbons and 25% water. The combined throughput time for the gases in the pyrolysis and oxidation sections is 3 to 4 seconds and the throughput time for the solid material is 5 to 15 minutes, depending on the volume of input and speed of kiln rotation. The afterburner can be fitted with a custom designed optional scrubbing system.

Power/Fuel Requirements - An electric power supply is required for the incinerator controls. The auxiliary burners are fired with natural gas (diesel or other fuels are optional).

Transportability - The units are highway-transportable. A transport truck is required to tow the mobile trailers.

CONTACT INFORMATION

Manufacturer

Bennett Environmental Inc.
Suite 200 - 1130 West Pender Street
Vancouver, British Columbia V6E 4A4

Telephone (604) 681-8828
Facsimile (604) 681-6825

REFERENCES

(1) Manufacturer's Literature.

BRULÉ INCINERATORS

No. 165

DESCRIPTION

A full range of portable, modular, spiral flow, rotary incinerators for treating solid and solid/liquid wastes.

OPERATING PRINCIPLE

Brulé systems use multiple chamber, high temperature technology to oxidize solids and solid/liquids.

STATUS OF DEVELOPMENT AND USAGE

Commercially available products used for remediation projects by industry, hospitals, other institutions, governments, and municipalities. Brulé specializes in custom engineering of systems using standard modules to meet specific requirements.

TECHNICAL SPECIFICATIONS

Physical

A full range of factory-completed, pre-engineered system modules is available for design flexibility. Remediation systems can be provided in portable or modular configurations for in-situ applications up to 450 kg/h (1,000 lb/h). Field systems are available for applications over 450 kg/h (1,000 lb/h).

Operational

Brulé incinerators are modular in design and can be custom designed for specific applications. The systems can process solids and solid/liquid combinations.

Power/Fuel Requirements

An electric power supply and fuel for the burners are required.

Transportability

The units are highway-transportable. The modular systems require minimal assembly on site.

CONTACT INFORMATION

Manufacturer

Brulé C.E. & E. Inc.
13920 S. Western Avenue, P.O. Box 35
Blue Island, IL 60406
U.S.A.

Telephone (708) 388-7900
Facsimile (708) 388-4372

OTHER DATA

Brulé also manufactures portable and modular liquid/vapour/fume incinerators (thermal and catalytic). Pollution control systems have proven performance beyond that required by the strictest codes. Services also available include field engineering supervision for project construction, startup/optimization technical assistance, and field maintenance repair, service, and parts.

REFERENCES

(1) Manufacturer's Literature.

CALCINER-BASED THERMAL PROCESSING SYSTEMS

No. 166

DESCRIPTION

Thermal processing systems, based on Bartlett-Snow rotary calciners, that separate volatile substances from solid granular materials, such as in reclaiming contaminated soil.

OPERATING PRINCIPLE

Soil is introduced into the rotary calciner and heat is indirectly applied from a furnace surrounding a rotating cylinder containing material to be processed. Organic contaminants are vaporized and driven off. Off-gas can then be incinerated in an afterburner or condensed and useful volatiles recovered.

STATUS OF DEVELOPMENT AND USAGE

Commercially available product in general use.

TECHNICAL SPECIFICATIONS

Physical - Systems are designed and built in many sizes and configurations, from stationary units for large permanent installations to skid- or trailer-mounted units for laboratory, pilot plant, or short-term applications.

Operational - These systems are effective in a range of cleanup and reclamation projects, including soil reclamation, metals recovery, tire pyrolysis, solvent recovery, refinery sludge treatment, drilling mud decontamination, and nuclear waste stabilization. Systems can be operated in several treatment atmospheres, including oxidizing, reducing, inert, or pyrolyzing. Particles sized down to fine clays can be easily handled. With multiple zone heating, temperature profiles along the length of the cylinder can be both varied and precisely controlled. Depending on construction materials, processing temperatures as high as 2,200°C (4,000°F) can be achieved. Volatilized organics can either be condensed and used as fuel for heating the calciner or incinerated in the afterburner. Soil discharged from the calciner is safe for return to the environment, although it may be necessary to add nutrients and moisture if revegetation is desired.

Power/Fuel Requirements - An electrical power supply is required for the controls, fuel for the calciner, and a water supply to cool the condensers. Specific requirements vary from unit to unit and with the application.

Transportability - The units are available as skid packages, mobile units, or housed in prefabricated buildings. Pilot plants are housed in highway-transportable trailers.

CONTACT INFORMATION

Manufacturer

ABB Air Preheater Inc.
Raymond® and Bartlett-Snow™ Products
650 Warrenville Road
Lisle, IL 60532

Telephone (708) 971-2500
Facsimile (708) 971-1076

Canadian Representative

ABB, ASEA Brown Boveri Incorporated
233 Colborne Street, Unit 103
Brantford, Ontario
N3T 2H4

Telephone (519) 756-0250
Facsimile (519) 756-9961

REFERENCES

(1) Manufacturer's Literature.

CLEANSOILS THERMAL DESORBER   No. 167

DESCRIPTION

A soil treatment system that uses a multi-phased thermal process to volatilize fuels, gasoline, oils, and similar hydrocarbons from soil particles and separately incinerate the hydrocarbons.

OPERATING PRINCIPLE

Screened soil is conveyed into the primary treatment unit, where heat is applied, causing moisture and hydrocarbons to volatilize from the surface of the soil. Gases exhausted from the primary treatment unit are filtered through a high-efficiency baghouse, and then incinerated in a secondary treatment unit. Fine particulate and dust from the baghouse are mixed with the processed soil and discharged. This is a permanent cleanup solution which leaves no residue.

STATUS OF DEVELOPMENT AND USAGE

Commercially available product developed specifically for soil remediation. CleanSoils Limited offers fixed-base soil remediation services at a number of locations across the United States. Such a site has recently been established in Hamilton, Ontario, serving south-central Ontario. A mobile system is also available for on-site remediation services.

TECHNICAL SPECIFICATIONS

Physical

The Thermal Desorber is a stand-alone, mobile processing unit. Three standard systems are available, rated at 18, 54, or 82 t/h (20, 60, or 90 tons/h). When set up, the system covers an area of about 46 × 30 m (150 × 100 ft).

Operational

This equipment is a stand-alone processing unit for treating solid material contaminated with jet fuel, gasoline, diesel fuel, heavy oils, and similar nonchlorinated hydrocarbons. Soil processed by the system typically contains total petroleum hydrocarbons below detectable limits. The equipment can be operated under different process conditions to remediate soil contaminated with crude oil, bunker oils, oily solids, and polyaromatic hydrocarbons with boiling points below 540°C (1,000°F). Soil throughput rates range from 30 to 70 t/h, depending on the size of the unit, type of soil, debris loading, moisture content, and contamination and cleanup criteria. Some soils require physical conditioning, such as screening, shredding, crushing, or pulverizing, before treatment. Stack emissions consist mostly of water vapour and carbon dioxide. Particulate emissions are low, averaging 700 $\mu g/m^3$ (0.025 gdscf) on stack tests. Operating test data indicate an average hydrocarbon destruction efficiency of 99.8% for soil containing 2,000 ppm of total petroleum hydrocarbons (TPH).

Power/Fuel Requirements

Existing electrical service can be accessed or the system can be powered by its own generator. The system also uses liquid propane or natural gas. About 2 L/s (30 gpm) of water are required to wet soils for dust control and workability.

Transportability

The system fits on four truck trailers and takes approximately two days to set up.

PERFORMANCE

CleanSoils have completed hundreds of successful remediation projects for major oil companies, manufacturers, government, the military, and others. In Hartford, Wisconsin, the system was used to clean up 21,000 t (23,000 tons) of soil contaminated with naphtha, gasoline, kerosine, heating oil, and lubricating oils. In mid-September, 1991, CleanSoils treated 7,700 t (8,500 tons) of soil contaminated with several thousand ppm of total petroleum hydrocarbons at the Eilson Air Force Base in Alaska. A mobile, trailer-mounted system was deployed on September 30, arriving on site October 8. The plant began remediation operations on October 14, completed treatment on October 30, and was demobilized on November 1. The system operated 24 hours a day in temperatures down to -26°C (-15°F).

CleanSoils opened a fixed-base site in Hamilton, Ontario in 1995 to receive large and small quantities of contaminated soil for remediation and disposal or reuse. The site, which is conveniently located off the Queen Elizabeth Way, operates 24 hours a day. On-site remediation services with a portable unit are also available anywhere in Ontario. The Ontario Ministry of Environment and Energy has granted certificates of approval to CleanSoils to allow the Thermal Desorber to be operated on an acceptable site anywhere in Ontario.

CONTACT INFORMATION

Manufacturer

CleanSoils Incorporated
2360 West County Road "C"
Roseville, MN 55113
U.S.A.

Telephone (612) 639-8811
Facsimile (612) 639-8813

Canadian Supplier and Remediation Service

CleanSoils Limited
225 Sheppard Avenue West
North York, Ontario
M2N 1N2

Telephone (416) 226-3838
Facsimile (416) 226-2931

OTHER DATA

CleanSoils can pretest soils using a bench-scale test apparatus. CleanSoils guarantees its work with laboratory analyses of processed soils in accordance with the company's quality control program.

REFERENCES

(1) CleanSoils Limited, "Statement of Qualifications - Remediating, Treating and Recycling Hydrocarbon Contaminated Soils by Low Temperature Thermal Desorption", CleanSoils Inc. (1992).
(2) Manufacturer's Literature.

CORECO ROTARY INCINERATORS   No. 168

DESCRIPTION

A line of five standard-sized mobile or modular rotary kilns for incinerating solid wastes and sludges and recycling metals.

OPERATING PRINCIPLE

Materials are fed into an indirectly heated rotary process kiln and heated to temperatures up to 870°C (1,600°F). This burns off volatile organic substances while the material cascades through a heated airstream for drying or oxidation. An internal afterburner destroys the volatile organic substances.

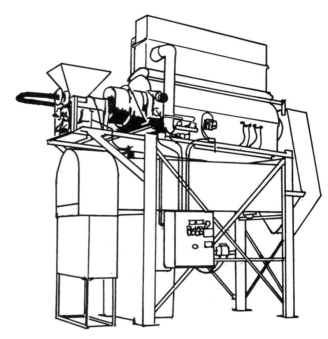

STATUS OF DEVELOPMENT AND USAGE

Originally developed and used to reclaim and detoxify foundry sands, this equipment is now being marketed for incineration of hydrocarbon-contaminated soils, oil spill cleanup materials, sludges, contaminated wastes, and recycling by-products. As of 1995, 26 machines were operating.

TECHNICAL SPECIFICATIONS

Physical - Five standard sizes are available with capacities from 450 kg/h (1,000 lb/h) to 6.4 t/h (7 tons/h).

Operational - All models are mobile or modular in design and material is continuously fed. Contaminated materials are fed into the motor-driven screw feeder, which is sealed against the rotary hearth of the process kiln and feeds material at a constant rate into the hearth. "Controlled temperature atmosphere incineration" technology is used in the incinerators to completely burn and afterburn the hydrocarbons with full use of the contained energy value and minimum auxiliary fuel demand. Oxygen is provided by air entering the rotary process kiln through the centre pipe of the screw feeder. The air is adjusted to provide 100% excess oxygen needed for incineration and afterburning. A high efficiency cyclone collector and a baghouse collect and separate the fines from the incinerated product. Incinerated material exits the rotary process kiln and enters a rotary cooler drum where it is cascaded repeatedly through a counterflow-induced draft air stream and exits at ambient temperature, cleaned and dust-free. Since the entire process takes place within a sealed, continuous flow rotary kiln, all of the product and effluents are captured for environmental control.

Power/Fuel Requirements and Transportability - An electric power supply is required and auxiliary fuel for the burners. The units are highway-transportable.

CONTACT INFORMATION

Manufacturer

CORECO - College Research Corporation
N116 WI6800 Main Street
P.O. Box 577
Germantown, WI 53022-0577, U.S.A.

Telephone (414) 255-4700
Facsimile (414) 255-528

OTHER DATA

CORECO maintains a pilot-plant facility which is available to customers for process evaluation.

REFERENCES

(1) Manufacturer's Literature.

HEYL & PATTERSON TRANSPORTABLE SOIL REMEDIATION UNIT

No. 169

DESCRIPTION

A portable remediation unit designed to thermally treat nonhazardous petroleum and other hydrocarbon-contaminated soil.

OPERATING PRINCIPLE

Contaminated soil is fed through a feed hopper and belt feeder into the soil treater where hot combustion gases drive off moisture and hydrocarbons. Clean soil is then removed from the treater and cooled in a discharge screw conveyor. Some exhaust gas is recycled to minimize fuel consumption and the rest is sent to the particulate collector where entrained soil fines are removed. Particulate-free gas goes through the exhaust fan to the thermal oxidizer which destroys hydrocarbon vapours by combustion at 760°C (1,400°F). Clean exhaust gas is then discharged.

STATUS OF DEVELOPMENT AND USAGE

Commercially available product used in soil reclamation projects.

TECHNICAL SPECIFICATIONS

Physical and Operational - The system is 4.7 m (16 ft) high and 3.5 m (12 ft) wide and is mounted on a 15 m (51 ft) "lowboy" trailer. The complete system weighs 39 t (43 tons). It treats soils to less than 10 ppm hydrocarbons. Hydrocarbons are thermally oxidized with a 95% destruction efficiency. Heyl & Patterson also manufactures a **Portable Indirectly Heated Soil Decontamination Unit** for thermally treating contaminated soil at remote locations.

Contaminated Soil Feed
Throughput   14 t/h (15 tons/h)
Moisture   8% (wt)
Hydrocarbons   0.8% (wt)

Clean Soil Discharge
Discharge rate   13 t/h (14 tons/h)
Moisture   < 3% (wt)
Hydrocarbons   < 10 ppm

Discharge temperature   93°C (200°F)
Soil retention time   3 to 15 minutes

Gas to Thermal Oxidizer
Particulate loading   0.07 g/m^3 (0.03 gdscf)
Outlet temperature   760°C (1,400°F)
Residence time   1 second
Destruction rate   95+%

Power/Fuel Requirements and Transportability - The burner is fired with No. 2 fuel oil and cooling water is required at a rate of 33 ft^3/h (246 gph). The system is mounted on a "lowboy" trailer which may require an oversized load permit for highway transport.

CONTACT INFORMATION

Manufacturer

Heyl & Patterson Incorporated
P.O. Box 36
Pittsburgh, PA 15230-0036
U.S.A.

Telephone (412) 788-6909
Facsimile (412) 788-6913

REFERENCES

(1) Manufacturer's Literature.

PORCUPINE PROCESSOR

No. 170

DESCRIPTION

Indirect contact dryer system with an integrated vapour recovery system used to dry soils or sludges while capturing VOCs driven off by the drying process.

OPERATING PRINCIPLE

Material is dried using a patented hollow-flight agitator design. Agitation, high heat transfer area, and particle contact with the heated agitator enhance evaporative drying of sludges, slurries, and soils. A vapour recovery unit captures all VOCs from the process for later treatment, disposal, or reuse.

STATUS OF DEVELOPMENT AND USAGE

Commercially available product used in many industrial applications and soil reclamation projects.

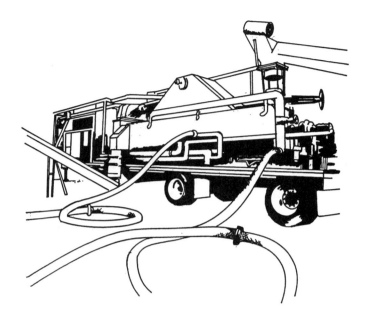

TECHNICAL SPECIFICATIONS

Physical

The dryer system features a trough, product inlet, rotor shaft and breaker bars, and a product outlet. It is available in stainless steel, carbon steel, nickel alloy, titanium, sanitary finish, and other custom alloys. The dryers are available in capacities from 45 kg/h (100 lb/h) to 36 t/h (40 ton/h). The vapour recovery system is available in capacities from 47 to 18,000 L/s (100 to 40,000 cfm).

Operational

The dryer operates in drying, cooling, heating, sterilizing, reacting, or low-temperature calcining modes, under atmospheric, vacuum, or pressure conditions to 5,280 kPa (765 psi). It operates continuously or in batch mode. Heating ranges from 27 to 370°C (80 to 700°F). Higher temperatures and below freezing temperatures are available upon special request. The heat transfer media can be water, steam, heat transfer oil, brine, or other heat and cooling transfer media. A separate boiler or heater is required. The vapour recovery system can reduce VOC emissions from the dryer to less than 10 ppm. The system is protected from organic acids by high alloy construction material. The Bethlehem Corporation maintains a test facility to evaluate the Porcupine Processor in relation to the customer's specific needs. Pilot plants can be rented for on-site testing.

Transportability

The units are skid- or trailer-mounted for transportability.

CONTACT INFORMATION

Manufacturer

The Bethlehem Corporation
P.O. Drawer 348
Easton, PA 18044
U.S.A.

Telephone (610) 258-7111
Facsimile (610) 258-8154

REFERENCES

(1) Manufacturer's Literature.

TRECAN SOLID WASTE INCINERATORS No. 171

DESCRIPTION

The **Solid Waste Incinerator** is a fully self-contained, skid-mounted, two-stage controlled air incinerator suitable for controlled combustion of a wide range of solid wastes. The **Transportable Pit Incinerator** is a basic modular incinerator designed for the combustion of general refuse, oil-contaminated sand, stones, water, seaweed, reeds, wood, and adsorbents.

OPERATING PRINCIPLE

Solid Waste Incinerator - Combustion occurs in two stages. The solid waste material is decomposed in the primary chamber, in a semi-pyrolytic operation (starved-air combustion). Smoke and gases are routed to a secondary chamber and completely oxidized under conditions of high temperature and turbulence with an additional supply of combustion air.

Transportable Pit Incinerator - Waste is fed into the incinerator through the open top of the combustion chamber. A remote fan then forces air through the top and bottom of the combustion chamber, which feeds the flames with the needed oxygen and provides a vortex that causes the flames to swirl, improving combustion and minimizing smoke.

STATUS OF DEVELOPMENT AND USAGE

Commercially available products. The **Solid Waste Incinerator** is marketed for a wide range of solid wastes including biomedical/pathological, industrial, low level radioactive, and municipal waste, as well as to recover precious metals. The Transportable Pit Incinerator was developed in cooperation with Environment Canada as part of the Arctic and Marine Oil Spill Program (AMOP) and is marketed primarily for disposing of oil-soaked debris in emergency situations.

TECHNICAL SPECIFICATIONS

Physical - Solid Waste Incinerator - Capacities are 1.4 to 14 m^3 (50 to 500 ft^3) for batch operation and 90 to 1,000 kg/h (200 to 2,200 lb/h) for ram feed operation. Ancillary equipment includes hydraulically operated ram feeders, on-line ash removal (wet or dry type), heat recovery, and gas scrubbers. The following are specifications for the six standard batch-feed incinerator models.

Model	Primary Chamber Volume	Incineration Rate Dry Waste	Incineration Rate Pathological Remains	Length	Width	Height*	Weight*
3	1.6 m^3 (55 ft^3)	45 kg/h (100 lb/h)	45 kg/h (100 lb/h)	2.5 m (8.3 ft)	1.6 m (5.1 ft)	2.0 m (6.7 ft)	6.8 t (7.5 tons)
7	2.4 m^3 (85 ft^3)	82 kg/h (180 lb/h)	73 kg/h (160 lb/h)	3.0 m (9.8 ft)	1.8 m (6.0 ft)	2.0 m (6.7 ft)	8.5 t (9.4 tons)
11	3.9 m^3 (138 ft^3)	125 kg/h (275 lb/h)	120 kg/h (260 lb/h)	3.6 m (12 ft)	2.1 m (7.0 ft)	2.0 m (6.7 ft)	11 t (12 tons)
21	6.3 m^3 (224 ft^3)	180 kg/h (400 lb/h)	160 kg/h (360 lb/h)	3.8 m (13 ft)	2.4 m (7.9 ft)	2.3 m (7.7 ft)	14 t (15 tons)
36	8.5 m^3 (300 ft^3)	230 kg/h (500 lb/h)	230 kg/h (500 lb/h)	4.8 m (16 ft)	2.5 m (8.1 ft)	2.6 m (8.7 ft)	16 t (18 tons)
52	14 m^3 (500 ft^3)	295 kg/h (650 lb/h)	320 kg/h (700 lb/h)	5.8 m (19 ft)	2.5 m (8.1 ft)	2.6 m (8.7 ft)	23 t (25 tons)

* excludes stack.

5.6 Treatment/Disposal - Solids Treatment Systems

Transportable Pit Incinerator - The system is made in sections, each weighing a maximum of 820 kg (1,800 lb). The complete system requires a 2.4 × 5.5 m (8 × 18 ft) area and weighs 9 t (10 tons). The combustion chamber is constructed of ten sections, each consisting of a steel frame with a 13-cm (5-in) thick cast refractory facing with a 5 cm (2 in) backup insulating block. The chamber is 3.1 m (10 ft) long × 2.5 m (5 ft) wide × 1.8 m (6 ft) high and has a hinged cleanout door at one end. The motor and fan unit have a 20 hp, 2 cylinder, air-cooled diesel engine.

Operational - **Solid Waste Incinerators** are suitable for the disposal of many types of solid waste and provide controlled and efficient combustion of combustible materials. The incinerators are fully self-contained and can be designed to meet specific emissions standards if all information about the various wastes is provided beforehand, and if some type of scrubbing or cleanup system is used. Adding such a system would, however, reduce the overall mobility of the incinerators. The flue gas cleanup equipment is mounted on a separate skid and may have to be protected from freezing. Trecan units can sometimes be offered in a different configuration with an auxiliary liquid burner in the secondary chamber to burn pumpable combustible liquids. When fitted with the liquid burner, no solid wastes can be burned. The unit is not designed to burn liquid and solid wastes simultaneously.

In the **Transportable Pit Incinerator**, the concept of both underfire air and overfire air ensures that the combustion is well aspirated to minimize smoke. As there is no form of secondary combustion, the unit cannot conform to specific emissions standards. The unit can be used, however, as an alternative method for disposing of bulk refuse generated by spill cleanup activities and may be acceptable at a temporary location in an emergency situation. The system can be fully assembled in two and a half hours without special tools or welding. Utility services are not required since all necessary power is supplied by an air-cooled diesel engine. Waste is loaded manually. Non-combustibles and combustion residue are removed through a cleanout door at one end of the combustion chamber.

Capacity   910 kg/h (2,000 lb/h) for domestic refuse rated at 1,400 J/g
                   450 kg/h (1,000 lb/h) for oil-soaked debris
Blower rating   160 m³/min at 2.5 kPa (5,700 cfm at 10 in WC)

Power/Fuel Requirements - The **Solid Waste Incinerator** requires an unspecified electric power supply, as well as an auxiliary fuel for the burners. The **Transportable Pit Incinerator** requires diesel fuel for the motor drive blower. Fuel consumption is rated at 7.5 L/h (2 gal/h).

Transportability - The **Solid Waste Incinerator** is skid-mounted and can be transported on a flatbed trailer. Smaller systems, complete with scrubbers, can be provided trailer-mounted. The **Transportable Pit Incinerator** is modular in design with each module weighing a maximum of 820 kg (1,800 lb). The entire unit can be transported by truck or helicopter and the separate modules can be slung with a medium-lift helicopter. If a helicopter is not available, lifting equipment is required for handling and assembly.

PERFORMANCE

The oil and gas industry in Alberta uses the **Solid Waste Incinerator** to treat oilfield waste. No detailed performance data were received. Use of the **Transportable Pit Incinerator** for hazardous materials has not been documented. Environment Canada conducted test burning on a mixture of 75% by weight oil, 10% water, 4% straw, 7% wooden logs, and 4% sorbents. The material was burned at a rate of about 450 kg/h (1,000 lb/h) with little or no visible smoke.

CONTACT INFORMATION

Manufacturer

MBB Trecan Inc.
1B Lakeside Industrial Park Drive
Lakeside, Nova Scotia
B3T 1M6

Telephone (902) 876-8213
Facsimile (902) 876-8492

Trecan Combustion Limited
#6, 4620 Manilla Road, S.E.
Calgary, Alberta
T2G 4B7

Telephone (403) 949-4378
Facsimile (403) 949-4378

Trecan also manufactures and supplies liquid waste incinerators, waste gas incinerators, and contaminated snow melters.

REFERENCES

(1) Manufacturer's Literature.

VULCANUS 400 MOBILE INCINERATOR — No. 172

DESCRIPTION

A basic modular incinerator for disposal of oil-contaminated sand, stones, water, seaweed, reeds, wood, and adsorbents.

OPERATING PRINCIPLE

Waste is fed into the incinerator through the open top of the combustion chamber. Air from a remote fan is then forced through the bottom of the combustion chamber beneath the fire and through the double-walled top. The air feeds the flames with the needed oxygen and provides a vortex that causes the flames to swirl, improving combustion and minimizing smoke.

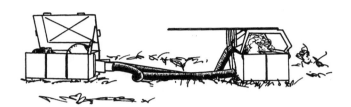

STATUS OF DEVELOPMENT AND USAGE

Commercially available product marketed primarily for disposing of oil-soaked debris in emergencies.

TECHNICAL SPECIFICATIONS

Physical - The incinerator consists of the combustion housing and the motor and fan unit. Two extra furnace grates and one set of heat-resistant aprons, gloves, and hood are included.

Motor and Fan Unit - Radial fan driven by a diesel engine, electric start, 12 VDC battery; 1.8 m (6 ft) long, 1.2 m (4 ft) wide, 1.1 m (3.6 ft) high, and weighing 700 kg (1,540 lb).

Combustion Housing - 1.8 m (6 ft) long, 1.4 m (4.7 ft) wide, 1.6 m (5.3 ft) high, and weighing 800 kg (1,760 lb).

Operational - As the incinerator is not designed for the combustion of hazardous wastes other than debris from oil spill cleanup, emissions may not be permissible in many jurisdictions. The maximum capacity of the incinerator is 400 kg/h (890 lb/h), which varies with the BTU value of the fuel.

Power/Fuel Requirements

Diesel fuel is required for the motor drive blower.

Transportability

The entire system can be transported by truck or medium-lift helicopter. Lifting equipment is required for handling.

CONTACT INFORMATION

Manufacturer

Roto Trading International AB
Violvaegen 4
S-914 33 Nordmaling
SWEDEN

Telephone +46 - 930 - 314 00
Facsimile +46 - 930 - 106 00

North American Distributor

Foss Environmental
7440 West Marginal Way South
Seattle, WA 98108-4141
U.S.A.

Telephone (206) 767-0441
Facsimile (206) 767-3460

REFERENCES

(1) Manufacturer's Literature.

INDEX - BY PRODUCT NAME

3L Filters Oil/Water Separators, **251**
Acid, Chemical Hoses - General Listing, **211**
Acid Leak Control Kit, **1**
Advanced Oxidation Water Treatment Systems, **252**
Airotech EcoSpheres, **65**
Air Plastics - Air Stripping Towers, **257**
Alberta Taciuk Process (ATP) System, **325**
Alfa-Laval Centrifuges, **233**
Algasorb Process, **258**
Amflow-Lift Peristaltic Hose Pump (formerly Mastr-Pump), **192**
Andco Mobile Treatment System, **260**
Andersen Mobile Soil Decontamination System, **323**
Andritz Mobile Dewatering Systems, **235**
Ansul Spill Treatment Kits and Applicators, **221**
ATL Port-A-Berm, **27**
Bennett MKIII Transportable Rotary Incinerator, **326**
Bentonite Soil Sealing Systems, **29**
BioAcceleration Treatment System, **262**
Breeze Air Stripping System, **264**
Brinecell Oxidizers, **265**
Brulé Incinerators, **327**
Busch Vacuum Pumps for Chemicals, **206**
Calciner-based Thermal Processing Systems, **328**
Calgon Activated Carbon Adsorption Treatment Systems, **267**
Calgon VentSorb and Sweetstreet Activated Carbon Adsorption Canisters, **311**
Canflex Sea Slug and Open-Top Reservoir, **155**
CAPSUR, MEXTRACT, and PENTAGONE Spill Agents, **224**
Carbtrol Diffused Air Strippers, **271**
Carbtrol L-1, L-4, and L-5 Activated Carbon Adsorption Canisters, **269**
Carbtrol Regenerative and Multi-phase Extraction Systems, **95**
Cartier Spill Response Kits, **226**
CDF Drum Liners, **184**
CEE Product Recovery Systems, **77**
Chemically Active Covers - General Listing, **53**
Chem-Tainer Polyethylene Spill Containment Pallet, **32**
Cherne Petro Plugs, **2**
Chlorine Institute Emergency Kits, **3**
Clark Spilstopper, **33**
CleanSoils Thermal Desorber, **329**
Contain-It Plus Polyethylene Spill Containment Pallet, **35**
Coppus Portable Ventilators, **58**
Coreco Rotary Incinerators, **331**
Crisafulli Dredges, **140**
Custom Metalcraft Transportable Storage Systems, **166**
Cyanide Destruction System, **273**

Denver Sand-Scrubber System, **236**
Devcon Zip Patch & Magic Bond, **4**
Dredging Equipment - General Listing, **131**
Dynamic Petro-Belt Hydrocarbon Skimmer, **79**
Edwards & Cromwell Response Kits, **5**
EG&G Rotron Regenerative Blowers for Environmental Remediation, **96**
Environment Canada's Reverse Osmosis Units, **274**
Environment Canada's Steam Stripping Units, **278**
Environment Canada's Microfiltration/Ultrafiltration Unit, **281**
Enviropack Emergency Spill Cleanup Kits, **73**
Enviropack Polyethylene Overpacks/Salvage Drums, **167**
Epoleon Odor-Neutralizing Agents, **228**
ESI Cascade LP500 Series Low Profile Air Stripper, **284**
ESI Mobile Recovery Trailer, **80**
Flexaust Hose and Duct, **214**
Flexible Containers - General Listing, **151**
Flo Trend Hydroclone Centrifugal Separator, Vibratory Screen, and Container Filter, **237**
Flygt Submersible Pumps, **194**
Foul-Up Hazardous Wet-Waste Solidifying Agent, **230**
Froth-Pak Portable Foam System, **36**
Fundamental Water Treatment Processes - General Listing, **243**
Furon (Bunnell) Teflon Flange and Valve Shields, **7**
Gast Vacuum Pumps, **209**
Gates Industrial Acid-Chemical Hoses, **216**
Geomembrane Liners - General Listing, **175**
Gilkes Portable Emergency Turbo Pump, **196**
GLE Slant Rib Coalescing (SCR) Oil/Water Separator, **286**
Global Vacuum Extraction and Vapour Liquid Separator Modules, **98**
Global Vapour Treatment Module - Chloro-Cat Catalytic Oxidizer, **313**
Goodall Chemical Hoses, **218**
Greif Steel Overpacks/Salvage Drums, **168**
H&H Dredges, **142**
Hazardous Waste Container Bag Liners, **168**
Hazelton Maxi-Strip Systems, **287**
Helios Flexible Bulk Containers, **157**
Heyl & Patterson Transportable Soil Remediation Unit, **332**
Hi Tech Berm, **38**
Holmatro Vacuum Leak Sealing Pad, **8**
Horizontal Directional Drilling for Trenchless Remediation, **100**
Hrubout In-situ Soil Processor, **101**
Hudson CS Series Coalescing Oil/Water Separators, **289**
Hydro-Stripper Air Stripper, **290**
ILC Dover DrumRoll, **10**

INDEX - BY PRODUCT NAME (cont.)

Induced Air Movement - General Listing, **56**
Inert Gas Blankets - General Listing, **60**
John Zink Flaring Systems, **315**
Kason Drum Sifter, **239**
Keck Product Recovery System, **82**
Koch Tubular Ultrafiltration (UF) System, **291**
Lamson Centrifugal Exhausters for Environmental Remediation, **103**
Lavin Centrifuges, **241**
Link-Seal Pipe Penetration Seals, **11**
Liquid-Miser Activated Carbon Adsorption Canisters, **293**
Lutz Drum Pumps, **197**
Matheson MESA (Mobile Emergency Scrubbing Apparatus), **13**
Max Vac Vacuum Systems, **115**
M-D Rotary Positive Displacement Blowers, **104**
Mechanical Covers - General Listing, **63**
Mechanical Skimmers - General Listing, **75**
Megator Sliding-Shoe Pump, **199**
Mercon Spill-treating Kits, **231**
Milsheff Spray-Stops Valve and Flange Covers, **15**
Minuteman International Vacuum Systems, **116**
ModuTank Modular Storage Tanks, **170**
Mud Cat Model 370 PDP Dredge, **144**
NAO Trailer-mounted Emergency Flaring Systems and Mobile VOC Wagon, **316**
NEPPCO PetroPurge Pump Systems, **84**
NEPPCO SoilPurge Soil Vapour Treatment System, **105**
Nilfisk Vacuum Systems, **118**
Nochar Solidifying Agents, **232**
Odor-Miser Activated Carbon Adsorption Canisters, **319**
ORS LO-PRO Air Strippers, **295**
ORS Scavenger Hydrocarbon Recovery Systems, **86**
Osmonics Crossflow Membrane Filtration Systems, **297**
PACE Oil/Water Separators, **299**
PACT Batch Wastewater Treatment System, **301**
Paratech Inflatable Leak Sealing Systems, **16**
Pego Soil Vapour Extraction Units, **106**
Permalon Drum Liners, **186**
Petersen Inflatable Pipeline Stopper Plugs, **18**
PetroTrap and SkimRite Free Product Skimmers, **89**
Pit Hog Auger Dredges, **145**
PLIDCO Pipeline Repair Fittings, **20**
Plug N'Dike Products and Plug Rugs, **22**
Poly-Spillpallet 6000 Polyethylene Spill Containment Pallet, **39**
Porcupine Processor, **333**
Portadam Barrier System, **51**
Port-A-Tank, **158**
Powervac Oil Recovery System, **120**
Protec Recip Pump and Rotary Pump, **91**
Pumps - General Listing, **189**

RETOX (Regenerative Thermal Oxidizer), **107**
Rigid Wall Containers - General Listing, **161**
Roots Blowers and Vacuum Pumps for Environmental Remediation, **109**
Ro-Vac Vacuum System, **121**
Sala Roll Pump, **201**
Sewer Guard Watertight Manhole Covers, **40**
ShallowTray Low Profile Air Strippers, **303**
Shields Welded Steel Containment Pallet, **42**
Skolnik Steel Overpacks/Salvage Drums, **173**
Solar-powered Pumping Systems, **202**
Spill Containment/Deflection Barriers - General Listing, **47**
Spill Sorbents - General Listing, **111**
Sprung Instant Structures, **66**
Suction Systems - General Listing, **113**
Syntechnics Track Collector Pan System, **43**
Terra Tank, **159**
Timco Isomega Recovery Pump, **93**
Tornado Vacuum Systems, **123**
Torvex Catalytic Oxidation Systems, **320**
Trac Pump Dredging System, **147**
Trans-Vac Oil Recovery Units, **125**
Trecan Solid Waste Incinerators, **334**
TriWaste Micro-flo Mobile Water Treatment System, **305**
Trouble Shooter Portable Containment Berms, **44**
Tub Scrub Units - Activated Carbon Adsorption Canisters, **321**
Vactagon Vacuum Systems, **127**
Vac-U-Max Vacuum Systems, **129**
Vacuum Pumps for Chemicals - General Listing, **205**
Vapour Suppressing Foams - General Listing, **68**
Vetter Systems Leak Sealing and Pipe Sealing Systems, **24**
VMI Mini Dredges and Piranha Auger Dredge Attachment, **148**
Vulcanus 400 Mobile Incinerator, **336**
Warren Rupp SandPIPER Pumps, **203**
Waste Container Bag Liners, **187**
Water Scrub Units - Activated Carbon Adsorption Canisters, **306**
Water Treatment Processes - General Listing, **243**
ZenoGem Process, **309**

INDEX - BY MANUFACTURER/DISTRIBUTOR

3L Filters Limited, **251**
3M Canada Incorporated, **71**
21st Century Container Limited, **162**
ABASCO, **152**
ABB Air Preheater Inc., **328**
ABB, ASEA Brown Boveri Incorporated, **328**
ABCO Industries Incorporated, **212**
A C Carbone Canada Inc., **54**
Aco-Asmann of Canada Limited, **162**
Adwest Technologies Incorporated, **108**
Aeromix Systems Incorporated, **264**
Aero Tec Laboratories Incorporated, **28, 153**
Affiliated-Dynesco, Division of Wajax Industries, **57, 59, 204**
Air Filter Sales and Service Ltd., **54**
Airguard Industries, Division of Delcon Filtration, **54**
Airotech Incorporated, **63, 65**
Air Plastics Incorporated, **162, 257**
Air Products Canada Limited, **62**
Alfa-Laval Separation Inc., **234**
AMFlow Master Environmental Inc., **193**
A.M.L. Industries Incorporated, **242**
Andco Environmental Processes, **261**
Andersen 2000 Inc., **324**
Anderson Water Systems Ltd., **54**
Andritz Ruthner, Inc., **235**
Ansul Fire Protection, **71**
Ansul Incorporated, **223**
Aqua-Guard Technologies Incorporated, **153**
Associated/I.R.P. Limited, **212, 215**
Beckland Equipment Ltd., **109**
Bennett Environmental Inc., **326**
Bethlehem Corporation, The, **333**
Bio Trol Incorporated, **263**
Bren-Ker Industrial Supplier, **17**
Breuer/Tornado Corporation, **124**
BrineCell Incorporated, **266**
Brulé C.E. & E. Inc., **327**
Burke Environmental Products, **177**
Busch Vacuum Technics Inc., **208**
Calgon Carbon Corporation, **54, 268, 312**
C.A. Litzler Co. Inc., **57**
Cameron-Yakima, Inc., **54, 308, 322**
Canadian Clay Products, **31**
Canbar Incorporated, **162**
Canflex Manufacturing Inc., **153, 156**
Can-Ross Environmental Services Ltd., **1, 23, 120, 163**
CAP International Inc., **202**
Carbtrol Corporation, **54, 95, 270, 272**
Cartier Chemical Ltd., **227**
Cassier Engineering Sales Ltd., **163, 166**
CDF Corporation, **177, 184**
Chemical Equipment Fabricators Ltd., **177**
Chem-Tainer Industries Inc., **32, 163**
Cherne Industries Inc., **2**
Chlorine Institute Inc., The, **3, 23**
Chubb National Foam, **71**
Clark Products Company, **34**
Class A Fire Equipment, **17**
Clean Environment Equipment, **78**
CleanSoils Incorporated, **330**
CleanSoils Limited, **330**
City Industrial Sales Corporation, **212**
C.M.P. Mayer Fire Equipment, **17**
Columbia Geosystems Limited, **177**
Cominco Limited, **62**
Containment Corporation, **39**
CORECO - College Research Corporation, **331**
Co-Son Industries, **196**
CSM Environmental Systems Incorporated, **320**
Custom Metalcraft Inc., **163, 166**
Cyanide Destruct Systems Incorporated, **273**
Deetag Limited, **212**
Delcon Filtration, **54**
DeMarco MaxVax Corporation, **115**
Donovan Enterprises Incorporated, **177**
Downs JJ Industrial Plastics Incorporated, **212**
Dresser Industries Inc., Roots Division, **109**
DRN Incorporated, **318**
Dynamic Process Industries, **79**
EG&G Rotron, Industrial Division, **97**
Ejector Systems Inc. (ESI), **81, 285**
Eldred Corporation, The, **153**
Ellicott Machine Corporation International, **144**
Empire Buff Limited, **178**
ENPAC Corporation, **39, 163**
ENVIROCAN, **141**
Envirocan Wastewater Treatment Equipment, **235**
Environmental Container Corporation, **74**
Environmental Protection Inc., **178**
Environment Canada, Emergencies Engineering Division, **277, 280, 283**
Enviropack Company, **163, 167**
Enviro Products Incorporated, **90**
Edwards and Cromwell Mfg., Inc., **6**
Epoleon Corporation of America, **229**
EPS Chemicals Incorporated, **231**
Fabrico Environmental Liners Inc., **178**
Fairplast Limited, **178**
Faltech AB, **201**
FARR Incorporated, **54**
Fleck Brothers Fire & Safety, **17**
Flexaust Company Inc., **212, 215**
Flex-Pression Limited, **212**
Flo Trend Systems, Incorporated, **238**
Fosroc Inc., **41**
Foss Environmental, **336**
Fred Cressman Sales Incorporated, **178**
Franklin Supply Company, **104**
Furon/Bunnell Plastics, **7**
Galbreath Incorporated, **163**

INDEX BY MANUFACTURER/DISTRIBUTOR (Cont.)

Gast Manufacturing Corporation, **210**
General Filtration Division of Lee Chemicals Ltd., **54**
Geogard Lining Incorporated, **178**
Geostructure Instruments, Inc., **200**
Gilkes Incorporated, **196**
Global Technologies Incorporated, **99, 314**
Goodall Rubber Company of Canada Ltd., **212, 220**
Great Lakes Environmental Incorporated, **286**
Greif Containers Incorporated, **164, 169**
GSE Lining Technology Inc., **179**
H&H Pump Company, **143, 147**
Harlock Industrial Incorporated, **179**
Hazelton Environmental, **288**
Helios Container Systems Inc., **153, 157**
Heyl & Patterson Incorporated, **332**
Hi Tech Berms (1993) Inc., **38**
HMT Canada, **179**
Holmatro Industrial Equipment, **9**
Hoyt Corporation, **54**
H.Q.N. Industrial Fabrics, **179, 187**
Hrubetz Environmental Services Inc., **102**
Hudson Industries, **289**
Hydro-Flex Incorporated, **212**
Hydro Group, Inc., **290**
(IMCO) Industrial Machine Co., **104**
IMCO Industriel Inc., **104**
Industrial Plastics Canada, **15, 179, 212**
Indian Springs Specialty Products Inc., **3**
ILC Dover Inc., **10**
Insta-Foam Products, Inc., **37**
Integrated Chemistries Incorporated, **225**
ITT Flygt, **195**
ITT Industries of Canada Ltd., **195**
ITW Devcon Environmental Products, **4**
J & L Industrial Tanks, **164**
J.V. Manufacturing Company, Inc., **54**
James Morton Ltd., **198**
John Zink Canada West, **315**
Kason Corporation, **240**
Keck Instruments Incorporated, **83**
Kepner Plastics Fabricators, Inc. **153**
Koch Membrane Systems Inc., **292**
Lamson Corporation, **103**
Levitt Safety Limited, **223**
Linatex Canada Incorporated, **212**
Liquid Carbonic Canada Inc., **61, 62**
Liquid Waste Technology Incorporated, **146**
Lutz Pumps, Incorporated, **198**
Lynum Engineering Sales, **318**
Machine Control B.A.A. Canada Inc., **130**
Mapleleaf Environmental Equipment, **99**
Marine, Industrial & Technical Sales, **104**
Mateson Chemical Corporation, **230**
Matheson Gas Products Canada, **14**
Matrix Photocatalytic Incorporated, **256**
May Fabricating Company, **164**
McDaniel Tank Manufacturing, **164**
MBB Trecan Inc., **335**
McIntyre Industries, **200**
M-D Pneumatics Division, Tuthill Corporation, **104**
Megator Corporation, **200**
Milepost Manufacturing Inc., **164**
Milsheff Inc., **15**
Minuteman International Inc., **117**
Minuteman Canada Inc., **117**
MK Plastics, **180**
ModuTank Inc., **164, 172**
Montgomery Environmental, **154, 180**
MPC Containment Systems Limited, **180**
Mud Cat, **144**
NAO Incorporated, **318**
National General Filter Ltd., **55**
Navenco Marine Incorporated, **122, 154**
NEPPCO, **85, 105**
Nilfisk Limited, **119**
Nochar, Inc., **232**
North East Environmental Products Inc., **304**
Northflex Manufacturing, **213**
Norwood Carbonation CO_2 Supply Company, **62**
N.R. Murphy Limited, **213**
Oil Pollution Environmental Control Ltd., **154**
Omniflex Industrial Sales Limited, **213**
Ontario Hose Specialties Limited, **213**
Ontario Rubber, **213**
ORS Environmental Systems, A Division of Sippican, Inc., **88, 296**
Osmonics Incorporated, **204, 298**
Packaging Research and Design Corp., **180, 185**
PAP Process Engineering Services, **238**
Patton Enterprises Inc., **92**
Patton Industries Ltd., **28**
Paratech Inc., **17**
Pego Systems Incorporated, **106**
Petersen Products Company, **19**
PLIDCO The Pipe Line Development Company, **21**
PND Corporation, **1, 23**
Polaris Fire & Safety, **17**
Poly-Flex, Incorporated, **180**
Poly Processing Co., **164**
Prolew-Scott Ltd., **109**
Protec, A Division of Patton Enterprises Inc., **92**
Q-Air Environmental Controls, **55**
Quickdraft, Division of C.A. Litzler Co. Inc., **57**
Reef Industries Incorporated, **180, 186**
Resource Conservation Company, **256**
Resource Management & Recovery, **259**
R.H. Woods Ltd., **100**
R.M.S. Enviro Solv Incorporated, **143, 193**
RNG Equipment Inc., **181, 182**
ROBCO Incorporated, **213**

INDEX BY MANUFACTURER/DISTRIBUTOR (Cont.)

Robertson White Engineering Ltd., **318**
Ro-Clean International A/S, **122**
Roto Trading International AB, **336**
RPM Mechanical Incorporated, **213**
SAI Engineering Sales Limited, **182**
Seaman Corporation, **182**
Scienco/Fast Divisions of Smith & Loveless Inc., **300**
SEI Industries Limited, **154, 159**
Separator Engineering Limited, **240**
Shadrack Engineering Limited, **130**
Shields Environmental Corporation, **42**
Sippican, Inc., **88, 296**
Skolnik Industries Incorporated, **165, 173**
Slickbar Products Corporation, **126**
Smith & Loveless Inc., **300**
SoilTech ATP Systems, Inc., **325**
Solarchem Environmental Systems, **256**
Sopers Engineered Fabric Products, **154, 158, 183**
Sprung Instant Structures Ltd., **67**
SRS Crisafulli Inc., **141**
Stady Oil Tools, **238**
Staff Industries, Incorporated, **183**
STANCHEM, Inc., **54, 268**
Stelfab Niagara Limited, **165**
Superior Safety Equipment, **17**
Svedala Pumps & Process, **236**
Syntechnics Inc., **43**
Taylor Fluid Systems Incorporated, **213**
Techstar Plastics Incorporated, **165**
Tes-Tech Engineering Systems Limited, **130**
ThermaFab Inc., **46**
Thomas Conveyor Company, **165**
Thunderline/Link Seal, **12**
Timco Manufacturing Inc., **93**
TOTE Systems, **165**
Trecan Combustion Limited, **335**
TriWaste Reduction Services Inc., **305**
Turner Equipment Company Inc., **165**
Tuthill Corp./Coppus Portable Ventilation Division, **57, 59, 104**
Ultrox, **256**
UMA Engineering Ltd., **325**
UMATAC Industrial Processes, **325**
Union Carbide Canada Inc., **61, 62**
Utility & Industrial Supply Limited, **213**
Ward Ironworks Limited, **115**
Warren Rupp Incorporated, **204**
Wes-Can Erosion Control, **38**
Westport Environmental Systems, **294, 319**
Wilden Pump & Engineering Company, **204**
Windsor Pump Company, **204**
Vactagon Pneumatic Systems, Inc., **128**
Vac-U-Max, **130**
Versatile Wales Limited, **213**
Vetter Systems, Inc., **26**
Vikoma International Limited, **120**
VMI Incorporated, **149**
Zedex Products Inc., **55**
Zenon Environmental Inc., **310**
Zimpro Environmental Inc., **302**